PERSPECTIVES ON PARTICLE PHYSICS

FROM MESONS AND RESONANCES TO QUARKS AND STRINGS

HIRONARI MIYAZAWA

PERSPECTIVES ON PARTICLE PHYSICS

FROM MESONS AND RESONANCES TO QUARKS AND STRINGS

In Commemoration of the Sixtieth Birthday of Professor II Miyazawa

Editors
S. Matsuda (Kyoto)
T. Muta (Hiroshima)
R. Sasaki

World Scientific
Singapore • New Jersey • London • Hong Kong

Published by

World Scientific Publishing Co. Pte. Ltd.
P O Box 128, Farrer Road, Singapore 9128

USA office: World Scientific Publishing Co., Inc.
687 Hartwell Street, Teaneck, NJ 07666, USA

UK office: World Scientific Publishing Co. Pte. Ltd.
73 Lynton Mead, Totteridge, London N20 8DH, England

Library of Congress Cataloging-in-Publication data is available.

PERSPECTIVES ON PARTICLE PHYSICS — FROM MESONS AND RESONANCES TO QUARKS AND STRINGS

ISBN 9971-50-589-4

Printed in Singapore by JBW Printers & Binders Pte. Ltd.

CONTENTS

CHAPTER 3 MODELS OF ELEMENTARY PARTICLES

CHAPTER 4 STRING AND CONFORMAL FIELD THEORY

CHAPTER 5 FIELD THEORY

Editors' Foreword

Professor Hironari Miyazawa of the University of Tokyo celebrated his sixtieth birthday on April 28 of 1987. In commemoration of this occasion was this volume planned.

Professor Miyazawa is well-known in the field of particle and nuclear physics for his works on nucleon magnetic moments and dispersion relations as well as his other important contributions to the field. The present commemoration volume includes recollections by his friends in connection with his activity in physics, and reviews by his freinds, former colleagues and ex-students on his original research contributions. Also included is the written version of talks presented at the Symposium of Perspectives on Particle Physics held on March 29 of 1988 at the University of Tokyo in commemoration of the sixtieth birthday of Professor Miyazawa. The symposium program and the list of participants may be found at the last section of the book.

It is always pleasant to remember the time when we were graduate students in the University of Tokyo working for Ph. D. degree under the guidance of Professor Miyazawa. It is needless to say that scientific achievements of Professor Miyazawa are innumerable and superb. At the same time it is unforgettable that he has been an outstanding teacher who has trained many students in particle physics. Many of those students are working in the field of particle physics at various places and made their contributions of original works to this commemoration volume.

We would like to thank all the contributors for their voluntary cooperation in preparing this volume.

November 1988

Satoshi Matsuda
Taizo Muta
Ryu Sasaki

PREFACE

My first encounter with Hironari Miyazawa occurred probably in 1947, in a classroom of the Physics Department of the University of Tokyo. As one of the research associate/assistants for the department, I was supposed to take turns with my colleagues to run exercise sessions in mathematical physics for first year students under the old university system. There he stood out immediately to my eyes, not only for his brilliance, but also for his serious and disciplined look with a short crew cut; there was an air of a military cadet or of an ascetic. These first impressions of mine about him were to persist for years to come as our paths crossed at various places and our relationship became closer.

Indeed the next occasion I remember is when I was at the Osaka City University, around 1951. In those days we indulged in a lot of self-criticisms, under the influence of such outspoken leaders as Sakata, Taketani and Watanabe, about our lack of a sense of direction, of close contact with experiment, of free interaction among different institutions, and so on. As for the last point, a suggestion was made that one should start the practice of "mushashugyo", an old custom of the feudal days whereby a free-lance samurai would travel from domain to domain, challenging other schools of samurai to test and improve his martial skills. I believe Hiro was the very first samurai to go on such a mission. Now freshly out of school, he stayed with us for a month or so, discussing physics by day and sleeping on a sofa by night in a makeshift wooden building of the department near the busy Umeda district. A few years later we would meet again, this time together as research associates at the University of Chicago under Gregor Wentzel, Maria Mayer, and Murph Goldberger, before he moved on to the Institute for Advanced Study in Princeton. In the Sixties he was frequently back to the U.S. I believe his second visit to the Institute was also his honeymoon trip. Later he spent some time at Argonne National Laboratory, Chicago, and Minnesota as a visiting professor.

So much for a brief description of Hiro's formative years viewed through my personal interactions with him. I would now like to recount some of his most significant scientific contributions. In the immediate post-War era, nuclear physics was still very much a part of elementary particle physics. The discoveries of strange particles, the pion, and the Lamb shift came one after the other, but the numerous hadron resonances had not been anticipated except in the context of nucleon isobars based on the strong coupling theory. The pion (or π meson) was thought to be Yukawa's mesotron, the new elementary particle that would once for all be responsible for the strong nuclear forces.

In the early Fifties, first indications of rich hadron dynamics began to emerge from accelerator experiments. The cross section for both charged and neutral pion production in $\gamma - p$ reactions from the electron synchrotron was unusually large. Soon another surprise showed up in ever increasing $\pi - p$ scattering cross sections at the Chicago cyclotron. On the basis of the $\gamma - p$ data and ahead of everybody else, Fujimoto and Miyazawa in 1950 made the bold suggestion that it is due to a nucleon isobar resonance, which was later to be confirmed and dubbed the 3-3 resonance (now the Δ resonance).

Around the same time, intriguing evidence for the V particles was accumulating, as people in Japan learned bits by bits through journals and private correspondence. Many were wondering about their true nature, but Hiro was among the several who actually proposed a model of V particles. He speculated that a new strongly interacting fermion pair was created, one of which formed a bound state with a nucleon and the other decayed into π and μ via a universal weak interaction. The two essential features which turned out right, namely associated production and weak decay, are there. It is also worth pointing out that all these model builders in Japan were in their twenties, and Hiro was the youngest at that.

Hiro's next important work concerns the deviation of nuclear magnetic moments from the Schmidt lines. Here he argued that the nucleon magnetic moment would be quenched in nuclear matter because the Pauli principle reduces the pion cloud. The idea preceded the systematic experimental work of Yamazaki and collaborators by two decades, at which time Hiro again got seriously involved in theoretical analysis.

In 1953 Hiro joined the Chicago theory group where Goldberger was developing the dispersion theory. On the experimental side, Fermi and his collaborators were busily exploring pion physics. Under these circumstances was born the classic paper by Goldberger, Miyazawa and Oehme on the dispersion relations for forward pion-nucleon scattering. It became the basis for the first successful application of dispersion theory to strong interactions that spurred the intense theoretical activity of the late Fifties.

As the SU(3) symmetry was developing into maturity in the mid-Sixties, Hiro got naturally interested, as many did, in higher symmetry schemes. But something few others would have dared was to combine both mesons and baryons into a supermultiplet. Because of the obvious difference in statistics, this did not make sense, or so I felt when he gave a seminar during his second stay at Chicago in 1967, quite apart from the question of whether the hadrons showed such supermultiplet structure or not. I simply did not anticipate the modern supersymmetry that started to blossom several years later. Reading his papers of 1966 and 1968, I appreciate in hindsight how Hiro had correctly grasped the mathematical problems involved.

Every theoretical physicist of stature has a distinct style. Thus many of Hiro's ideas are characteristically simple and down to earth; his arguments may even sound naive. Actually he is at his best and on his own when he reasons from physical intuition rather than from mathematical sophistication. Although he did sometimes get involved in the latter, he also kept sight of the forest of the physical world. His excursions into quark confinement and the string model since the Seventies show that he is primarily concerned with qualitative and realistic questions rather than the mathematical or philosophical grandeur of a theory. In fact it was a revelation for me when I recently happened to read in the same week his 1979 paper in which he discussed rearrangement of quark bonds in dense nuclear matter, and P. W. Anderson's resonating valence bond theory of high T_c superconductivity. It has been with a sense of rediscovery, admiration and delight for me to trace Hiro's footprints and write this preface dedicated to him.

Y. Nambu

Chapter 1 MESONS AND NUCLEI

H. Miyazawa, My Personal Memories and in the History of Sciences

H. Morinaga
Physik-Department, E17, Technical University of Munich
8046 Garching, W.-Germany

I. Miyazawa, My Personal Memories

The first time I met H. Miyazawa, the little prince from the war-striken Physics-Department of the University of Tokyo, must have been around 1952 in Chicago either at Nambu's home or at the Institute of Metallurgy of the University of Chicago which was not so far from the famous foot-ball ground under which the first Fermi's atom reactor became critical.

The first time I heard of him, however, was a little time before. Professor R. Sagane, my teacher, was already back to Berkeley after the war. He used to pick up Japanese physicists who arrived at the San Francisco pier with his car. Prof. Sagane told me personally about his meeting with Miyazawa, whom he took to the Stanford University in the next following days. "All the high power physicists showed great respect to this little boy from Japan and followed him about." I still remember Prof. Sagane's happy and proud expression quite vividly.

What made this young physicist so famous was the interpretation of the 3-3 resonances and the reduction of the nuclear magnetic moments due to mesonic effects. I meet the 3-3 resonances once in a while in my work recently on the study of cosmic ray backgrounds for solar neutrino detection, but the mesonic effects in nuclei has often been at the center of nuclear structure physics. Indeed, as I met Miyazawa for the first time, there was no overlap between his physics and my physics, - until then I had worked mainly on high frequency techniques and accelerators. However, gradually as my interest shifted toward nuclear structure and nuclear spectroscopy, this pioneering thinking of Miyazawa appeared often on my horizon, on the other side of which a lot of mysteries were hidden.

Miyazawa resulted in further unravelling of deep mysteries of elementary particle physics. But his very pioneering work which was done just before the separation of nuclear and elementary particle physics, had a great impact on the nuclear physics community of Japan. The work of Arima and Horie to explain the reduction of the magnetic moments of nuclei with the help of configuration mixing, a classical work which also is a starting point of much further nuclear structure work, once threw doubt on the necessity of the Miyazawa effect. But this must also be considered to have appeared to some extent excited by the initial Miyazawa's work. I also was involved with some small investigation to prove experimentally that the g factor for neutrons is almost zero. But the final decisive work, to prove the existence of the mesonic current of nucleons in the nucleus experimentally as Miyazawa pointed out more than a quarter century before, was done by Yamazaki. He showed through an ingenious and convincing way that the g_1 factor for neutrons in nucleus is ~0.1, that is much smaller than 1 but finite. The theoretical work had been pushed forward already by Hara and Karoboshi.

Elucidation of nuclear magnetism became a highlight of the postwar Japanese nuclear physics. The latter was damaged by politically minded members. Miyazawa kept himself out of this unfortunate tendency.

Back to Chicago. At that time I was in Ames, Iowa. Gyo Takeda, my classmate from Tokyo was in Madison, Wisconsin. We often arranged to meet each other in Chicago. But after all, we could not do much in the city and ended up calling upon Nambu's family, where we knew that Mrs. Nambu treated us well, if we listened to her problem as to whether they should go back to Japan or not. Miyazawa came then, and we played Majong together. Miyazawa was always by far the best. We accumulated the points and Miyazawa's gain could never be surpassed.

I believe that there is a certain correlation between the type of scientists or more in general profession and hobby, game or sports. It is well known that among theoretical physicists there are good "go" players. N. Fukuda or G. Takeda are examples. But Majong is a little different. I have, however, no explanation why Miyazawa was so good in Majong. I have not enough knowledge of theoretical physics to classify his particular type.

After I came back to Sendai in 1957 I often met him in Tokyo and saw his performances in meetings, which are known to be direct, even, with for example, Prof. Tomonaga. I was proud of him and it has been a great pleasure to know this brilliant theorist.

My academic life in Sendai was quite pleasant. Indeed it was one of my most productive times. But after two intensive years, naturally dissatisfaction accumulated in different ways. It then happened that Miyazawa played an important role in my life.

It must have been sometime in 1959. As he came to Sendai, he asked me to apply for a new position in the Physics Department of the University of Tokyo. It took a long time, but finally I was offered a position, and from April 1960 I came back to my Alma Mater, until I left it again in 1968 to take a chair at the Technical University of Munich. The same month I married my wife. In my wedding I arranged Gyo Takeda to be the official go-between, and two of my important teachers, Prof. Kaya and Prof. Sagane, as honored guests and asked Miyazawa to act as the ceremony chief. This was more than what is normally enough to satisfy our mothers and myself!

So my life started close to Miyazawa. But actually in the 8 years at the Physics Department, I had not as much contact with him as before. However even now, every time I go back to Tokyo, I visit him regularly at his office. In some sense he has not changed much from the time when he was in Chicago. I have the feeling that he has gradually become quieter. But one thing is clear, one by one, brilliant new theorists have grown up near him, - during a fairly long time span. I know Miyazawa's young students personally rather well and this was probably one of the most fruitful consequences of the time I spent in Hongo -the Physics Department of the University of Tokyo. "How nice it is that in particle theory there is always something new and exciting." This has been my greeting to him over the last 25 years. And I am so glad that I could have chats with him off and on so as to feel the greatness of physics in between my practical and often fruitless time.

II. Miyazawa in the History of Physics

In order to place H. Miyazawa in the history of physics, I must first describe the history of science, or especially of physics, as I sec it.

It is usually accepted that 1600, the time Gallileo Gallilei discovered his physics, as the starting point of the basic way of thinking, or in another word the motivation, of the present age of science. Philosophers call this way of thinking "positivism". It is suggestive that this great event of human history took place exactly 1600 (=400 4) years after O.A.D., the most important point in the western - probably also in the whole-world. 400 years is what Toynbee identifies as the active life time of a civilization. Indeed one could identify certain important events in the development of human society at about n 400 years like the fall of the Roman Empire (n=1). Establishment of middle age order in Europe (n=2). Rise of Arabic sciences (n=3).

It would not be necessary to go into this part of history of science in order to discuss Miyazawa, a modern field theorist. But the development of physical science after 1600 must be reviewed before I shed a spotlight on this ever young physicist.

The development of physics and technological world after 1600 can be clearly divided in four periods of 100 years each as in Fig. I. Society follows with 100 years time lag.

In Fig. II is shown that the development of Modern Physics in the 20th century is again divided in four periods, as I showed in the summary talk at the Tokyo

Conference on Nuclear Structure in 1977.

Now, ten years later, one could make some more detailed comments on this figure. What was not quite so clear but what started to be certain is a half century phase lag of the development of modern elementary particle physics to nuclear physics - which was at that time elementary particle physics. It is more than just interesting to point out that the Gell-Mann-Zweig's quark appeared almost exactly 50 years after the Bohr's theory of atoms. The establishment of Q.C.D. for which one of our old Majong friend played the most fundamental and decisive contribution, came again exactly 50 years after the establishment of quantum mechanics. If the development of the string theory corresponds to the development of the quantum field theory and the renormalization theory the answer should still take some years.

This remarkable success in dividing a period by four in the history of science suggests that we might divide the 25 years period by a further four. Is there any observable which changes every 6.25 years? The answer seems to be yes. This observable is the personal motivation to do science (more narrowly physics).

The question why one person does physics professionally has never been treated as a problem of physics. Even not explicitly in the history of science (physics), a branch accepted in sciences (physics), although the subject deserves great attention. It is not difficult to develop such a branch because the science itself, general and social history, as well as the biography of scientists are well known. Non-personal motivation to push sciences can be studied even easier because there are megatons of documents and proposals stored in the governments and private organizations supporting sciences.

Before a big wave of computerization comes to this field also, I should like to try a first order fit, which is necessary in order to place my friend Miyazawa in the proper place. In the scheme which is schematized in Fig. III, the time is the time when a professional physicist decides to choose physics as his life profession. (The time when one physicist decides to study physics is a few years before that and the motivations are surprisingly similar. For him physics was easy and clear and that physics is fascinating was the important reason. Also dislike of engineering is an important factor.) I have chosen the time from 1925 to the present, in which I know many of the physicists personally. Since it is a personal story I shall not mention any names, but readers are invited to apply this scheme to their teachers, friends and pupils. You will definitely find a large amount of truth in it, of course, there are exceptions. This should interprete only as a tendency. On accepting the truth in this theory one sees immediately a social influence to personal motivation. It is not necessarily a social requirement. It is rather a social influence on the individual mind. (The motivations for scientific research of organizations are more directly social requirement. About this point Japanese readers could refer to the author's letter to the Journal Baturi Oct. 1985 issue titled "Motivations in Physics").

Some comments should be made to Fig.III. Actually it is assumed that readers are familiar with the present history, which is briefly in the last column. In the period of 1931.25 to 1937.5 come physicists who had direct contact with the founders of modern physics and directly contributed in giving a final touch to the completion of quantum mechanics (Weisskopf, Peierls, Yukawa, Tomonaga, Fusimi). They were often ardent peace-movement supporters. As they started to work in physics, the very basis of science, the human freedom had begun to be in danger. In the period of 1937.5 to 1943.75, where the military dominated and the war went on, the only way to protect themselves was to use in someway or other the bureaucratic machinery, which was even stronger than the military order. It is useful to know that the people who went through this time knew this.

This period is followed by the period of samurais. External excitation and spirit of physics encourages the creation and a number of individual wolfs. I, belonging to this generation, must confess that many of my good friends did not receive a proper university education. Air battles were much more exciting (not only interesting) than the lectures. Physics had spine tingling power based on the human capacity. Physicists in this generation are often society or philosophy conscious on an individual basis, and rather strict with themselves.

The next generation grew up in the post-war poverty - especially in Japan and in most of the European countries. The peace and the rest, was the main thing. It is a well known fact that Tomonaga allowed anybody to attend his seminar on the condition that he would not arrange a job for them. Physics was to be enjoyed and also to establish ones identity, but never to be made ones life or carrier. I have noticed that many physicists of this generation maintain their academic activities longer and find their own reason of existence only in scientific contributions. (academists)

The economic recovery brought, together with the

social recognition of science, especially nuclear physics, a new aspect to the physics community forming the next generation. Career physicists were together with economical animals, the great contributors to the country. Nobel prizes became a great social issue. An American physicist known for his great activity said: "There are four respectable professions in the society. Politicians and military men who save a country, medical doctors who save human life, priests who save men's soul and nuclear physicists who save mankind."

The next generation, the angry young men could not stand this dirty new world. Fortunately, there was still an anti-thesis, represented by Mao in New China. It is somewhat surprising that the motivation of work for young people changes so quickly. I had contact with many of these young students (many of them were students of Miyazawa). In spite of their immaturity and chaotic behaviour my feeling was sympathetic to them. It is a spiritually high time. I left the University of Tokyo in the middle of this chaos and moved to Munich, where changes were also going on.

The time of disenchantment which follows this time is actually a good time. But the following generation reflects the difficulties of the social condition of the time.

One sees, after going through the change of physics of the last half century divided into sub-period of 6.25 years, an intriguing tendency which is given in the last column. This is the alternative of three spiritual state of mind, high spirited, good and bad. Of course, good and bad could be a somewhat subjective judgment. The very problematic time of 1956.25 to 1968.5 is in a sense a flourishing time and the economic wondor was founded there. But it is my feeling that this time is a rather unfortunate time for the scientists.

The sudden appearance of 3 in the scheme of dividing the history by 4, is also interesting. It has been stated by meteorologists that there seems to be a period of 20 years which is close to 18.75 years (6.25 years 3). There might be something to it. But anyway it seems as if there is a feeling of attraction between people whose age is 18.75 years apart. Both neighbouring generations are not friends. The former generation which is challenged, never feels happy. Maybe this is the mechanism by which the number 3 comes about. Readers (who are mainly theorists) are encouraged to do the Gedankenexperiments to this point.

I hope, I have explained the history of physics in sufficient detail to place H. Miyazawa in it. He finished his physics degree in 1950, just between the time of Samurais and Academists. The most free thinking little prince still lives in buddha's palm, the fate of the development of physics. In history, from the separation time of elementary particle physics from nuclear physics and as a human being as a Samurai who purely loved physics.

Received March 29, 1988

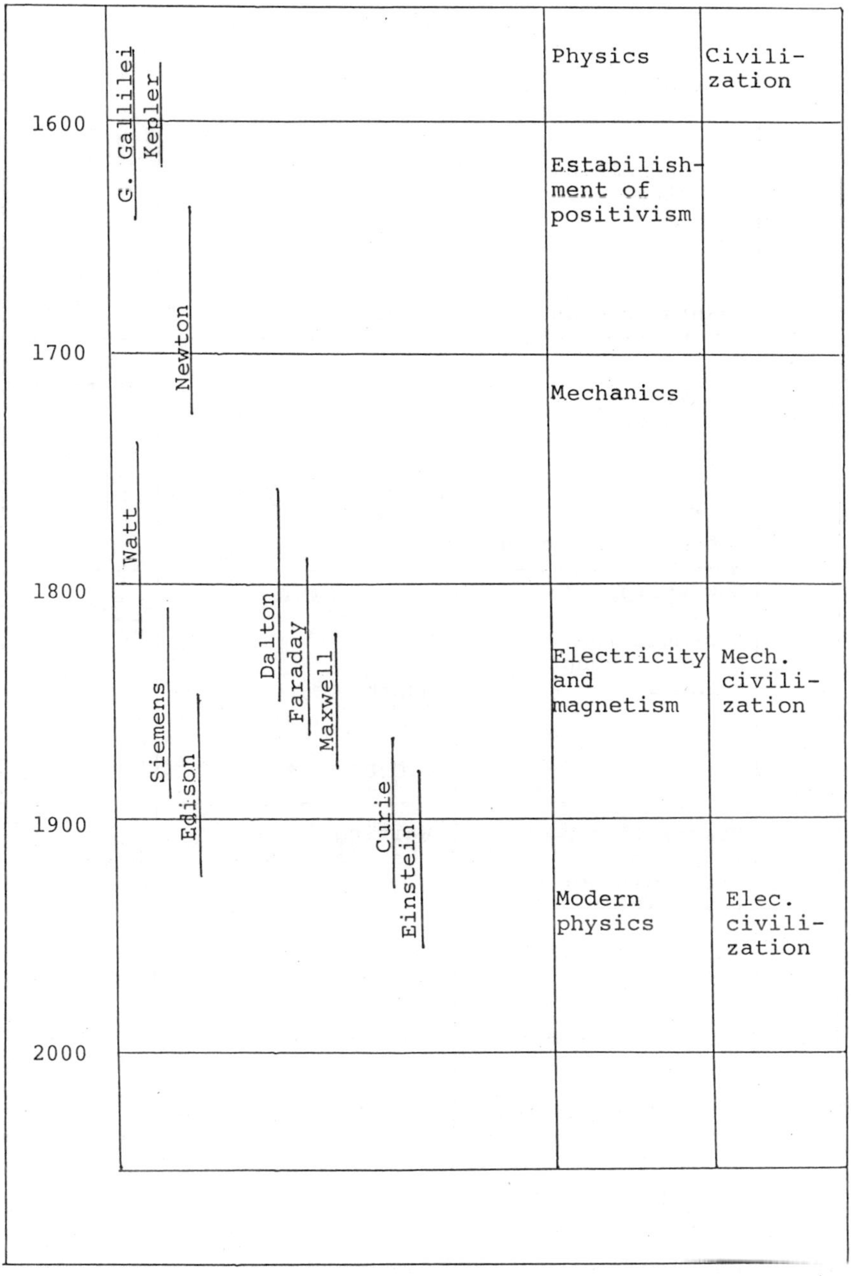

Fig. I Development of physics and civilization in the last 400 years.

Year			Leaders (typical)
	Beginning of Modern Physics		
1900	Surprising discoveries		England Rutherford
	Quantum Mechanics		
1925	Victory of new theories		Germany Heisenberg
	Social Recognition		
1950	Low energy Accumulation of vast amount of data	High energy Zoo to quark Q.C.D	U.S.A. AEC
1975	Preparation to become the basis for next civilization	Waiting for the final theory?	
2000			

Fig. II Development of modern physics between 1900 and 2000

	Mentality	Motto of the time	
1925	nobilities	No comment	high
1931.25	old liberalists	Is the united states of the World possible?	good
1937.5	bureau-crats	You can not take me to the army. I have tuberculosis.	bad
1943.75	samurais	I could have died that time.	high
1950	academists	No matter what you do, you can not have good life. Why not enjoy academic life.	good
1956.25	opportun-ists	Science is a good profession.	bad
1962.5	'68 students	I do physics, but not in order to write papers.	high
1968.75	disenchant-ment	What a sense to make such a pose?	good
1975	fear	Physics might be useful any time.	bad
1981.25	new generation	(Please make the motto of this time by yourself!)	(high?)
1987.5			

Fig. III Change of motivation from 1925 to 1987.5.

PHENOMENOLOGICAL THEORIES OF THE ELECTROMAGNETIC STRUCTURE OF NUCLEAR MATTER

Robert G. Sachs

It is an honor and a pleasure to have this opportunity to contribute to the commemoration of Professor Hironari Miyazawa's important work in nuclear and particle physics. Professor Miyazawa's work has covered a broad range of topics and has been characterized by, among other things, its bearing on timely issues of physics. He has dealt with the relationship between the fundamental theoretical concepts as they have developed over the years and the physical phenomena as determined by experimental measurements at the time. In this he has made a valuable contribution to the progress of theoretical physics toward the objective of formulating fundamental theories that can meet the test of successful confrontation with experiment.

One thread that has run through much of his work has been concerned with the way in which measured electromagnetic properties of nuclei and of the nucleon may provide information both about their structure and about the origins of the nuclear interactions. Since a parallel thread has run through much of my own earlier work and our efforts in this direction had much in common, I have chosen this occasion to present a reminiscence about work on this subject in which we had some common interest. Although many of the specific results presented here have given way to an interpretation in terms of quark models, the methods used and their motivations continue to be useful.

I shall present a few examples of the way in which these methods bear on current problems to illustrate the point.

1. PHENOMENOLOGICAL THEORIES OF ELECTROMAGNETIC PROPERTIES

Two distinct approaches have characterized attempts to compare theories of nuclear structure with experiment. One of them makes use of a specific theory to calculate the property to be measured. In this approach to nuclear problems, the calculation must be based on some approximation method and on a nuclear model. Therefore it is not always clear whether the measurement of the calculated quantity provides a test of the model, a test of the approximation used, or a test of the specific features of the theory.

The other approach requires only the most general assumptions about the theory, for example, that the nuclear system is described by quantum mechanics and is subject to the usual conservation laws and boundary conditions. The calculation of the quantity to be measured is then carried out in terms of the most general wave function or state vector that is consistent with these conditions. Then the comparison of the theory with measurements is used to place constraints on acceptable wave functions or state vectors. This process, which I refer to as a "phenomenological theory," defines a whole class of states of the system any one of which is consistent with the measurement. It does not involve a dynamical calculation of the state but it provides a constraint on what must be the outcome of any dynamical calculation in order that it lead to agreement with the measurement, so that it can be a very useful tool for testing specific theories and models.

In some cases it may turn out that the phenomenological theory shows the measured quantity to be dependent only on the grossest features of a model, such as the distribution of spin and orbital angular momentum. Then any model that is consistent with these features will lead to agreement with the measurements so that the measurements do not provide a useful test of a specific model.

An example of the way in which agreement with a specific model can be misleading may be taken from the early days of nuclear physics when the independent particle ("Hartree") model was used to fit data on nuclear ground states without any clear justification. The success of the model in obtaining a reasonably good fit to measured magnetic moments encouraged this use. But magnetic moments depend primarily on the distribution of angular momentum and not on other details of the model and it turned out that the alpha particle model, corresponding to an entirely different form of wave function, could give just as good an account of the magnetic moments.[1]

Of course later developments made it possible to overcome some of this ambiguity; strong support for the Mayer-Jensen shell model was provided by the Schmidt diagram of nuclear moments. However the shell model in its pure form was not capable of accounting for deviations of the measured moments from the Schmidt lines and other properties of the nuclear states had to be considered.[2] In one of his earliest papers,[3] Miyazawa contributed to this subject by showing that the effect of the exclusion principal on the calculation of the moments would tend to shift them in the right direction.

An early example of the general constraints that may be placed on a wave function by the phenomenological theory occurred as a result of the discovery of the quadrupole moment of the deuteron. The implication was that the ground state deuteron wave function must be a mixture of S and D states. No model that was limited to only central forces between the neutron and proton could lead to such a function. Therefore it was necessary to introduce the tensor force as an important contributor to the theory of nuclei. But application of the mirror theorem to the measured magnetic moment (see below) made it possible to determine the amount of D state without reference to this interaction. This was one of many useful applications of the principal underlying the mirror property which is typical of phenomenological theories.

The measured electromagnetic moments of nuclei are good examples for application of phenomenological theory because the electromagnetic interactions are so much smaller than the specific nuclear (strong) interactions that their effect on the nuclear state is negligible. Therefore the measurement does not change the characteristics of the nuclear states that are being determined.

The phenomenological theory and the more specific dynamical theory are of course not mutually exclusive. Miyazawa has been particularly adept at using the combination of the two as a means for testing models of nuclear forces based on pseudoscalar meson theory.

2. MIRROR THEOREMS FOR MAGNETIC MOMENTS

The mirror theorems display the broadest features of phenomenological theory. For nuclear systems they are based on the concept of charge symmetry of nuclear forces, that is, the notion that the forces are not altered by interchanging neutrons and protons. Charge symmetry implies that a wave function of a nucleus made up of Z protons and N neutrons is transformed into the corresponding wave function of the "mirror" nucleus having N protons and Z neutrons by simply identifying the Z(N) coordinate and spin variables of the protons (neutrons) in the one with the neutron (proton) variables in the other. Another way of saying this is in the language of isotopic spin I: the wave functions for given I and for $I_3 = \pm|I_3|$ have the same space and spin dependence. The effect on the wave function of coulomb repulsion between protons is being neglected here, so that the symmetry can be expected to apply only to light nuclei.

If the magnetic moment operator (observable) is the sum of orbital and spin moments of the nucleons, this mirror symmetry implies that the isoscalar magnetic moment, the sum of the moments of the mirror pair, depends only on a known linear combination of the quantities $|C(J,L,S)|^2$, where C(J,L,S) is the amplitude for a state of total orbital angular momentum L and total spin angular momentum S in the wave function of a nucleus of total angular momentum J.[4] Thus from the sum of the measured moments it is possible to determine a linear combination of these probabilities, providing a severe constraint on the wave function that must apply to every nuclear model.

A particularly fruitful case is that of the self-mirrored nuclei, those with $Z = N$, because application of the mirror theorem in that case requires the measurement of only one magnetic moment. The best known case is that of the deuteron for which the known (measured) value of $J = 1$ so that only the $^3S_1$ ($L = 0$, $S = 1$) and $^3D_1$ ($L = 2$, $S = 1$) states are accessible. From the measured magnetic moment the mirror theorem yields the well-known D-state probability, $|C(1,2,1)|^2 \simeq 0.04$.

The mirror theorem is exact only in the non-relativistic limit. Since the average kinetic energy of a nucleon in a nucleus may be of the order of a few percent of the nucleon mass, substantial corrections to the non-relativistic results may be anticipated. That relativistic corrections to the magnetic moment of a Dirac particle moving in a fixed central field can be calculated by use of phenomenological theory was shown as early as 1928 by Gregory Breit[5] who also pointed out that in making such corrections in general one must be careful to specify the properties of the central potential under Lorentz transformations. Margenau[6] and Caldirola[7] applied these methods to estimate the corrections for the deuteron but their treatment tacitly assumed that the nuclear potential arises from a vector field and they also neglected important recoil effects. Breit[8] and Sachs[9] extended the phenomenological methods to take recoil effects (two body system versus one body in a fixed field) into account and to nuclear potentials arising from fields that are either scalars or vectors under proper Lorentz transformations. Miyazawa[10] considered another correction to the mirror theorem, that resulting from the change in the isoscalar anomalous moment of one nucleon due to the pion field produced by the other nucleon. His calculation was carried out by applying perturbation methods to the lowest non-vanishing order of a cutoff pseudoscalar meson theory. Since the same methods applied to the calculation of the anomalous moments of free nucleons were unsuccessful, there remained some question about how to interpret Miyazawa's results.

As it happens, at about the same time I attempted to develop a phenomenological meson theory of nucleon structure by considering the nucleon state to be a linear combination of states of the pion field including 0, 1 and 2 pions.[11] From angular momentum conservation it follows that the pion field must be in either an S state (L = 0) or a P state (L = 1). The assumption of isotopic spin invariance again leads to a mirror theorem for the sum of proton and neutron moments and therefore determines the P-state probability from the isoscalar anomalous moment. This probability is about nine percent.

One can then ask whether it is possible to account for the isovector anomalous moments within this limitation on the pion fields. The answer is that a nucleon state containing at least two pions is required to account for the isovector moment, the nine percent probability for a one pion state (a single pion state must be a P state to conserve parity) is much too small to produce the anomalous moments. These two pions must be highly correlated since their important contribution to the moments arises from $\pi^+\pi^-$ annihilation. It was only later, when the existence of the ρ meson was suggested by Frazer and Fulco[12] on other grounds, that the possible connection of this correlation with the ρ was recognized.

Again Miyazawa[13] examined the basis of the mirror theorem as it applies to the isoscalar nucleon moment and pointed out that deviations could be expected due to the contribution of nucleon-antinucleon (virtual) pair states that are to be associated with the proper field of a nucleon. He found that the corrected value of the maximum one pion state probability is about 15 percent, which is an improvement but is still not large enough to account for the isovector moment without introducing the two pion state.

These results explain why the lowest order finite source pseudoscalar meson theory does not account for the isoscalar moment: Only the one pion state is produced in this approximation and it occurs with much too high a probability because of the strong coupling.

Miyazawa[14] showed that a semi-phenomenological method of Chew and Low relating the theoretical calculation of the moments in the meson theory to measurable pion-nucleon scattering cross sections leads to good agreement with the isovector moment but not with the isoscalar moment. It has been pointed out[15,16] that the agreement with the isovector moment is probably fortuitous because nucleon recoil effects that have been neglected can be shown to be large. The results in any case are subject to the limitation that the finite source theory is not gauge invariant in the form that was used. The modifications required[16] to establish gauge invariance modify the form of the magnetic moment operators, a subject to be discussed later.

Except for its methodological value, the above discussion of the nucleon moment is highly academic relative to the current view based on the quark model of nucleons. However in the non-relativistic constituent quark model a mirror theorem still applies to the isoscalar nucleon moment[17] because the proton and neutron are analogous to the mirror nuclei ^{3}He and ^{3}H with protons and neutrons replaced by up and down quarks respectively. In this case the allowed angular momentum states are, in addition to the dominant ^{2}S state, a ^{2}P, a ^{4}P and a ^{4}D state. Therefore the observed isoscalar nucleon moment simply establishes a linear relationship among the probabilities of these three states, which is not very restrictive.

This approach has been extended to the treatment of the magnetic moments of the baryon octet,[18] seven of which have been measured. Comparison with the most recent measurements has led Brekke[19] to conclude that the only modification of the SU(6), S state dominated quark model that is required to fit the data is the introduction of small anomalous quark moments. The significance of this success of the nonrelativistic quark model in accounting for the baryon moments is that the moments themselves are not sensitive to the details of the basic theory. Distinguishing fine points of the theory will require much higher precision in the measurements of the magnetic moments as well as other high precision information concerning the structure of at least some of the baryons, such as their electromagnetic form factors.

3. EXCHANGE MOMENTS AND GAUGE INVARIANCE

The application of the mirror theorem to the mirror nuclei ^3H and ^3He provided the first experimental indication that the magnetic moments of the nucleons in nuclei are not simply additive one-particle operators but that the magnetic moment operator includes two-body or many-body terms. The allowed ground states of these spin 1/2 nuclei are the ^2S, ^2P, ^4P and ^4D states. The reasonable expectation is that the ^2S state is dominant, because of the higher kinetic energy associated with the other states. If the nuclei are in the simplest (lowest kinetic energy) of two possible ^2S states and the moments are additive, then their values for ^3H and ^3He should be the same as those of the proton and neutron respectively. Their measured values differ from these values by only a few percent. Application of the mirror theorem provides a measure of the amount of other states needed to account for this small difference. It turns out that a four percent probability for the ^4D state, as might be expected from experience with the deuteron, is all that is required. However it also turns out that a D state probability of this magnitude leads to a large discrepancy between the calculated and measured isovector moments of the ^2H, ^3He system. In fact the agreement between isoscalar and isovector moments can be improved only by including the ^2P and ^4P states with very large probabilities, of the order of 20 percent for each. This seems so unreasonable from the point of view of nuclear physics that it suggested the existence of multibody contributions to the magnetic moments.[20]

Many years earlier Siegert and then Pais[21] had suggested the possible existence of such contributions on the basis of the assumption that exchange forces between nucleons were due to the exchange of some kind of charged field. Of course it was immediately clear that Yukawa's theory would lead to an exchange moment and that the measured moments of ^3H and ^3He might provide both evidence for the existence of mesons and a test of the various forms of the theory extant at the time. It is to be kept in mind that the pion had not yet been observed so that its existence was still in question at the time these moments were measured. F. Villars[22] was the first to show that the lowest order pseudoscalar meson theory leads to an exchange moment of the correct order of magnitude. In carrying out such calculations, the assumption of isotopic spin invariance implies that the exchange moments are isovector in character and therefore do not contribute to the mirror relations.

Before the ^3H or ^3He moment had been measured I had the opportunity to discuss with Fermi the question of whether their precise measurement might be a way to establish the existence of mesons in nuclei. He suggested that such evidence would not be decisive because he suspected that the mere existence of an exchange potential between neutrons and protons would give rise to an exchange moment, independently of the existence or non-existence of mesons.

Fermi's remark led me to embark on a study of the effect of exchange potentials on electromagnetic interactions.[23] I found that the requirement of electromagnetic gauge invariance on the nuclear hamiltonian did indeed lead to the existence of exchange moments to be associated with a phenomenological exchange potential. This procedure formed the basis for the development of a phenomenological theory of the electromagnetic properties of nuclei, providing a minimal approach to the interpretation of the data. By "minimal" I mean an approach requiring the least detailed assumptions about the underlying theory. One of its implications is that the study of these phenomena cannot in themselves provide detailed information about the theory at a basic level. It can provide information on the consistency of electromagnetic phenomena with the effective (phenomenological) potentials used to account for other nuclear properties.

This phenomenological theory based on the requirement of gauge invariance led to many other interesting results applicable not only to nuclear physics but also to atomic and molecular physics.[24]

Among the most interesting general result is that which we called the "Siegert Theorem" because a particular example of it was suggested by Siegert in his discussion of exchange currents.[21] The statement of the theorem is as follows: The calculation of the matrix elements of the electric multipole moments of a system depends only on the form of the wave functions. This means that, in contrast to the case of magnetic moments, the details of the nature of the forces is irrelevant except insofar as they determine the form of the wave functions.

From the viewpoint of phenomenological theory this is a very powerful tool. It means that from measurements of electric multipole moments and transitions it is possible to determine general charateristics of the wave functions which describe the structure of the system. Thus this is a large extension of the kind of opportunity provided by the mirror theorem for magnetic moments.

4. ELECTROMAGNETIC FORM FACTORS OF THE NUCLEON

In addition to the measurement of the magnetic moments, the measurement of the interaction of slow neutrons with electrons was an early source of information concerning the structure of the nucleon. This interaction is related directly to the mean square radius $<R^2>$ of the neutron charge distribution.

The phenomenological theory of nucleon structure[11] suggests that $<R^2>$ determined in this way is just the mean square radius of the distribution of the negative pion state in the virtual field of a neutron fixed at the origin. However, Foldy[25] developed a relativistic phenomenological theory of the neutron-electron interaction by making use of an effective, relativistically covariant Dirac Hamiltonian. The resulting interaction included an effect of the Pauli magnetic moment term which had been inserted to account for the anomalous neutron moment. He found, surprisingly, that by inserting the known magnetic moment he could account for almost all of the measured electron-neutron interaction and concluded that $<R^2>$ must be quite small.

Calculations[26] of $<R^2>$ for the pion cloud by means of lowest order pseudoscalar meson theory in the static approximation led, however, to a value of $<R^2>$ that was much too large. An attempt was then made[27] to determine the recoil corrections to this static approximation by making use of the same semi-phenomenological method that Miyazawa had used to calculate the magnetic moments. The resulting value of $<R^2>$ was still much too large.

The origin of this discrepancy turned out to be a misinterpretation of Foldy's result. If one could calculate correctly the mean square radius of the charge, including both the charge on the π^- and the charge on the recoiling proton, the Foldy term proportional to the anamolous moment would be included. This was demonstrated[16] by making use of Salzman's[26] generalization of Foldy's phenomenological neutron-electron interaction Hamiltonian to express its effect in terms of two electromagnetic form factors, a Dirac form factor $F_1(q^2)$, proportional to the minimal electromagnetic coupling of a Dirac particle, and a Pauli form factor $F_2(q^2)$, proportional to the Pauli anomalous moment. Here, q is the four momentum transfer between electron and nucleon. It was shown that the Fourier transform of the charge distribution is determined by a linear combination of $F_1(|\vec{q}|^2)$ and $F_2(|\vec{q}|^2)$ measured in the Breit reference frame for which $q_4 = 0$. A correct mesonic calculation would yield a value of $<R^2>$ based on this combination, which includes the Foldy term. Of course, an approximation to pseudoscalar meson theory that does not yield a correct value of the anomalous moment cannot be expected to yield a correct value of $<R^2>$ under these circumstances.

The form factors became a powerful means for phenomenological analysis of the elastic scattering of high energy electrons from nucleons and the resulting measurements for high values of q^2 provide detailed information concerning the electric charge and current structure of the nucleon.[28]

Before leaving the subject, I would like to comment on one more source of information concerning the "static" electromagnetic structure of the nucleon, that is the mass difference between proton and neutron. Under the assumption that this mass difference is electromagnetic in origin, it provides another measure of charge and current distributions. In fact, the phenomenological theory leads to very simple results. If one assumes that the nucleon structure involves only the one pion state it follows that the proton must be heavier than the neutron because the dominant contribution to the electromagnetic energy is the electrostatic attraction in the neutron between the virtual π^- and the recoiling proton. This is in agreement with the usual classical argument that the charged body must be the heavier because of the associated electrostatic energy.

However, if we impose the requirement that there be two pions in the field, as suggested by the phenomenological theory of the nucleon magnetic moment, then the observed mass difference can easily be explained. It is only necessary to use the same electromagnetic annihilation of the two pion state that accounts for the magnetic moment to obtain a contribution to the self-energy of the proton that lowers its mass.[29] As mentioned earlier, the π-π correlation required to obtain this result is equivalent to the introduction of the ρ meson.

It is interesting that, by using an entirely different approach, Miyazawa, Oi, and Suzuki[30] also found that the contribution of the ρ meson to the electromagnetic energy is needed to account for the neutron-proton mass difference within the context of the phenomenological theory.

5. ELECTRIC DIPOLE MOMENT OF THE NUCLEON

The fact that the Fourier transform of the charge density is given by a linear combination

$$F_{ch}(|\vec{q}|^2) = F_1 - (q^2/2M)F_2$$

in the Breit frame implies that all of the electric moments of the nucleon may be obtained from tensors formed by taking derivatives of F_{ch} with respect to the components of $i\vec{q}$ and evaluating at $\vec{q} = 0$.[31] Note that this statement is based on the use of the Siegert theorem: The electric moment tensors are given by the expectation value (average over the charge distribution) of the sum of tensors formed from the coordinates of each particle multiplied by its charge. There are no exchange contributions to electric moments. The quantity $|R|^2$ is one such moment, it is the trace of the tensor of second rank and is equal to

$$-[\nabla^2 F_{ch}(|\vec{q}|^2)]_{|\vec{q}|^2=0} = -6\, F'_{ch}(0),$$

where the prime denotes differentiation with respect to $|\vec{q}|^2$.

The electric dipole moment of the nucleon, especially the neutron, is of considerable interest because it provides a measure of the violation of time reversal (T). One can ask, in the spirit of the phenomenological theory, what information concerning the structure of the nucleon could be obtained from the measurement of a non-zero dipole moment?

A calculation based on F_{ch}, given above, will always yield a vanishing electric dipole moment because the form factors F_1 and F_2 are those allowed by the requirements of parity (P) conservation and T invariance, as well as by Lorentz invariance and gauge invariance.[32] In order to obtain a finite result it must be assumed that, due to P and T violation of the weak interactions, there exists another form factor for the nucleon which may be called G_2^5. It has the form of the axial vector weak interaction form factor that is disallowed by T invariance.[33]

Since G_2^5 is due to changes in the charge distribution caused by the P and T violating aspects of the weak interactions it is expected to be very small compared to F_{ch}. Therefore the electric dipole moment, which turns out to be just $G_2^5(0)$, is also expected to be very small. However its measurement would provide, in the manner of the phenomenological theory, an indication of the way in which the wave function of the nucleon must be modified by the combined P and T violation of the weak interactions in order to account for the observed value (if it is not zero).

References

1. R.G. Sachs, Phys. Rev. 55, 825 (1939).
2. R.G. Sachs, Nuclear Theory, Addison-Wesley Publishing Co. (1953) pp. 254-57.
3. H. Miyazawa, Progr. Theor. Phys. 6, 263 (1951).
4. Reference 2, p. 247.
5. G. Breit, Nature 122, 649 (1928).
6. H. Margenau, Phys. Rev. 57, 383 (1940).
7. P. Caldirola, Phys. Rev. 69, 608 (1946).
8. G. Breit, Phys. Rev. 71, 400 (1947). There is an error in this paper that is corrected in G. Breit and R.M. Thaler, Phys. Rev. 89, 1177 (1953).
9. R.G. Sachs, Phys. Rev. 72, 91 (1947).
10. H. Miyazawa, Progr. Theor. Phys. 7, 207 (1952).
11. R.G. Sachs, Phys. Rev. 87, 1100 (1952).
12. W.R. Frazer and J.R. Fulco, Phys. Rev. Lett. 2, 365 (1959).
13. H. Miyazawa, Phys. Rev. 97, 1413 (1955).
14. H. Miyazawa, Phys. Rev. 101, 1564 (1956).
15. J.P. Walecka, Nuovo Cimento 11, 821 (1959).
16. F.J. Ernst, R.G. Sachs, K.C. Wali, Phys. Rev. 119, 1105 (1960).
17. R.G. Sachs, Phys. Rev. D23, 1148 (1981).
18. L. Brekke and R.G. Sachs, Phys. Rev. D28, 1178 (1983).
19. L. Brekke, paper submitted to Phys. Rev. D (1988).
20. For further details see Sections 8.3, 9.2 and 9.3 of reference 2.
21. A.J.F. Siegert, Phys. Rev. 52, 787 (1937). Siegert used the adiabatic approximation to the beta decay theory available to him at the time. A. Pais, Kgl. Danske Videnskab. Selskab Mat.-fys. Medd. 20, No. 17 (1943) made use of Yukawa's meson field theory.
22. F. Villars, Helv. Phys. Acta 20, 476 (1947).
23. R.G. Sachs, Phys. Rev. 74, 433 (1948).
24. R.G. Sachs and N. Austern, Phys. Rev. 81, 705 (1951). See also N. Austern and R.G. Sachs, Phys. Rev. 81, 710 (1951).

25. L.L. Foldy, Phys. Rev. 83, 688 (1951); 87, 688 (1952); 87, 693 (1952).
26. G. Salzman, Phys. Rev. 99, 973 (1955).
27. S.B. Treiman and R.G. Sachs, Phys. Rev. 103, 435 (1956).
28. The story of the measurement of these form factors by R. Hofstader is presented in Pions and Quarks: Particle Physics in the 1950's, L. Brown, M. Dresden and L. Hoddeson, editors, Cambridge University Press, 1988. Somewhat more detailed comments on the history of their interpretation by R.G. Sachs and K.C. Wali also appear there.
29. W.G. Holladay and R.G. Sachs, Phys. Rev. 96, 810 (1954).
30. H. Miyazawa, Y. Oi and M. Suzuki, Prog. Theo. Phys. Supplement Extra Number, 436 (1965).
31. See Appendix II of R.G. Sachs, Phys. Rev. 126, 2256 (1962).
32. See Section 8.4 of R.G. Sachs, The Physics of Time Reversal, University of Chicago Press (1987).
33. See Section 8.5 of reference 31.

THE NUCLEONS AND MESONS

Yasuo Hara

Institute of Physics, University of Tsukuba, Tsukuba,
Ibaraki 305
Japan

ABSTRACT

Theoretical investigations on the nucleons and their isobars, based on the static strong coupling model, static moderately-weak coupling model, nonrelativistic quark model, Skyrme model and Monte Carlo simulation of lattice QCD, are reviewed.

1. FROM THE INVENTION OF THE PIONS TO THE CONFIRMATION OF Δ(1232)

In 1935 charged pions (π^+ and π^-) were invented by Yukawa as particles which mediate the nuclear force.[1] In addition the neutral pion, which decays into photons, was predicted[2] in order to explain charge independence of the nuclear force.[3]

The meson theory was applied to various problems; the nuclear force, the magnetic moments of the nucleons, pion-nucleon scattering, cosmic ray phenomena etc. As approximation methods the strong coupling model,[4),5)] the intermediate coupling model[6] (which includes both weak and strong coupling approximation as limiting cases) as well as the weak coupling method were used.

It should be mentioned that in the strong and intermediate coupling models it is essential to introduce a cutoff factor, i.e., the source function $\rho(x)$, into the pion-nucleon interaction Hamiltonian,

$$H_{int} = (f/m_\pi) \int \rho(x)\psi^\dagger(\vec{x})\, \tau_\alpha \vec{\sigma}\psi(\vec{x})\vec{\nabla}\, \phi_\alpha(\vec{x})d^3x. \qquad (1)$$

The physical meaning of the factor was not clear at that time.

It should also be mentioned that in the static limit of the strong coupling model with the interaction Hamiltonian, nucleon isobars, i.e., excited states of the nucleons with I=J=3/2, 5/2, 7/2, ··· and with mass $m=I(I+1)/2\lambda=J(J+1)/2\lambda$, was predicted by Pauli and Dancoff.[5] The rotational spectra is due to the fact that the isospin and spin of the nucleon are frozen in the directions of the charge and momentum of the pion which is strongly coupled to the nucleon, respectively, and the fact that they rotate like tops with the moment of inertia, λ, which is $O(1/f^2)$.

However, quantitative understanding of the pion-nucleon interaction was not obtained since the muons which were discovered[7] in 1936 were misidentified as the pions.

The charged pions were discovered in cosmic rays[8] in 1947, and the charged and neutral pions were produced by a synchrocyclotron at Berkeley in 1948[9] and 1950,[10),11] respectively.

As soon as candidates for the neutral pions which decayed into two photons were produced, by the synchrotron X ray beam of 330 MeV maximum energy, with cross sections similar to those for the charged pions and with an excitation function flatterning off for photon energies above 250 MeV,[10),11] Miyazawa and Fujimoto[12] pionted out that these experimental results suggested the existence of a proton isobar, i.e., an excited state of the proton, and showed that these results were reproduced by identifying the isobar with the I=J=3/2 state predicted by the strong coupling theory.

The I=J=3/2 resonance now called Δ(1232) was not confirmed until pion-nucleon scattering experiments[13),14] and phase-shift analysis of these experimental results[13),15] were carried out. However, there were still ambiguities in the phase-shift analysis.

In 1955 the dispersion relation, which was originally derived by Kroning[16] and Kramers[17] between the real and imaginary parts of the forward scattering amplitude for the scattering of light on atoms from the causality condition that the scattered wave should have zero amplitude until the incident wave reaches the scatterer, was extended to the scattering of massless spin 0 bosons on fermions with spin 1/2 by

Goldberger.[18)]

The derivation of the dispersion relation used the formalism of quantum field theory and the causality condition that the commutator of two Heisenberg operators for the boson fields taken at space-like points should vanish.[18)]

The boson-fermion dispersion relations were cast into a form suitable for comparison with the pion-nucleon scattering experiments by Goldberger, Miyazawa and Oehme.[19)] By making use of the dispersion relations, the following results were obtained;

(1) the magnitude of the πN coupling constant,[20)] $f^2/4\pi=0.08$,

(2) correct values for the s-wave phase-shifts[21)] δ_1 and δ_3,

(3) removal of phase-shift ambiguity and confirmation of the I=J=3/2 resonance.[21)]

2. THE STATIC MODEL ON P-WAVE PION-NUCLEON SCATTERING AT LOW ENERGIES

A successful theory of low energy pion-nucleon interaction was proposed by Chew and Low.[22)] In this static theory, in which the nucleon recoil is negledcted, the amplitude for $\pi(p)+N \to \pi(q)+N$ scattering is expressed as

$$T_q(p) = - \sum_n \left[\frac{T_p^\dagger(n) T_q(n)}{E_n - \omega_p - i\varepsilon} + \frac{T_q^\dagger(n) T_p(n)}{E_n + \omega_p} \right] , \tag{2}$$

where

$$T_k(n) = (\Psi_n^{out}, \, i f^{(0)} (\vec{\sigma}\cdot\vec{k}) \tau_k \rho(k)/(m_\pi \sqrt{2\omega_k}) \Psi_0). \tag{3}$$

If we use the one-pion approximation in which only "one-nucleon" and "one-nucleon and one-pion" intermediate states are retained, we can express low energy pion-nucleon scattering amplitudes in terms of the pion-nucleon coupling constant f and the cutoff factor $\rho(k)$ using Eq.(2) and the unitarity relation. In this theory it is possible to obtain rough quantitative agreement with experimental data on p-wave pion-nucleon scattering[22)] and photo-pion effect[22)] for $E_\pi \lesssim 300$ MeV and the isovector part of the anomalous magnetic moments of the nucleons[23)] if the coupling constant $f^2/4\pi$=0.08 and the cutoff momentum $K \approx 6m_\pi$ are used.

It is possible to use the amplitudes (2), which reproduce p-wave pion-nucleon scattering experiments for $E_\pi \leq 300$ MeV fairly well, as off-mass-shell pion-nucleon scattering amplitudes.[24] By making use of the amplitudes, Miyazawa and collaborators derived two-pion-exchange potential,[24],[25] though contributions from two-pion resonances were not included.

The off-mass-shell amplitudes (2) were also used for the evaluation of two-pion exchange contributions to the absorptive parts of the nucleon-nucleon forward scattering amplitudes in the unphysical region.[26] The dispersion relations for nucleon-nucleon forward scattering amplitudes[27] with these absorptive parts were used for the analysis of low and intermediate energy nucleon-nucleon scattering and good agreement was obtained if $f^2/4\pi$=0.08±0.01 was used.[26]

Therefore, low energy pion-nucleon system and two-nucleon system were satisfactorily understood based on the amplitude (2), though there was an important exception, i.e., low energy s-wave pion-nucleon scattering. To understant this process we have to include the contribution from the ρ-mesons exchanged in the t-channel.[28]

In the moderately-weak coupling static model the bootstrap condition between πNN coupling constant $3f^2 = \gamma_{11}$ and $\pi\Delta N$ coupling constant γ_{33},

$$\gamma_{IJ} = \sum_{I'J'} \alpha_{II'} \beta_{JJ'} \gamma_{I'J'} , \tag{4}$$

was derived by solving the unitarity relation and Eq.(2) using N/D method in which the driving force was due to the nucleons and $\Delta(1232)$ exchanged in the u-channel and in which the D-functions were approximated by linear functions of the pion energy.[29] In Eq.(4) α's and β's are the crossing matrices for spin and isospin, respectively.

In the static limit of the strong coupling model the bootstrap condition (4) and similar conditions among nucleon isobars were obtained[30],[31] from the unitarity relation and the fact that the mass differences among baryons Δm were $O(1/f^2)$. The bootstrap conditions in this case are the spectra generating algebra of isobars.[31]

From the bootstrap condition we can predict that expectation values of σ_j and τ_α between the nucleons vanish.

Analogous bootstrap conditions between SU(3) multiplets to which the nucleons and Δ(1232) belong were derived by Dashen[32] and Hara[33),34] and used to derive qualitative predictions on various properties of the nucleons and Δ(1232).[32-35] It was found that in the static model the agreement between experimental and theoretical results improved if the driving force due to the exchange of mesons in the t-channel was included.

The static model of pion-nucleon interaction cannot be applied to higher nucleon resonances such as N*(1520) and N*(1680) since momenta of pions which produce these resonances $p_\pi > m_N/2$ and the recoil of nucleons cannot be neglected. Though there were proposals[36] that N*(1520) and N*(1680) were formed by strong attractive force due to ρN intermediate states, no quantitative justification was given.

From the studies on pion-nucleon interaction reviewed in this section it is clear that various properties of the nucleons and Δ(1232) are qualitatively understood in the static model. However, for quantitative understanding we have to include vector-mesons and other mesons into the model. As for higher nucleon resonances, even qualitative understanding was not obtained until 1964.

3. THE QUARK MODEL AND QCD

In 1964 the quark model was invented by Gell-mann[37] and Zweig.[38] In the quark model baryons are composed of three quarks and mesons are composed of a quark and an antiquark.

In the same year an SU(6) symmetric model, in which nonrelativistic quarks were assumed to interact with each other through central force independent on both spin and unitary spin, was proposed by Gürsey and Radicati[39] and Sakita.[40] In this model qualitative understanding of the mass spectra[39] and that of d/f ratios of Yukawa couplings of mesons and baryons were obtained,[41] and the ratio of the magnetic moments of the nucleons $\mu_p/\mu_n \approx -3/2$ was explained.[42]

In the SU(6) model even-parity baryons are SU(6) 56-plets and their radially and orbitally excited states with even L, and odd-parity baryons are SU(6) 70-plets with odd L, while mesons are SU(6) 35-plets + singlet and their orbitally and radially excited states.[43] These hadrons are considered to lie on Regge trajectories.[44]

The color degree of freedom was attached to the quarks in order to reconcile the Fermi statistics obeyed by the quarks with the totally symmetric wave functions for baryons in the SU(6) model.[45]

In is interesting to notice that the spectrum of the nucleon isobars contained in the totally symmetric quark model is given by

$$I = J = 1/2,\ 3/2,\ \cdots,\ N/2, \tag{5}$$

if we extend the SU(3) group associated with the color to SU(N) and if the orbital excitations are neglected.[46),47] (In SU(N) case the baryons are composed of N quarks.) Thus, the spectrum in the N → limit is the same as that in the strong coupling limit of the pseudoscalar meson theory.

The discovery of the SU(3) symmetry associated with the color degree of freedom, combined with the appeal of local gauge theories in which interaction is uniquely fixed, led to the proposal of quantum chromodynamics,[48] an asymptotic free local gauge theory[49] with gluons as gauge-bosons.

Though quarks and gluons have not been detected, jet-events have convinced us the existence of quarks and gluons confined inside hadrons.

Quark confinement has been proved through Monte Carlo simulation[50] of lattice QCD,[51] though it has not yet been proved analytically.

The masses of light hadrons have been calculated by using the Monte Carlo simulation of lattice QCD. By calculating the propagators of hadrons,

$$A(\tau) = \sum_{n} \langle P(n) P^{\dagger}(0) \rangle\ ,\ [\, n(\vec{n},\ \tau),\ 0(\vec{0},\ 0)\,] \tag{6}$$

in the quenched approximation to lattice QCD with a renormalization group improved lattice SU(3) gauge action and Wilson's quark action on a $16^3 \times 48$ lattice at $\beta(=6/g^2)=2.4$, Iwasaki, Yoshie and Itoh calculated the masses of the pion, ρ-meson, nucleon and Δ(1232).[52] In (6) P(n) is an annihilation operator of a hadron at site n. In this calculation hadron masses are expressed in terms of the hopping parameter K, which is related to the bare quark mass, and lattice spacing a. If K and a are determined by using m_π and m_ρ as inputs, the following values are obtained, m_N=900(180) MeV and m_Δ=1290(210) MeV, which should be compared with experimental values 940 and 1232 MeV, respectively. Though scaling of the results has not been checked, the value of the coupling constant =2.4 is believed to lie in the scaling region.

The success of the calculation in the quenched approximation in which quark loops are neglected may be related to OZI rule.[53]

However, it is still impossible to calculate the masses of higher resonances and matrix elements of various operators between hadrons because of the ability of available computors. As for the matrix elements, only decay constants responsible for leptonic decays[54),55)] such as $\pi\to\mu\nu$ and $\rho\to e^+e^-$ are calculable at present.

Therefore, for estimation of the masses of higher resonances and hadronic matrix elements, we have to use a nonrelativistic quark model at present.[56),57)]

4. THE SKYRME MODEL

In the previous section I have mentioned that there is a correspondence of the spectra of nucleon isobars between the strong coupling limit of the static pseudoscalar meson theory and the large N limit of the symmetric quark model.

In the large N limit, QCD becomes equivalent to an effective field theory of mesons,[58] and at low energies this theory reduces to a nonlinear sigma model of spontaneously broken chiral symmetry. The simplest choice of the model is the Skyrme model,[59]

$$L = -(1/4)F_\pi^2\mathrm{Tr}(\partial_\mu U\partial_\mu U^\dagger) + (1/32e^2)\mathrm{Tr}[(\partial_\mu U)U^\dagger,\ (\partial_\nu U)U^\dagger]^2, \qquad (7)$$

where F_π=93 MeV is the pion decay constant, e is a dimensionless parameter, and U is an SU(2) matrix,

$$U(\vec{x}) = \exp[i\phi_i(\vec{x})\tau_i/F_\pi] \ . \qquad (8)$$

In the Lagrangian (7) contributions from the ρ-mesons and other mesons are neglected.

In the Skyrme model the baryons are viewed as the solitons.[59),60)] The lowest energy solution with the topologically nontrivial U field configuration can be written as

$$U_0(\vec{x}) = \exp[iF(r)\hat{r}\cdot\vec{\tau}]. \qquad (9)$$

Since the soliton solution U_0 is not invariant to the action of the isospin $\vec{I}$ and the spin $\vec{J}$, we can obtain a family of soliton solutions,[61)] all degenerate with the solution U_0, by rotating U_0, $U = AU_0A^{-1}$, where $A=a_0+i\vec{a}\cdot\vec{\tau}$ with $a^2+|\vec{a}|^2=1$ is an arbitrary constant SU(2) matrix acting on $\vec{\tau}$ in U_0.

Now, A is treated as a set of collective coordinates, and the Lagrangian (7) and all physical variables are expressed in terms of a time dependent A(t). When the Lagrangian which is equal to $-M_0 + 2\lambda\,\Sigma(\dot{a}_\alpha)^2$ with the constraint $\Sigma\, a_\alpha^2=1$ is quantized, a series of nucleon isobars with I=J=1/2, 3/2, 5/2, ···, and with masses,

$$M = M_0 + (1/2\lambda)J(J+1) \ , \qquad (10)$$

are obtained. The mass spectra (10) is the same as that in the strong coupling limit of the pseudoscalar meson theory.[5)]

Independent on the Skyrme model, it has been shown that, in general, QCD in the large N limit is the same as the static limit in the strong coupling model by showing that the exact large N equations are identical to the bootstrap conditions in the static strong coupling model if the correct strong coupling model is chosen.[46)]

The collective coordinates A of the Skyrme model with the Lagrangian (7) have been shown[62)] to be proportional to the square-roots of the fields of pions bound to the bare nucleons,

$$\phi_{\alpha k} = \int \phi_\alpha(\vec{x})\partial\rho(x)/\partial x_k d^3x \ , \qquad (11)$$

in the strong coupling model of the pseudoscalar mesons.[5)]

Since resonances with the same internal quantum numbers are associated with a Regge trajectory ($|\Delta\vec{J}|=2$) and since no exotic resonances, such as resonances with (I,J)=(5/2,5/2), (7/2,7/2) etc., have been observed, apparently the collective coordinates $A=a_0+i\vec{a}\cdot\vec{\tau}$ are not good ones for higher nucleon resonances.

It has been found that we cannot improve the situation by including the ρ- and ω-mesons into the Skyrme model since the collective coordinates are still $A=a_0+i\vec{a}\cdot\vec{\tau}$ in these cases.[63)] We have to look for a soliton model in which excitations with $|\Delta\vec{I}|=0$ appear, i.e., we have to look for a strong coupling model which contains I=0 mesons such as f-mesons with J=2.

Through the investigation of the quark model and Skyrme model, the source function ρ(x) has been found to be closely related to the distribution of the quarks or that of the pions in the nucleon which emits and absorbes mesons.

5. CONCLUSIONS

In the previous sections we have reviewed theoretical investigations on the nucleons and their isobars based on the static strong coupling model, static moderately-weak coupling model, nonrelativistic quark model, Skyrme model and Monte Carlo simulation of lattice QCD. Through these investigations we have understood the static properties of the nucleons and Δ(1232). For example, their masses can be calculated by Monte Carlo simulation of lattice QCD, and the ratios of the magnetic moments and those of the meson-baryon coupling constants have been explained in the quark model.

Relations among the static strong coupling model, the large N limit of the symmetric quark model and the Skyrme model have been clarified, and now we have several alternative view points on the nucleons and Δ(1232).

As for higher nucleon resonances, we have only nonrelativistic quark model at present.

REFERENCES

1. Yukawa, H., Proc. Phys.-Math. Soc. Japan 17, 48 (1935).

2. Sakata, S. and Tanikawa, Y., Phys. Rev. 57, 548 (1940).

3. Yukawa, H., Sakata, S. and Taketani, M., Proc. Phys.-Math. Soc. Japan 20, 319 (1938),
Kemmer, N., Proc. Camb. Phyl. Soc. 34, 354 (1938),
Yukawa, H., Sakata, S., Kobayashi, M. and Taketani, M., Proc. Phys. -Math. Soc. Japan 20, 720 (1938).

4. Wentzel, G., Helv. Phys. Acta 13, 209 (1940); 14, 633 (1941),
Oppenheimer, R. and Schwinger, J., Phys. Rev. 60, 150 (1941).

5. Pauli, W. and Dancoff, S., Phys. Rev. 62, 85 (1942).

6. Tomonaga, S., Sci. Pap. I. P. C. P. 39, 247 (1941),
Miyazima, T. and Tomonaga, S., Sci. Pap, I. P. C. P. 40, 21 (1942).
See also, Tomonaga, S., Prog. Theor. Phys. 1, 83, 109 (1946); 2, 6, 63 (1947).

7. Anderson, C. D. and Neddermeyer, S., Phys. Rev. 50, 273 (1936).

8. Lattes, C.M.G., Occhialini, G.P.S. and Powell, C.F. , Nature 160, 453, 486 (1947).

9. Gardner, E. and Lattes, C.M.G., Science 109, 270 (1948).

10. Steinberger, J., Panofsky, W.K.H. and Steller, J., Phys. Rev. 78, 802 (1950).

11. Bishop, A.S., Steinberger, J. and Cook, L.J., Phys. Rev. 80, 291 (1950).

12. Fujimoto, Y. and Miyazawa, H., Prog. Theor. Phys. 5, 1052 (1950).

13. Anderson, H.L., Fermi, E., Martin, R. and Nagel, D.E., Phys. Rev. 91, 155 (1953).

14. Ashkin, J. et al., Phys. Rev. 93, 1129 (1954).

15. de Hoffmann, F., Metropolis, N. and Alei, E.F., Phys. Rev. 95, 1586 (1954), Beth, H.A. and de Hoffmann, F., "Meson and Fields", vol.2 (Row, Peterson, Evanston, 1955).

16. Kronig, R., J. Opt. Soc. Am. 12, 547 (1926).

17. Kramers, H.A., Atti. congr. intern. fisici Como 2, 545 (1927).

18. Goldberger, M.L., Phys. Rev. 99, 979 (1955).

19. Goldberger, M.L., Miyazawa, H. and Oehme, R., Phys. Rev. 99, 986 (1955).

20. Haber-Schaim, U., Phys. Rev. 104, 1113 (1956).

21. Anderson, H.L., Davidon, W.C. and Kruse, U.E., Phys. Rev. 100, 339 (1955).

22. Chew, G.F. and Low, F.E., Phys. Rev. 101, 1570, 1579 (1956).
See also, Chew, G.F., Phys. Rev. 95, 1669 (1954); Low, F.E., Phys. Rev. 97, 1392 (1955); Wick, G.C., Rev. Mod. Phys. 27, 339 (1955);

Chew, G.F., Goldberger, M.L., Low, F.E. and Nambu, Y., Phys. Rev. 106, 1337, 1345 (1957).

23. Miyazawa, H., Phys. Rev. 101, 1564 (1956).
24. Miyazawa, H., Phys. Rev. 104, 1741 (1956).
25. Konuma, M., Miyazawa, H. and Otsuki, S., Prog. Theor. Phys. 19, 17 (1958).
26. Hara, Y. and Miyazawa, H., Prog. Theor. Phys. 23, 942 (1960).
27. Matsuyama, S. and Miyazawa, H., Prog. Theor. Phys. 18, 328 (1957); 19, 517 (1958).
28. Weinberg, S., Phys. Rev. Lett. 17, 616 (1966), Tomozawa, Y., Nuovo Cimento 46A, 707 (1966).
29. Chew, G.F., Phys. Rev. Lett. 9, 233 (1962).
30. Abers, E.S., Balazs, L.A.P. and Hara, Y., Phys. Rev. 136, B1382 (1964).
31. Cook, T., Goebel, C.J. and Sakita, B., Phys. Rev. Lett. 15, 561 (1965).
32. Dashen, R.F., Phys. Lett. 11, 89 (1964).
33. Hara, Y., Phys. Rev. 135, B1079 (1964), Hara, Y., Phys. Rev. 137, B1553 (1965), Hara, Y., Phys. Rev. 144, 1241 (1966), Hara, Y., Phys. Rev. 164, 1844 (1967).
34. Goebel, C.J., Phys. Rev. Lett. 16, 1130 (1966).
35. Cutkosky, R. and Tarjanne, P., Phys. Rev. 132, 1355 (1963), Dashen, R.F. and Frautschi, S., Phys. Rev. Lett. 13, 497 (1964); Phys. Rev. 137, B1331 (1965).
36. Itabashi, K., Kato, M., Nakagawa, K. and Takeda, G., Prog. Theor. Phys. 24, 529 (1960).
37. Gell-Mann, M., Phys. Lett. 8, 214 (1964).
38. Zweig, G., CERN Preprint 8182/Th.401; 8419/Th.412 (1964) (unpublished).
39. Gürsey, F. and Radicati, L.A., Phys. Rev. Lett. 13, 173 (1964).
40. Sakita, B., Phys. Rev. 136, B1756 (1964).
41. Pais, A., Phys. Rev. Lett. 13, 175 (1964), Gürsey, F., Pais, A. and Radicati, L.A., Phys. Rev. Lett. 13, 299 (1964).
42. Beg, M.A.B., Lee, B.W. and Pais, A., Phys. Rev. Lett. 13, 514 (1964), Sakita, B., Phys. Rev. Lett. 13, 643 (1964).
43. As a review, see, for example, Rosner, J.L., Phys. Rep. 11c, 189 (1974).
44. Regge, T., Nuovo Cimento 14, 951 (1959); 18, 947(1960),

Chew, G.F. and Frautschi, S., Phys. Rev. 123, 1478 (1961).

45. Greenberg, O.W., Phys. Rev. Lett. 13, 598 (1965),
Miyamoto, Y., Prog. Theor. Phys. Suppl, extra no., 187 (1965),
Nambu, Y. and Han, M.Y., Phys. Rev. 139, B1006 (1965).

46. Gervais, J.-L. and Sakita, B., Phys. Rev. Lett. 52, 87 (1984); Phys. Rev. D30, 1795 (1984).

47. Bardakci, K., Nucl. Phys. B243, 197 (1984),
See, also, Manohar, A.V., Nucl. Phys. B248, 19 (1984).

48. Fritsch, H., Gell-Mann, M. and Leutwyler, H., Phys. Lett. 47B, 365 (1973).

49. Gross, D.J. and Wilczek, F., Phys. Rev. Lett. 30 1343 (1973),
Politzer, H.D., Phys. Rev. Lett. 30, 1346 (1973),
t'Hooft, G., (1972) (unpublished).

50. Creutz, M., Phys. Rev. D21, 2308 (1980).

51. Wilson, K.G., Phys. Rev. D10, 2445 (1974).

52. Itoh, S., Iwasaki, Y. and Yoshie, T., Phys. Rev. Lett. 183B, 351 (1987).

53. Okubo, S., Phys. Lett. 5, 165 (1963),
Zweig, G., ref. 38,
Iizuka, J., Prog. Theor. Phys. Suppl. 37-38, 21 (1966).

54. Itoh, S., Iwasaki, Y., Oyanagi, Y. and Yoshie, T., Nucl. Phys. B274, 33 (1986).

55. Hara, Y., Itoh, S., Iwasaki, Y. and Yoshie, T., Phys. Rev. D34, 3399 (1986).

56. De Rujule, A., Georgi, H. and Glashow, S.L., Phys. Rev. D12, 147 (1975).

57. Isgur, N. and Karl, G., Phys. Rev. D18, 4187 (1978); D19, 2653 (1979).

58. 't Hooft, G., Nucl. Phys. B72, 461 (1974): B75, 461, (1974).

59. Skyrme, T.H.R., Proc. Roy. Soc. A260, 127 (1961); Nucl. Phys. 31, 556 (1962).

60. Witten, E., Nucl. Phys. B223, 433 (1983).

61. Adkins, G.S., Nappi, C.R. and Witten, E., Nucl. Phys. B228, 552 (1983).

62. Hara, Y., Phys. Rev. D (to be published).

63. See, for example, Adkins, G. and Nappi, C., Phys. Lett. 137B, 251 (1984); Igarashi, Y., et al., Nucl. Phys. B259, 721 (1985); Fujiwara, T., et al., Prog. Theor. Phys. 73, 926 (1985).

PIONS IN NUCLEI: FROM VIRTUAL-PION EXCHANGE TO REAL-PION TRANSFER

Toshimitsu Yamazaki

Institute for Nuclear Study, University of Tokyo
Tanashi, Tokyo 188, Japan

Tracing the work of Miyazawa on nuclear magnetic moments, we discuss possible experimetal ways to see whether a real pion exists in nuclei or not. While virtual pions are known to play an important role in nuclei, as clarified experimentally from anomalous orbital g factors of nucleons in nuclei, nearly nothing is known for the behavior of real pions in nuclei. We have shown that deeply bound hybrid states of π^- are expected to exist in heavy nuclei, which can be populated by "pion transfer" reactions.

1. INTRODUCTION

The pion was originally predicted by Yukawa as a particle mediating the nuclear force. The pion was actually discovered and is produced abundantly in large accelerators. The pion-nucleon interaction, and then the pion-nucleus interaction have been studied extensively. The understanding of nucleon-nucleon interactions in terms of pions has been developed. While pions play important roles also in nuclei, it was widely believed that their role is implicit; namely, the role of pions is burried in the nuclear force so that nuclear physics does not need pions explicitly. Miyazawa, however, put a milestone in the development of nuclear physics, showing that nuclear magnetic moments are affected by pions in an explicit way [1], but this aspect had been discarded for many years until a strong experimental evidence was obtained in 1970 [2].

Nowadays, the pion is believed to be a composite particle, a (u,d) quark-antiquark pair. It is also believed to be a Nambu-Goldstone boson. The question as to what is the pion does not seem to be settled down. To solve this question we want to embed pions deep inside nuclei to study their behaviors experimentally and address the following questions; how do pions keep their free identities in nuclei?

In the present paper the author would like to begin with reviewing the Miyazawa effect on nuclear magnetic moments and then to proceed to discuss a new aspect of pion bound states and a new way to study them experimentally.

2. MIYAZAWA EFFECT ON NUCLEAR MAGNETIC MOMENTS

In 1951 Miyazawa published a paper on nuclear magnetic moments [1]. It had been known that the observed magnetic moments deviate from the Schmidt lines [3], and a dedicated effort was made by Miyazawa to explain this fact in terms of meson exchange effect.

The famous Schmidt estimate is the expectation value of the magnetic moment operator

$$\vec{\mu} = (g_\ell \vec{l} + g_s \vec{s})\mu_N \tag{1}$$

assuming the following two independent assumptions:

(A) single-particle model ($\vec{j} = \vec{\ell} + \vec{s}$) for the nuclear states of interest

and

(B1) free-nucleon g factors,

$$g_s(\mathrm{p}) = +5.586,$$
$$g_s(\mathrm{n}) = -3.826,$$

together with

(B2) $g_\ell(\mathrm{p}) = 1$ and $g_\ell(\mathrm{n}) = 0$.

Miyazawa studied how the assumptions (B1) and (B2) are modified in nuclei through the exchange of pions in nuclear environment.

The physics of nuclear moments is described in all text books of nuclear physics. For instance, one of the famous books on nuclear moments, written by Kopfermann [4], says,

"The experimentally observed μ_I values do not lie on, but between the Schmidt lines. This could be explained by assuming that the intrinsic (spin) magnetic moments of nucleons are smaller inside the nucleus than in the free state. (The g factors of the orbital momenta behave normally throughout atomic physics, so that one would not like to admit anomalies in the nucleus, such as deviations from the values $g_\ell = 1$ for protons and $g_\ell = 0$ for neutrons.)"

As represented in this description, a possible anomaly for the spin g factors was expected, because the free-nucleon magnetic moments themselves are already anomalous supposedly due to pion cloud (at that time this was believed to be only one source of the anomaly; there was no concept of quark constituents), but none for the orbital g factors. Miyazawa's famous paper was often referred to as a literature in which the effect of pion exchange on the spin g factors was studied. However, the empirical g_s factors deduced by assuming (A) vary from one nucleus to another, which appeared to be in contradiction with the meson exchange picture. A few years later, Arima and Horie [5] proposed a totally different explanation based on configuration mixing in nuclear wavefunctions; in other words, the assumption (A) was questioned and modified. This theory turned out to be very successful. Although, in principle, some meson exchange effect on the g_s factors might exist, this is "eaten up" by a more pronounced "nuclear core polarization" effect. So, people believed that the global feature of nuclear maganetic moments is explained by configuration mixing without introducing meson exchange effect. However, the most striking case, the ^{209}Bi ground-state moment, remained puzzling, though the shell-model wavefunction of this state was supposed to be best known.

A new era emerged after 20 years. First, the development of experimental methods made it possible to produce high-spin nuclear excited states which are often metastable so that their magnetic moments can be measured to high precision [6]. Among them the most striking was the magnetic moment of the 11^- isomeric state of 24 nsec half life in ^{210}Po [2]. Since this state has a relatively pure configuration of $h_{9/2}i_{13/2}$ of two protons coupled to the "stretch angular momentum" ($J = j_1 + j_2 = 11$), its magnetic moment is essentially an orbital magnetic moment. Specifically,

$$g(11^-) = g_\ell + \frac{1}{121}(g_s - g_\ell). \tag{2}$$

Thus, a measured magnetic moment should yield a g_ℓ value for proton nearly independent of others. The observed value, $g(11^-) = 1.107 \pm 0.019$, showed a significant anomaly in g_ℓ.

A more surprising fact was that such an anomaly in g_ℓ,

$$g_\ell(\mathrm{p}) = 0.1 \text{ and } g_\ell(\mathrm{n}) = -0.1, \tag{3}$$

had been predicted in Miyazawa's paper of two decades ago. The anomaly in g_ℓ is a rather unique effect of pion exchange current, as depicted in Fig.1. This produces a positive anomaly, $\delta g_\ell^{\mathrm{meson}}$, for proton and a negative anomaly for neutron. Since the pion is lighter than the nucleon, it contributes to an additional magnetic moment

$$\delta\mu = \frac{M_{\mathrm{P}}}{m_\pi}\mu_{\mathrm{N}} = 6.7\ \mu_{\mathrm{N}} \tag{4}$$

while it undergoes orbital motion. (The same enhancement is responsible for the anomalous magnetic moments of nucleons in the pion exchange model.) The g_ℓ anomaly is a genuine nuclear phenomena that involves exchange of pions among nucleons.

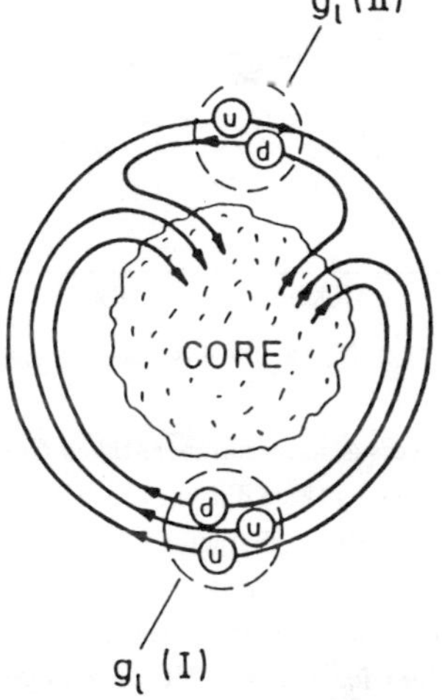

Fig.1 Quark diagram for orbital motion of proton. A proton is exchanged by a core neutron emitting a positive pion, which produces an additional δg_ℓ factor, $\delta g_\ell^{(\mathrm{II})}$, or $\delta g_\ell^{\mathrm{meson}}$. This diagram also shows the contribution of a possible deconfined quark cluster in a nucleus, whose electromagnetic coupling produces an effective nuclear magneton, or a correction $\delta g_\ell^{(\mathrm{I})}$.

One may suspect this interprotation, because nuclear states may not be described by simple shell-model wavefunctions. It was, however, shown by Nagamiya and Yamazaki [7] that higher-order impurities in the wavefunctions should contribute to decrease of "proton orbital moments" and to increase of "neutron orbital moments", thus opposite to the meson exchange effect.

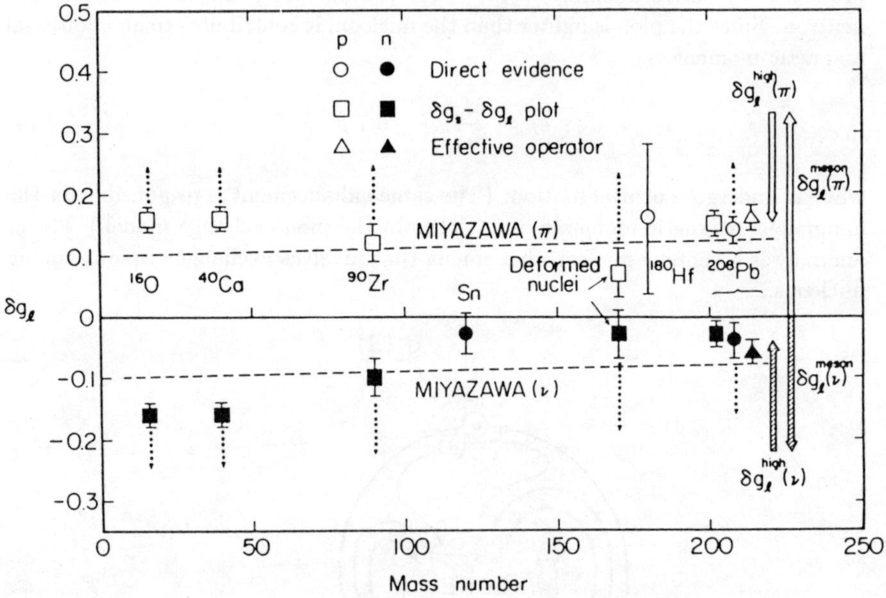

Fig.2 Empirical data of δg_ℓ versus mass number, taken from Yamazaki [8]. The deduction of δg_ℓ factors was made by three diffrent procedures; a) directly from two-particle states with dominantly orbital character (circles), b) from δg_ℓ - δg_s plots combining two or more single-particle moments (squares), and c) from fitting single-particle moments to the effective M1 operator (triangles). Detailed procedures are described in ref.[8]. The values for ^{16}O and ^{40}Ca are obtained after the higher-order correction of Shimizu *et al.* [14]. Otherwise, the directions after higher-order corrections are indicated by dotted arrows. The Miyazawa prediction after the N/Z correction is shown by broken lines.

The discovery of anomalous orbital g factors was followed by subsequent investigation of the problem by Nagamiya and Yamazaki [7], who showed that the g_ℓ anomaly is a general phenomena. They analyzed empirical data all over the nuclear mass region. The latest compilation of the experimental data [8] is presented in Fig.2. The experimental data show that the orbital g factors of proton deviate to the positive side, while those of neutron to the negative direction, which are expected from the meson exchange effect. In heavy nuclei the neutron number N exceeds the proton number Z, and the one-pion exchange current brings the following corrections,

$$\delta g_\ell(\mathrm{p}) = \frac{N}{A}\xi$$
$$\delta g_\ell(\mathrm{n}) = -\frac{Z}{A}\xi \tag{5}$$

where ξ is a constant (Miyazawa's estimate is 0.2). These modified Miyazawa values are depicted by broken lines in Fig.2.

Another evidence for the presence of meson exchange effect was shown by Riska and Brown [9] in the enhancement of the deuteron photodisintegration cross section. Furthermore, Fujita and Hirata [10] pointed out that the hitherto known enhancement factor κ in photonuclear reaction is related to the anomalous g_ℓ factor as

$$\tau_3 \cdot \kappa = 2\delta g_\ell. \tag{6}$$

The observed enhancement factor κ was as much as 0.5-1 [11], which in the light of the Fujita-Hirata relation (6) indicates the δg_ℓ might be larger than the Miyazawa value. What is the genuine g_ℓ^{meson} arising from meson exchange effect? To answer this we have to consider effects coming from nuclear wavefunctions.

Nagamiya and Yamazaki derived a general rule that the higher-order correction for proton is negative, while that for neutron is positive. Fulrtheremore, Yamazaki [12] took into account the difference between the proton number and the neutron number, and gave the following expression,

$$\delta g_\ell^{\mathrm{high}}(\mathrm{p}) = -\frac{N}{A}\zeta$$
$$\delta g_\ell^{\mathrm{high}}(\mathrm{n}) = +\frac{Z}{A}\zeta, \tag{7}$$

where ζ is the mixing ratio of higher-order impurities or the fraction of time in which the nucleon is combined with other nucleons [13]. This infunitive estimate is consistent with a realistic calculation given by Shimizu *et al.* [14].

The genuine meson exchange contribution $\delta g_\ell^{\mathrm{meson}}(\mathrm{p})$ is an observed value less the above correction, which yields $\delta g_\ell^{\mathrm{meson}}(\mathrm{p})$ always larger than $\delta g_\ell^{\mathrm{obs}}(\mathrm{p})$. The orbital g factors of both proton and neutron combined with a recent precise measurement of κ by Nolte *et al.* [15] are summarized in Fig.3. Yamazaki

[12] pointed out that the magnitude of $\delta g_\ell^{\rm meson}({\rm n})$ appears a little too small as compared with $\delta g_\ell^{\rm meson}({\rm p})$ in the light of the relation (5) and argued that the nuclear magneton itself, $\mu_{\rm N} = e\hbar/2M{\rm c}$, in the expression of nuclear magnetic moments may be different from the free nuclear magneton and enhanced by 10%, which can be interpreted as giving an effective "electromagnetic mass" of proton, $M = 0.9 \times M_{\rm N}^{\rm free}$. Since the nucleons have composite structure, such an anomaly is expected, as symbolically indicated by the $\delta g_\ell^{({\rm I})}$ in Fig.1. Whether this is ascribed to deconfinement of quarks or to relativistic effect [16] is an open problem; anyway, the effect is not very large even in the heavy nuclei.

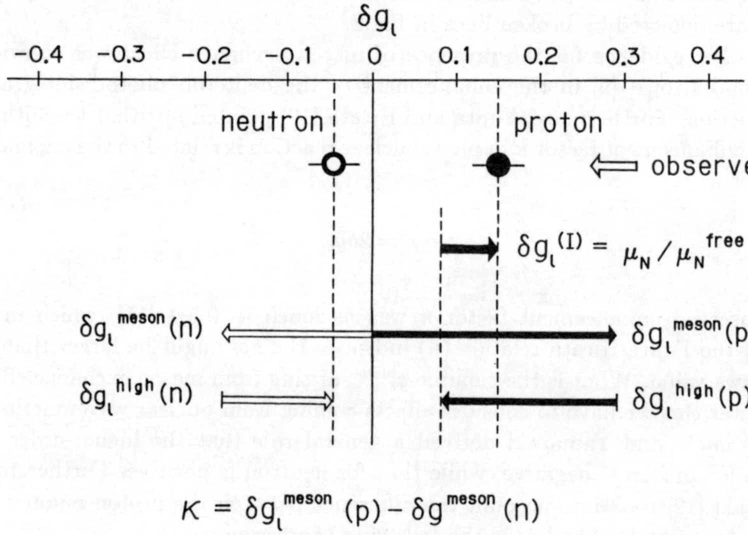

Fig.3 Summary of the δg_ℓ anomalies in the ^{208}Pb region. The observed values are decomposed into $\delta g_\ell^{\rm meson}$ and $\delta g_\ell^{\rm high}$, where the ratio between proton and neutron is assumed to follow eqs.(5) and (7), and the $\delta g_\ell^{\rm meson}$'s are adjusted so as to satisfy $\delta g_\ell^{\rm meson}({\rm p}) = (N/A)\kappa^{\rm obs}$, the latter being measured by Nolte *et al.* [15]. The resultant discrepancy, $\delta g_\ell^{\rm meson(I)} = 0.08$, is interpreted by Yamazaki [12] as an enhancement of the effective nuclear magneton.

3. REAL PIONS IN NUCLEI

The important role of pion exchange current in nuclei has thus been demonstrated. In the meson exchange vertex a virtual pion is exchanged. Could a real pion be bound and produced in nuclei? In this section we address this question. Before going into this, let us mention a very interesting case of the lightest pion-nucleus system, namely, πNN. Among the T=2 πNN multiplet the π^-nn and π^+pp systems would not be able to decay by the strong interaction, provided that they were bound; only the decay channel would be via the weak interaction and thus the lifetime would be as long as 10 nsec. Thus, it would be fantastic, if the system were really bound. One could determine its magnetic moment, which would show the value (5) as far as the pion keeps its identity and moves in a p orbital. A dedicated effort to find such bound states from d($\pi^\pm, \pi^\mp$) reactions is in progress at Los Alamos [17].

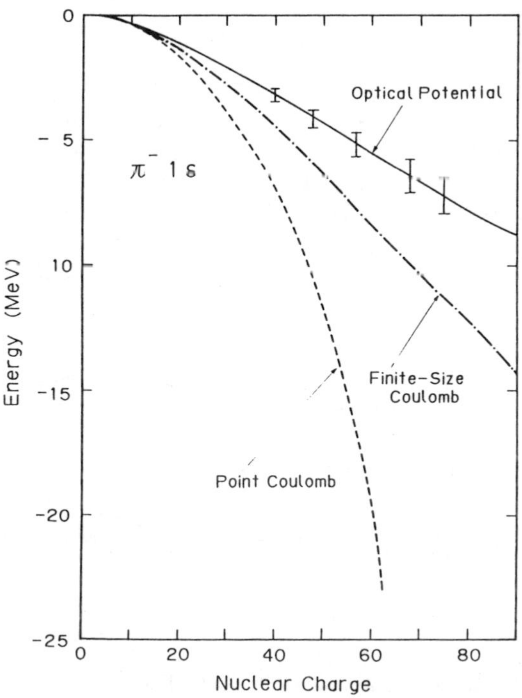

Fig.4 The Z dependence of the 1s levels of π^- calculated with the pion-nucleus optical potential (solid curve), the finite-nuclear-size Coulomb potential (dash-dotted curve) and the point Coulomb potential (dashed curve). The widths are indicated by vertical bars for representative Z. Here, $N = Z$ is assumed. From Toki and Yamazaki [19].

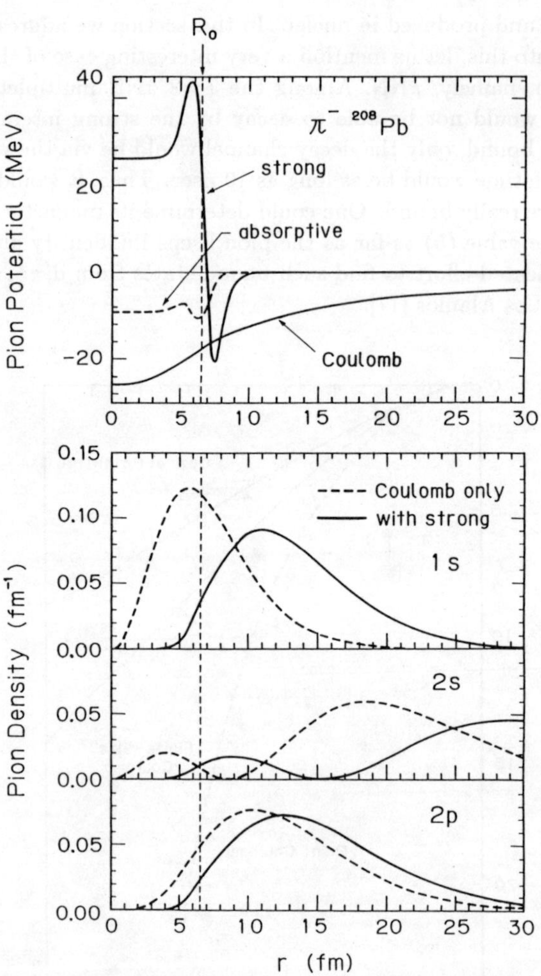

Fig.5 (Upper) The Coulomb potential and the optical potential for π^- in ^{208}Pb. (Lower) The π^- densities of the 1s, 2s and 2p states calculated by Hayano [21] are shown. The nucleus serves as a soft repulsive core, and pushes out s-state pions, thus forming narrow-width hybrid states.

Now, let us look into pion bound states in heavy nuclei. In general, a negatively-charged particle forms Coulomb bound states in nuclei, usually orbiting oustide the nucleus. In heavy nuclei, the Bohr radius is close to the nuclear radius so that the particle moves inside the nucleus. A negative hadron feels a strong-interaction potential as well. No matter whether the potential is strong or weak, the negative hadron can be bound in an orbital $(n\ell)$ by the nucleus, assisted by the large Coulomb attraction. The particle resides partially outside and partially inside, thus the state formed is of "hybrid" character, lying somewhere between "atomic state and "nuclear state". Thus, the strong-interaction absorption width is somewhat attenuated,

$$\Gamma \ = \ -2\int W(r)\rho_{n\ell}(r)\mathrm{d}r = \ -2W_0\int_0^{R_0}\rho_{n\ell}(r)\mathrm{d}r \tag{8}$$

where $\rho_{n\ell}(r) = 4\pi r^2|\Phi_{n\ell}(r)|^2$ is the radial density of π^- in an orbital $(n\ell)$, and $W(r)$ is the imaginary part of the optical potential with a depth W_0. Such hybrid states are studied in connection with Σ^- hypernuclei [18].

Motivated by the above arguments Toki and Yamazaki [19] studied deeply bound pionic states theoretically. Assuming the conventional pion-nucleus potential of Ericson-Ericson type [20] they calculated bound states of inner shells in heavy nuclei. The result appears surprising. The energy and width of the 1s bound state versus Z are shown in Fig.4. Even for heaviest nuclei the width is still "narrow" compared with the level spacing, indicating that the particle undergoes orbital motion more than ten times before it dies away! Why is the width so narrow?

To understand this, let us show the density distributions of the deeply bound states in ^{208}Pb. The 1s energy level is pushed up by switching on the strong interaction, because the S-wave potential is repulsive. Namely, there is a soft repulsive core inside the nucleus. For this reason, the pion is pushed away from the center of the nucleus, as shown in Fig.5. The 1s, 2s and 2p states as shown in the figure are all "hybrid". The fraction of the pion density inside is reduced so as to yield a narrow width.

The level structure calculated is shown in Fig.6 for the cases of ^{90}Zr and ^{208}Pb. Nobody knows if these bound states be described by the known optical potential. They have to be examined experimentally. A precise determination of such a level structure, if possible, would yield entirely new information on the pion-nucleus interaction as seen from the bound states. Hitherto, the optical potential has been obtained experimentally only from pion scattering and very shallow pionic atom data. It would be much better to determine the potential directly from deeply bound states. How can we study deeply bound pionic states experimentally? This is the problem we are going to discuss in the next section.

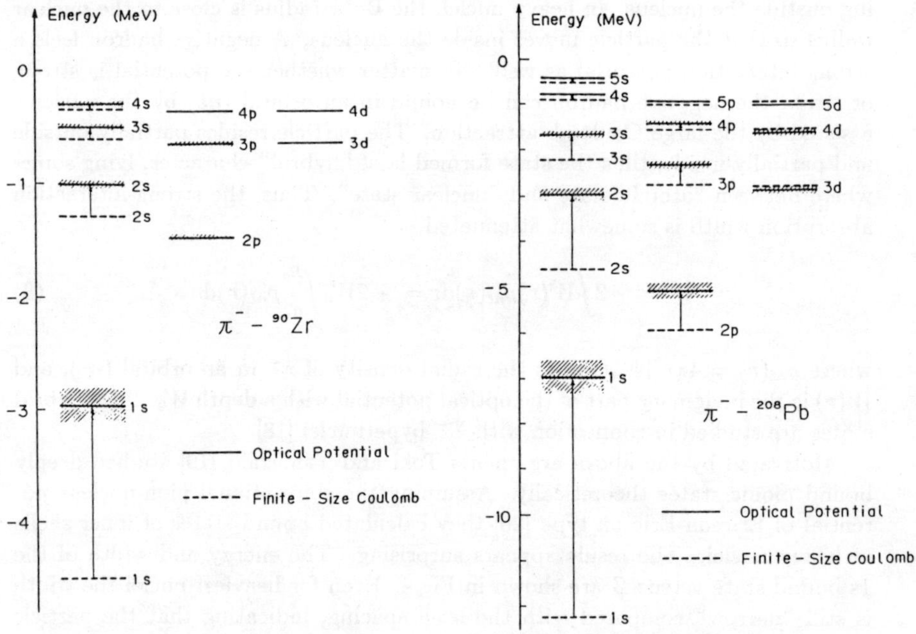

Fig.6 The calculated energy levels of $\pi^- - {}^{90}$Zr and $\pi^- - {}^{208}$Pb [19]. The results with the use of the optical potential are denoted by solid bars, while dashed bars are those with the finite-size Coulomb potential only. The widths are indicated by hatched areas.

4. PION TRANSFER REACTION

Deeply bound pionic states are guarded by a so called "last orbital", where the pionic atom cascades terminates because of the overwhelming pion absorption process compared with the x-ray cascade process. Here, the level width is in the region of 1-10 keV. The fact that inner bound states cannot be reached by mesic x ray cascade often misleads people who tend to believe that there are no observable states inside the last orbital. Whether a discrete state exists or not is a matter of its width, namely, whether it is smaller or larger than the level spacing (or, in other words, how many times the particle undergoes orbital motion within its lifetime.). Here, a width as much as 1 MeV can be allowed for a discrete level. This is two or three orders of magnitude larger than the width of the "last orbital", which is essentially the intervening strong absorption width in competition with the radiative width around 10 keV.

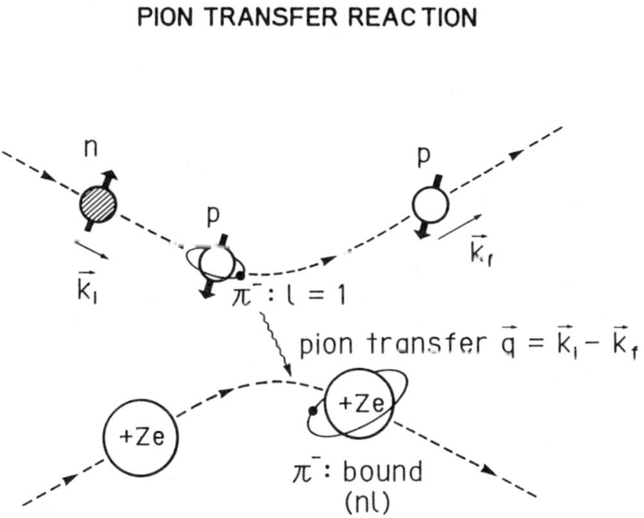

Fig.7 Diagram of the (n,p) pion-transfer reaction to populate deeply bound pionic states in heavy nuclei, where the p-wave π^- is transferred to a nuclear bound state. The intrinsic spins of the nucleons are also indicated.

In anology with the strangeness exchange reactions in the case of hypernuclei, we can populate pionic states with a "pion transfer" reaction. The best way is to use (n,pπ^-) reactions. This is just the Yukawa pion vertex, where the π^- is a real pion trapped by a target nucleus (Fig.7). In other words, the π^-cloud in the incoming neutron is transferred to a real pion bound state in a nucleus.

The cross section was given by Toki and Yamazaki [19] as follows.

$$\frac{d\sigma}{d\Omega} = \frac{M_P}{(2\pi)^2}\frac{k_f}{k_i}\left(\frac{f}{m_\pi}\right)^2 \frac{q^2}{m_\pi}|\Psi_{n\ell}(q)|^2. \tag{9}$$

where $\Psi_{n\ell}(q)$ is the Fourier transform of the pionic wave function of the $(n\ell)$ orbit, $\vec{k}$ and $\vec{k}_f$ the incoming and outgoing nucleon momenta, respectively, $f(=1)$

is the πNN coupling strength, and $\vec{q} = \vec{k}_i - \vec{k}_f$ is the momentum transfer. The cross section is proportional to the square of the Fourier transform of the pion wavefunction at the momentum transfer around $q = 200$ MeV/c which is the case in the (n,p) reaction above $T_n = 400$ MeV. This means that the cross section is larger for a more tightly bound π^-.

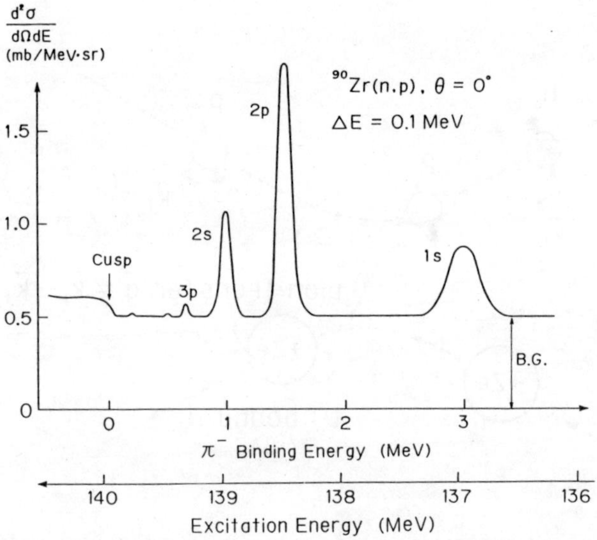

Fig.8 Expected spectrum of ^{90}Zr(n,p) reactions around the pion production threshold at $\theta = 0°$ and $E_n = 400$ MeV [19]. A background cross section of 0.5 mb/sr/MeV is assumed. The spectrum assumes an experimental resolution of $\Delta E = 0.1$ MeV.

The calculated cross section for the 1s state is 0.12 mb/sr for ^{90}Zr and 0.8 mb/sr for ^{208}Pb. An expected spectrum for ^{90}Zr(n,p) is shown in Fig.8. Recently, a medium-resolution (n,p) reaction spectroscopy has been established at TRIUMF using the 200-460 MeV proton beam for the purpose to study Gamow-Teller nuclear resonance states [22]. Though the excitation region around 140 MeV has not been exploited, the background cross section seems to be below 0.5 mb/sr. Therefore, detection of the 1s state in ^{208}Pb appears quite feasible. In the future, high-resolution spectroscopy ($\Delta E = 100$ keV) will also be possible at the forthcoming new cyclotron at the Research Center for Nuclear Physics of Osaka University.

So, we expect a "clean" pion-nucleus spectroscopy to come in the near future, which will yield rich information on the behavior of real pions in nuclei. In the future, based on experimental grounds, we hope to answer hitherto unsolved questions such as: "what is pion?", "does pion keep its free identify in nuclei?,.......

Acknowledgement

This article is dedicated to the 60th birth day of Professor H. Miyazawa. The authors would like to thank Professor Miyazawa for the stimulating discussions and encouragement over 20 years. The author is indebted to Professors H. Toki and R.S. Hayano for the collaborative work on the pion bound states.

References

[1] H. Miyazawa, Prog. Theor. Phys. **6** (1951) 801.
[2] T. Yamazaki, T. Nomura, S. Nagamiya and T. Katou, Phys. Rev. Letters **25** (1970) 547.
[3] T. Schmidt, Z. Phys. **106** (1937) 358.
[4] H. Kopfermann, "Nuclear Moments" (Academic Press, 1958)
[5] A. Arima and H. Horie, Prog. Theor. Phys. **12** (1954) 623.
[6] T. Yamazaki and E. Matthias, Phys. Rev. **175** (1968) 1476;
T. Yamazaki, Int. School of Physics "Enrico Fermi", Course XL
(ed. M. Jean and R. Ricci, Academic Press, New York, 1969) p.791.
[7] S. Nagamiya and T. Yamazaki, Phys. Rev. **C4** (1971) 1961.
[8] T. Yamazaki, in "Mesons in Nuclei" (ed. M. Rho and D. Wilkinson, North-Holland Pub. Co., Amsterdam, 1979) Vol.II, Ch.16.
[9] D.O. Riska and G.E. Brown, Phys. Letters **32B** (1970) 662.
[10] J.I. Fujita and M. Hirata, Phys. Letters **37B** (1971) 237.
[11] J. Ahrens, H. Borhert, H.B. Eppler, H. Cimm, H. Gudrun, P. Richn, G. Sita Ram, A. Ziegler, M. Kröning and B. Ziegler, Proc. Int. Conf. on Nuclear Structure Studies using Election Scattering and Photoreaction (ed. K. Shoda and H. Ui, Sendai, 1972) p.213.
[12] T. Yamazaki, Phys. Letters **160B** (1985) 227.
[13] G.F. Bertsch, Phys. Letters **28B** (1968) 302
[14] K. Shimizu, M. Ichimura and A. Arima, Nucl. Phys. **A226** (1974) 282.
[15] R. Nolte, A. Baumann, K.W. Rose and M. Schumacher, Phys. Letters **173B** (1986) 388.
[16] B. Serot and J.D. Walecka, "Advances in Nuclear Physics" (Plenum

Press, New York, 1986) Vol.16, p.1.
[17] D. Ashery *et al.*, private comm. (1988).
[18] T. Yamazaki, R.H. Hayano, O. Morimatsu and K. Yazaki, Phys. Letters 207B (1988) 393.
[19] H. Toki and T. Yamazaki, Phys. Letters B, to be published.
[20] M. Ericson and T.E.O. Ericson, Ann. Phys. **36** (1966) 496.
[21] R.S. Hayano, private communication (1988).
[22] M.C. Vetterli *et al.*, Phys. Rev. Letters **59** (1987) 439.

HADRON-HADRON INTERACTION WITH STRING FLIP-FLOP

Koichi Yazaki

Department of Physics, Faculty of Science, University of Tokyo
Bunkyo-ku, Tokyo 113, JAPAN

ABSTRACT

The string flip-flop model of the interaction among quarks proposed by H. Miyazawa is applied to a simple case of two meson systems. Scattering and bound state problems have been solved for the case of harmonic oscillator confinement. The symmetric channel allows a loosely bound state and threshold resonances while the antisymmetric channel exhibits a strong short range repulsion due to the Pauli principle for quarks. The structure of the bound state is discussed in some detail. The importance of the quark exchange effect in hadronic interactions is emphasized.

1. INTRODUCTION

The non-relativistic potential model for quarks has been successful in describing static properties and excitation spectra of single hadrons.[1] Although the theoretical basis of the model is very fragile for light quarks, it provides us with a convenient starting point to treat hadronic interactions in terms of the quark degrees of freedom. I will discuss here a possible extension of the model to multi-hadron systems based on the idea of string flip-flop proposed by H. Miyazawa.[2] The full account of the work has been given in ref. 3).

Let us start from a system of a single meson consisting of a quark (1) and an antiquark ($\bar{1}$). The hamiltonian for the system is given by

$$H = K_1 + K_{\bar{1}} + V_{1\bar{1}} \tag{1}$$

where K_1 and $K_{\bar{1}}$ are the kinetic energies of 1 and $\bar{1}$, i. e.

$$K_1 = \frac{\vec{P}_1^{\,2}}{2m}, \quad K_{\bar{1}} = \frac{\vec{P}_{\bar{1}}^{\,2}}{2m} \tag{2}$$

and $V_{1\bar{1}}$ is the interaction potential for 1 and $\bar{1}$, i. e.

$$V_{1\bar{1}} = v(r_{1\bar{1}}), \tag{3}$$

The confinement implies $v(r) \to \infty$ as $r \to \infty$.

For a system of two mesons consisting of two quarks (1, 2) and two antiquarks ($\bar{1}$, $\bar{2}$), the hamiltonian can be expressed as

$$H = K_1 + K_2 + K_{\bar{1}} + K_{\bar{2}} + V \tag{4}$$

where V is the interaction among quarks. We require the exchange symmetry i. e. V should be symmetric with respect to the interchange of 1 and 2 and/or $\bar{1}$ and $\bar{2}$. A straightforward extension of the two-body interaction $V_{1\bar{1}}$ leads

$$V = V_{1\bar{1}} + V_{1\bar{2}} + V_{2\bar{1}} + V_{2\bar{2}} \tag{5}$$

which confines the whole system and does not allow asymptotically free two mesons. A possible cure is provided by adding quark-quark and antiquark-antiquark interactions, i. e.

$$V = V_{1\bar{1}} + V_{1\bar{2}} + V_{2\bar{1}} + V_{2\bar{2}} + V_{12} + V_{\bar{1}\bar{2}} \tag{6}$$

with $V_{12} = -v(r_{12})$ and $V_{\bar{1}\bar{2}} = -v(r_{\bar{1}\bar{2}})$. The added interaction cancels the confining interaction between two mesons on the average but the residual fluctuation gives rise to a long range force between the two mesons which is the analogue of the Van der Waals force between two neutral atoms. This can be seen in the following way. Let us consider the case where $(1, \bar{1})$ and $(2, \bar{2})$ form well-separated two mesons, i.e. $r_{1\bar{1}}, r_{2\bar{2}} \ll y$ with

$$\bar{y} = \frac{1}{2}(\vec{r}_1 + \vec{r}_{\bar{1}} - \vec{r}_2 - \vec{r}_{\bar{2}}). \tag{7}$$

The relevant residual interaction to the lowest order is

$$V_{1\bar{2}} + V_{\bar{1}2} + V_{12} + V_{\bar{1}\bar{2}} \simeq (\vec{r}_{1\bar{1}}\cdot\vec{\nabla}_y)(\vec{r}_{2\bar{2}}\cdot\vec{\nabla}_y)\ (y), \tag{8}$$

The second order perturbation theory gives the "Van der Waals" potential

$$U_{vdw}(y) \simeq -\frac{\langle r^2\rangle^2}{\Delta}\frac{1}{9}\sum_{ij}[\nabla_y^i\nabla_y^i v(y)]^2, \tag{9}$$

where $\langle r^2\rangle$ is the mean square radius of the single meson and Δ is the average excitation energy. For a power law potential $v(r) \propto r^\alpha$, the long range interaction becomes

$$U_{vdw}(y) \propto -\ y^{2\alpha-4} \tag{10}$$

The power is different from that of the "color Van der Waals" force discussed by H. Miyazawa and others[4)] because the color degrees of freedom have not been included in the present case. For the "colored" quarks, the confining interaction between the "colored" mesons dominates the internal excitation energy and Δ in eq. (9) becomes proportional to to $v(y)$, which leads to the usual power law for $U_{vdw}(y)$, i. e.

$$U_{vdw}(y) \propto -\ y^{\alpha-4} \tag{11}$$

There has been no experimental evidence for the long range strong interaction between hadrons and, therefore, the confining interaction does not seem to be expressible in terms of a superposition of two-body potentials. In the next section, we show that the idea of string flip-flop solves the problem. The scattering problem for two mesons is formulated and solved in §3. The bound state properties are studied in §4. Concluding remarks are given in §5.

2. FLIP-FLOP CONFINEMENT

Let us continue the discussion about the interaction, V, for the system of two quarks (1, 2) and two antiquarks ($\bar{1}$, $\bar{2}$). We want to have

$$V = u_a \equiv V_{1\bar{1}} + V_{2\bar{2}} \tag{12}$$

in the case where (1, $\bar{1}$) and (2, $\bar{2}$) form well separated two clusters i. e. $r_{1\bar{1}}$, $r_{2\bar{2}} << y$ and

$$V = u_b \equiv V_{1\bar{2}} + V_{2\bar{1}} \tag{13}$$

for $r_{1\bar{2}}$, $r_{2\bar{1}} << z$ with

$$\vec{z} = \frac{1}{2}(\vec{r}_1 + \vec{r}_2 - \vec{r}_{\bar{2}} - \vec{r}_{\bar{1}}) \tag{14}$$

This can be achieved by choosing

$$V = \theta(u_b - u_a)u_a + \theta(u_a - u_b)u_b \tag{15}$$

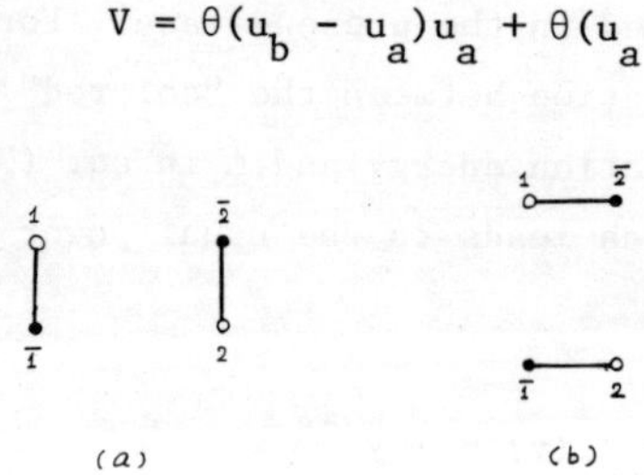

Fig. 1. The string configurations for two mesons. ∘(•) denotes a quark (antiquark) and the solid lines describe strings.

This choice actualizes the idea of string flip-flop since $u_a(u_b)$ is the potential energy for the string configuration (a)((b)) as illustrated in Fig. 1 and the system makes transitions between the two choosing the energetically favored one. The interaction, V, given by eq. (15), satisfies the exchange symmetry and has no problem of the Van der Waals force since there is no residual interaction between well separated two mesons.

Let us now take a quadratic confining potential, i. e.

$$v(r) = ar^2 \tag{16}$$

and study the two meson system in some detail. As will become clear in the following, this choice offers the maximal separation of variables and allows an almost exact treatment of the four-body problem. We choose the variables, $\vec{R}_{cm}$ and $\vec{x}$, as

$$\vec{R}_{cm} = \frac{1}{4}(\vec{r}_1 + \vec{r}_2 + \vec{r}_{\bar{1}} + \vec{r}_{\bar{2}}) \tag{17}$$

$$\vec{x} = \frac{1}{2}(\vec{r}_1 + \vec{r}_2 - \vec{r}_{\bar{1}} - \vec{r}_{\bar{2}}), \tag{18}$$

together with $\vec{y}$ and $\vec{z}$ defined by eqs. (7) and (14), respectively. In terms of these variables, the potentials, u_a and u_b, can be expressed as

$$\begin{aligned} u_a &= a(r_{1\bar{1}}^2 + r_{2\bar{2}}^2) = 2a(x^2 + z^2) \\ u_b &= a(r_{1\bar{2}}^2 + r_{2\bar{1}}^2) = 2a(x^2 + y^2) \end{aligned} \tag{19}$$

and thus the interaction, V, becomes

$$V = 2a\{x^2 + z^2\theta(y^2 - z^2) + y^2\theta(z^2 - y^2)\}. \tag{20}$$

The kinetic energy takes the form

$$K = K_1 + K_2 + K_{\bar{1}} + K_{\bar{2}} = K_{cm} + K_x + K_y + K_z \quad (21)$$

with

$$K_{cm} = \frac{\vec{P}_{cm}^{\,2}}{8m}, \quad K_x = \frac{\vec{P}_x^{\,2}}{2m}, \quad \cdots \quad . \quad (22)$$

The variable, $\vec{x}$, as well as the center-of-mass variable, $\vec{R}_{cm}$, are thus separated and the hamiltonian can be written as

$$H = K_{cm} + K_x + 2ax^2 + h \quad (23)$$

where the reduced hamiltonian, h, for the variables, $\vec{y}$ and $\vec{z}$, is given by

$$h = K_y + K_z + V(y, z),$$
$$V(y, z) = 2a[z^2\theta(y^2 - z^2) + y^2\theta(z^2 - y^2)]. \quad (24)$$

The exchange of quarks, $1 \leftrightarrow 2$, corresponds to $\vec{y} \leftrightarrow -\vec{z}$ while that of antiquarks, $\bar{1} \leftrightarrow \bar{2}$, corresponds to $\vec{y} \leftrightarrow \vec{z}$ and the exchange symmetry is reflected in the invariance of h with respect to $\vec{y} \leftrightarrow \pm \vec{z}$. We also note that the potential, V(y, z), depends only upon the magnitudes of y and z and, therefore, the angular momenta $\vec{l}_y$ and $\vec{l}_z$ are separately conserved. We can thus make the partial wave decomposition and for each partial wave, specified by l_y and l_z, the problem becomes that of two dimensions. The potential, V(y, z), is illustrated in Fig. 2. One sees that the diagonal line, y = z, plays a speical role in this problem.

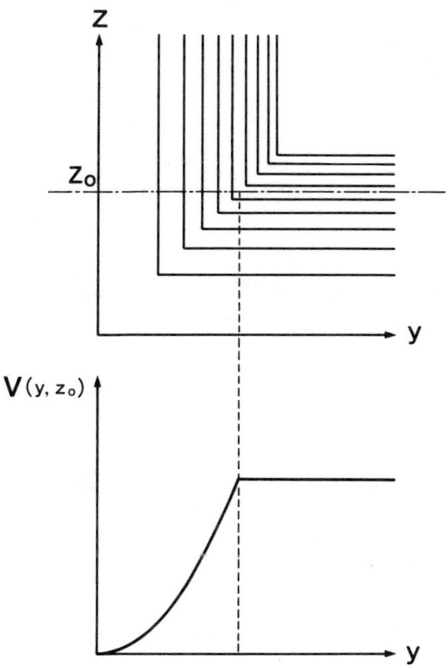

Fig. 2. The equi-potential surface for V(y, z) given by eq. (24). V(y, z) as a function of y for a fixed z is also shown.

The complete separation of the variable, $\vec{x}$, and the angular parts of $\vec{y}$ and $\vec{z}$ is a special feature of the quadratic confining potential and greatly simplifies the problem. We take here this speical form for the potential and examine the scattering and bound state properties of the system by solving the problem almost exactly, rather than discussing the more general form of the potential which requires a complicated numerical treatment and thus obscures the important features of the system.

3. SCATTERING PROBLEM

The meson-meson scattering in the present case is described by a wave function which has the following asymptotic form.

$$\Psi \sim e^{i\vec{k}_{00}\vec{y}} {}_0(\vec{r}_{1\bar{1}})\ {}_0(\vec{r}_{2\bar{2}}) + \sum_{\mu,\nu} f_{\mu\nu}(\hat{y}) \frac{e^{ik_{\mu\nu}y}}{y} \varphi_\mu(\vec{r}_{1\bar{1}})\varphi_\nu(\vec{r}_{2\bar{2}})$$

$$+ \sum_{\mu,\nu} g_{\mu\nu}(\hat{z}) \frac{e^{ik_{\mu\nu}z}}{z} \varphi_\mu(\vec{r}_{12})\varphi_\nu(\vec{r}_{2\bar{1}}), \tag{25}$$

where φ_μ describes the internal state of a meson and satisfies the Schrödinger equation for a harmonic oscillator, i.e.

$$(- \frac{1}{m} \Delta_r + ar^2)\varphi_\mu(\vec{r}) = \varepsilon_\mu \varphi_\mu(\vec{r}), \tag{26}$$

$k_{\mu\nu}$ is related to the total energy in the c.m. system as

$$E = \frac{k_{\mu\nu}^2}{2m} + \varepsilon_\mu + \varepsilon_\nu \tag{27}$$

and $f_{\mu\nu}$ and $g_{\mu\nu}$ are the scattering amplitudes for the direct and exchange channels, respectively. The exchange symmetry implies that only the particular combinations of $f_{\mu\nu}$ and $g_{\mu\nu}$ appear in the observed cross sections. In eq, (25), the two mesons are assumed to be initially in their ground states. Using the variable $\vec{x}$, one can write

$$\varphi_0(\vec{r}_{1\bar{1}})\varphi_0(\vec{r}_{2\bar{2}}) = \phi_0(\vec{x})\phi_0(\vec{z}) \tag{28}$$

where $\{\phi_\mu\}$ is another set of harmonic oscillator wave functions with the same frequency as $\{\varphi_\mu\}$, i. e. ϕ_μ satisfies the equation

$$(- \frac{1}{2m} \Delta_x + 2ax^2)\phi_\mu(\vec{x}) = \varepsilon_\mu \phi_\mu(\vec{x}). \tag{29}$$

Since x is separated from y and z, the wave function, Ψ, in eq. (25) can be expressed as

$$\Psi \sim \phi_0(\vec{x})\{e^{i\vec{k}_0\vec{y}} + \sum_\mu F_\mu(\hat{y}) \frac{e^{ik_\mu y}}{y} \phi_\mu(\vec{z}) + \sum_\mu G_\mu(\hat{z}) \frac{e^{ik_\mu z}}{z} \phi_\mu(\vec{y})\}, \tag{30}$$

where k_μ is related to E as

$$E = \frac{k_\mu^2}{2m} + \varepsilon_\mu + \varepsilon_0. \tag{31}$$

The amplitude, $f_{\mu\nu}$, can be expressed by F_μ as

$$f_{\mu\nu} = \sum_\sigma \langle\mu\nu|0\sigma\rangle\, F_\sigma \tag{32}$$

where the sum is over all σ satisfying

$$\varepsilon_\mu + \varepsilon_\nu = \varepsilon_0 + \varepsilon_\sigma \tag{33}$$

and $\langle\mu\nu|0\sigma\rangle$ is the Moshinsky bracket for the two harmonic oscillators. The amplitude, $g_{\mu\nu}$, can also be expressed by G_μ in the same way. The amplitudes, F and G, are obtained by the partial wave decomposition with respect to the angular momenta, $\vec{\ell}_y$ and $\vec{\ell}_z$. We will not describe here the details of the procedure since we are mostly interested in the low energy scattering where the partial wave with $\ell_y = \ell_z = 0$ is dominant.

The Schrödinger equation for this partial wave ($\ell_y = \ell_z = 0$) is

$$\{-\frac{1}{2m}(\frac{\partial^2}{\partial y^2} + \frac{\partial^2}{\partial z^2}) + V(y, z)\}\psi(y, z) = \bar{E}\psi(y, z) \tag{34}$$

where V(y, z) is given by eq. (24) and $\bar{E}$ is the reduced energy, i.e.

$$\bar{E} = E - \varepsilon_0 \tag{35}$$

The exchange symmetry is reflected in the invariance of eq. (34) under the interchange of y and z. We can thus assume ψ(y, z) to be either symmetric, $\psi^s(y, z)$, or antisymmetric, $\psi^a(y, z)$, i. e.

$$\psi^s(y, z) = \psi^s(z, y), \ \psi^a(y, z) = -\psi^a(z, y), \tag{36}$$

and restrict ourselves to the region $y \geq z$, for example, with the following boundary conditions.

a) $\psi(y, 0) = 0$ (37)

b) $\psi^a(t, t) = 0, \ (\partial_z - \partial_y)\psi^s(y, z)|_{y=z=t} = 0$ (38)

c) $\psi_n(y, z) \sim \Phi_n^-(y, z) - \sum_m S_{mn} \Phi_m^+(y, z)$ for large y (39)

where $\Phi_n^{\pm}$ is defined by

$$\Phi_n^{\pm}(y, z) \equiv e^{\pm ik_n y}\phi_n(z)$$

with ϕ_n being an s-wave solution of eq. (29) and

$$\bar{E} = \frac{k_n^2}{2m} + \varepsilon_n, \ \varepsilon_n = (2n + \frac{3}{2})\omega, \ \omega \equiv 2\sqrt{\frac{a}{m}} \tag{40}$$

We have introduced here the scattering wave functions with the internal excitations in the initial state.
Introducing the functions along the diagonal as

$$\xi_n^{\pm}(t) \equiv \Phi_n^{\pm}(t, t), \ \eta_n^{\pm}(t) \equiv (\partial_z - \partial_y)\Phi_n^{\pm}(y, z)|_{y=z=t} \tag{41}$$

$$\omega_n(t) \equiv -\frac{1}{2ik_n}\psi_n^s(t, t), \ \theta_n(t) \equiv \frac{1}{2ik_n}(\partial_z - \partial_y)\psi_n^a(y, z)|_{y=z=t} \tag{42}$$

and using the Green theorem, we can derive the following relations.

$$(\eta_m^+, \omega_n) = (\xi_m^+, \theta_n) = \delta_{mn} \tag{43}$$

$$\frac{k_m}{k_n} S_{mn}^s = (\eta_m^-, \omega_n), \ \frac{k_m}{k_n} S_{mn}^a = (\xi_m^-, \theta_n) \tag{44}$$

where the inner product (f, g) is defined by

$$(f, g) \equiv \int_0^\infty dt \ f(t)g(t) \tag{45}$$

We can use eq. (43) to obtain $\{\omega_n\}$ and $\{\theta_n\}$ as biorthogonal sets to $\{\eta_n^+\}$ and $\{\xi_n^+\}$, respectively, and then use eq. (44) to get the S-matrix element, S_{mn}. Further details about the method of solution have been described in ref. 3).

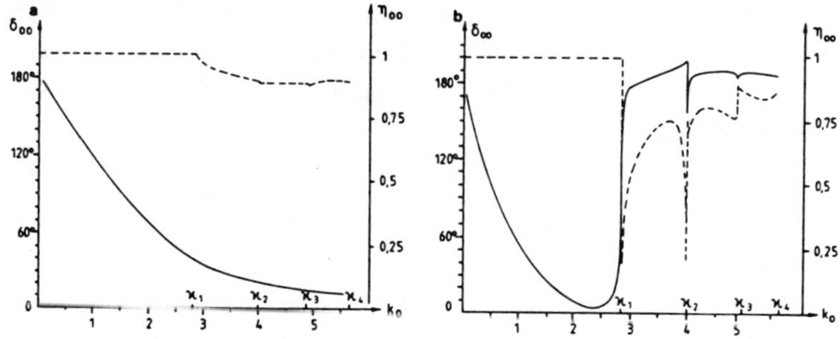

Fig. 3. The S-matrix, S_{00}, for antisymmetric (a) and symmetric (b) scatterings. The wave number, k_0, is given in unit of $2\sqrt{m\omega}$, and κ's are inelastic thresholds.

Fig. 3 shows the calculated elastic S-matrix elements, S^s_{00} and S^a_{00}, for symmetric and antisymmetric scatterings, respectively. They are given in the form $S_{00} = \eta_{00} \exp(2i\ \delta_{00})$. One sees that there are a bound state and sharp resonances near the thresholds for the symmetric case. The corresponding energies for the lowest three states are

$$E_0 = (\frac{3}{2} - 0.04)\omega$$

$$E_1 = (\frac{7}{2} + 0.015 - i0.015)\omega \qquad (46)$$

$$E_2 = (\frac{11}{2} + 0.008 - i0.019)\omega$$

The low energy scattering is dominated by the loosely bound state and the scattering length, a, and the effective range, r_0, are

$$a^s = 4.98\ b,\ r_0^{\ s} = 2.14\ b,\ b \equiv \frac{1}{\sqrt{m\omega}} \qquad (47)$$

In contrast to the symmetric scattering, the antisymmetric scattering has much weaker structures and the threshold effects are barely seen in the figure. The scattering is essentially that of a hard core and the low energy parameters are

$$a^a = 1.57\ b,\quad r_0^{\ a} = 1.00\ b \qquad (48)$$

Such a repulsive interaction is a direct consequence of the boundary condition for the antisymmetric scattering which requires the wave function to vanish along the diagonal y = z. In a somewhat different approach using the resonating group method, this can also be interpreted as due to the appearance of the "Pauli-forbidden" state which corresponds to the $(0s)^2(0s)^2$ configuration in the harmonic oscillator representation of the $q^2\bar{q}^2$ system.

Loosely bound states, resonances near thresholds and repulsive cores are characteristic features of the observed hadron-hadron interaction and it is interesting to see that the present simple model with the flip-flop confinement already shows these features.

4. BOUND STATE PROPERTIES

The bound state for the symmetric scattering is of a purely quantum mechanical nature and has no classical analogue. The potential, V(y, z), given by eq. (24) has clearly no classical stable point and a particle in the potential moves away either along the y-axis or the z-axis. In quantum mechanics, however, there is an effective restoring force since the particle, moving in the direction of increasing y, gets more and more completely confined in the z-direction, and thus the energy available for the motion in the y-direction decreases. Since such a restoring force is not so strong,

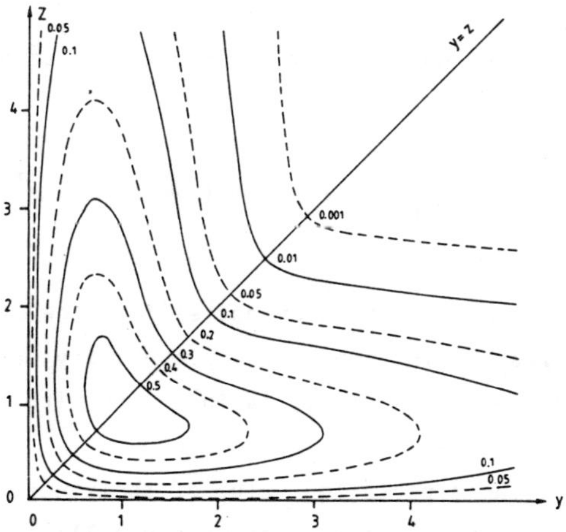

Fig. 4. The bound state wave function, $\psi_B(y, z)$. y and z are given in unit of $2\sqrt{b} = \sqrt{2/m\omega}$.

the binding energy is very small and the wave function has long tails both in the y- and z-directions as one sees in Fig. 4.

Given the bound state wave function, $\psi_B(y, z)$, one can examine how the internal structures of the mesons are modified in this composite system. Let us first calculate the probability of the two mesons being in their ground states, by introducing an appropriate projection operator, P_0. Denoting, in general, the projection operator for the internal state, φ_n, by P_n, we require that P_n's satisfy the orthogonality, $P_n P_{n'} = \delta_{nn'} P_n$ and the exchange symmetry. Unfortunately, there seem to be no simple operators satisfying both requirements. An obvious choice which satisfies the first requirement is

$$P_n(y, z; y', z') = \varphi_n(z)\varphi_n^*(z')\delta(y - y') \tag{49}$$

Because of the lack of exchange symmetry, however, the resulting probability for the ground state is $<\Psi_B|P_0|\Psi_B> \simeq 0.70$, which is considerably smaller than what one expects from the loosely bound state. A more reasonable choice would be

$$P_n(y,z;y',z')=\theta(y-z)\varphi_n(z)\varphi_n^*(z')\delta(y-y')+\theta(z-y)\varphi_n(y)\varphi_n^*(y')\delta(z-z'). \tag{50}$$

which gives $<\Psi_B|P_0|\Psi_B> = 0.993$ but does not satisfy the orthogonality. There is, thus, no clear definition of the probability, although the result of the second choice, eq. (50), is in accord with our expectation.

We next consider the expectation value of the kinetic energy. We note that the virial theorem for a harmonic oscillator holds for the present interaction, since it is based on the scale transformation property of the hamiltonian and the θ-functions contained in $V(y, z)$ (eq. (24)) are invariant under this transformation. Including the x-part to get the full four-body bound state wave function, Ψ_B, i. e.

$$\Psi_B(x, y, z) = \phi_0(x)\psi_B(y, z). \tag{51}$$

we obtain

$$<\Psi_B|K|\Psi_B> = \frac{1}{2} <\Psi_B|H|\Psi_B> = \frac{1}{2}(3 - 0.04)\omega \tag{52}$$

which is slightly smaller than the sum of the internal kinetic energies for separated two mesons. Since the kinetic energy for the relative motion also contributes to $<\Psi_B|K|\Psi_B>$ in eq. (52), the internal kinetic energy of a meson is definitely decreased in the bound state which suggests that the mesons there have a more extended structure than what they have in the free space.

In order to examine such a possibility further, we now calculate the momentum distribution of quarks and antiquarks in the bound state. The distribution function, $n(\vec{P})$, is defined by

$$n(\vec{P}) \equiv \frac{1}{4} \langle \Psi_B | \{\delta(\vec{P}_1 - \vec{P}) + \delta(\vec{P}_2 - \vec{P}) + \delta(\vec{P}_{\bar{1}} - \vec{P}) + \delta(\vec{P}_{\bar{2}} - \vec{P})\} | \Psi_B \rangle, \quad (53)$$

which can be expressed by the reduced wave function, ψ_B, as

$$n(\vec{P}) = \left(\frac{b^2}{\pi}\right)^{\frac{3}{2}} \int d\vec{P}_1 d\vec{P}_2 \, e^{-\frac{b^2}{2}(2\vec{P}-\vec{P}_1-\vec{P}_2)^2} |\tilde{\psi}_B(\vec{P}_1, \vec{P}_2)|^2 \quad (54)$$

where $\tilde{\psi}_B$ is the Fourier transform of ψ_B. We compare it with the corresponding momentum distribution for the case where the quark exchange is neglected and, therefore, the wave function, Ψ_B, becomes a product of the internal wave functions for two mesons times the wave function for the relative motion, i. e.

$$\Psi_B^{(h)}(x, y, z) = \phi_0(x)\phi_0(z)\chi(y) \quad (55)$$

where $\chi(y)$ is generated by a potential which simulates the two-meson interaction in the present model. The resulting distribution function, $n_h(\vec{P})$ is

$$n_h(\vec{P}) = \left(\frac{b^2}{2\pi}\right)^{\frac{3}{2}} \int d\vec{q} \, e^{-\frac{b^2}{2}(\vec{P} - \frac{\vec{q}}{2})^2} |\tilde{\chi}(q)|^2 \quad (56)$$

where $\tilde{\chi}$ is the Fourier transform of χ. The momentum distribution of a single meson is

$$n_0(\vec{P}) = \left(\frac{b^2}{2\pi}\right)^{\frac{3}{2}} e^{-\frac{b^2}{2}P^2} \quad (57)$$

Fig. 5a shows $n(P)$ and $n_h(P)$. In order to emphasize the difference, we compare in fig. 5b the relative deviation from the single meson distribution, δ and δ_h, defined by

$$\delta(\vec{P}) \equiv \frac{n(\vec{P}) - n_0(\vec{P})}{n_0(\vec{P})}, \quad \delta_h(\vec{P}) \equiv \frac{n_h(\vec{P}) - n_0(\vec{P})}{n_0(\vec{P})}. \quad (58)$$

Fig. 5b also shows δ_h for the case when the size parameter of a meson is slightly increased. One sees that the quark exchange gives rise to softening of the momentum distribution and the effect can be approximately simulated by the size change of the constituents (mesons in the present case). A similar effect of the quark exchange is recently noted in connection with the EMC effect.

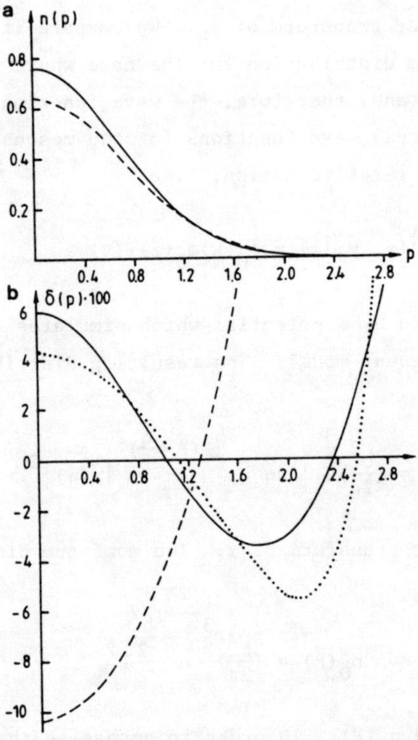

Fig. 5. The quark momentum distribution n(P) for the bound state (a) and its relative deviation, $\delta(P) \equiv (n(P) - n_0(P))/n_0(P)$, from the quark momentum distribution $n_0(P)$ for an isolated meson. The solid lines correspond to the full calculation while the dashed lines show the results, $n_h(P)$ and $\delta_h(P)$, of an approximate calculation without the quark exchange. The dotted line in (b) is $\delta_h(P)$ with a modified meson size parameter, $\bar{b} \equiv 1.055b$.

5. CONCLUDING REMARKS

We have studied here the interaction between two mesons based on the idea of the string flip-flop within the framework of the non-relativistic quark model. For the harmonic oscillator confinement, the problem has been reduced to that of a two-dimensional system with the confining potential acting on the smaller coordinate. (See eq. (24) and Fig. 2). It is instructive to note an analogy between the present system and the electron-hydrogen system. The latter is described by a hamiltonian of the form

$$H = \frac{P_1^2}{2m} + \frac{P_2^2}{2m} - \frac{e^2}{r_1} - \frac{e^2}{r_2} + \frac{e^2}{r_{12}} \tag{59}$$

where the two electrons are denoted by 1 and 2 and the proton recoil is neglected. In the monopole approximation, where one takes an average over the angular variables of the two electrons, the hamiltonian becomes

$$H \simeq \frac{P_1^2}{2m} + \frac{P_2^2}{2m} - \frac{e^2}{r_1}\,\theta(r_2-r_1) - \frac{e^2}{r_2}\,\theta(r_1-r_2) \tag{60}$$

The interaction is similar to that given by eq. (24) except that the confining potential is replaced by the Coulomb potential. It is well known[5)] that the system has a loosely bound state, H^-, and the threshold resonances in the spin-singlet (spatially symmetric) channel while it exhibits the hard core scattering in the spin-triplet (spatially antisymmetric) channel. Using the virial theorem, one can again show that the average kinetic energy for the two electrons in H^- is less than twice of that for a single electron in the hydrogen atom. Thus, one sees that the electron-hydrogen system shares many important features with the present two-meson system. The essential ingredient responsible for these features is the quantum effect of particle exchange.

We have restricted ourselves to the case where quarks have no internal degrees of freedom. Although a straightforward extension of the model can incorporate the flavor and the spin, the inclusion of color is not trivial and requires further sophistications as have been done in ref. 3) and in the applications to two-baryon systems.[6] There is also a conceptual problem in assigning colors to quarks and antiquarks at different space-time points, since such an assignment is not compatible with the local gauge invariance of QCD. The problem has recently been discussed in ref. 7). Further efforts in this direction are required to clarify the basis of the model.

It is a great honor for me to be able to contribute this paper for celebrating the sixtieth birthday of Professor H. Miyazawa, who has been guiding me in physics since I started my research.

REFERENCES

1. De Rújula, A., Georgi, H. and Glashow, S. L., Phys. Rev. D12, 147 (1975).
Isgur, N. and Karl, G., Phys. Lett. 72B, 109 (1977); Phys. Rev. D19, 2563 (1979); D20, 1191 (1979).

2. Miyazawa, H., Phys. Rev. D20, 2593 (1979).

3. Lenz, F., Londergan, J. T., Moniz, E. J., Rosenfelder, R., Stingl, M. and Yazaki, K., Ann. of Phys. 170, 65 (1986).

4. Matsuyama, S. and Miyazawa, H., Prog. Theor. Phys. 61, 942 (1979).
Greenberg, O. W. and Lipkin, H. J., Nucl. Phys. A370, 349 (1981).

5. McGowan, J. W., Phys. Rev. 156, 165 (1967); Chen, J. C. Y., in "Case Studies in Atomic Physics III" (McDaniel, E. W. and McDowell, M. R. C., eds.), North-Holland, New York.

6. Oka, M. and Horowitz, C. J., Phys. Rev. D31, 2773 (1981)
Koike, Y., Nucl. Phys. A454, 509 (1986).

7. Morimatsu, O., Contribution to the 11th Int. Conf. on Particles and Nuclei (PANIC '87), Kyoto, 1987, No.b12.

THE MEAN FIELD THEORY AT FINITE TEMPERATURE FROM NUCLEAR PHYSICS TO HIGH-T_c SUPERCONDUCTIVITY

Kosai Tanabe

Department of Physics, Saitama University
Urawa, Saitama 338
JAPAN

ABSTARACT

A brief review is given of the thermal Hartree-Fock-Bogoliubov (THFB) theory which is a statistical extension of the HFB theory and of the thermal random phase approximation (TRPA). The TRPA formalism can be generalized to the extended TRPA (ETRPA) by applying the thermo field dynamics. Usefulness of these formalisms is exemplified by some applications to highly excited nuclear collective states. It is shown that these theoretical methods are extensively applicable to the superconductivity in association with local spin fluctuation (LSF). New type of order parameter is introduced within the framework of the extended THFB theory.

1. INTRODUCTION

Recent measurements of γ-ray spactra after formation of compound nucleus by heavy-ion collision have provided detailed informations of nuclear levels and intrinsic properties in high accuracy. Such studies are concerned with details of the mechanism realizing collective excitations of the finite nuclear system and the interplay between single-particle behaviour and collective motion. It is well-known that the pairing correlation between nucleons plays an important role to realize superconducting state in the lowlying nuclear states, in which nucleons composing a Cooper pair do not contribute to the moment of inertia of the total system. Mottelson and Valatin[1)] argued that the pairing field disappears at a certain rotational frequency ω_c in a deformed rotating nucleus. This is the proposed mechanism known as the Mottelson-Valatin effect caused by the Coriolis antiparing (CAP) effect. However, no clear evidence of such a structure change had been found until very recent experiment which discovered the super-deformed rotational states in rare-earth nuclei[2,3]. The backbending behaviour of the rotational frequency ω with increasing spin I can be understood rather in terms of the rotational alignment (RA) mechanism proposed by Stephens and Simon[4]. In the RA paired nucleons with large Ω ($= j_z$) located near the Fermi surface are dissociated, and their spins are aligned to the direction of rotation as in a gyroscope. Correspondingly the pairing gap Δ for relevant orbitals decreases with increasing angular frequency, so that the moment of inertia gains certain contribution from the decoupled pair.

From theoretical point of view all of the above phenomena occuring in the self-sustained nucleon system must be described in terms of the complete self-consistent calculation based on the single-particle space and the microscopic two-body interactions. With the mean field theory we attempt to establish the precise relation between experimantal data and the assumed phenomenological interaction, and to describe all the above nuclear properties on the unified microscopic fashion.

In sect. 2, we will see that the cranked HFB theory is capable of reproducing simultaneously the backbending and g-factors[5] and also describing the superdeformed high spin states in ^{132}Ce[6]. In sect. 3, the HFB theory will be extended to the THFB theory with the statistical recipe[7,8]. Furthermore, the TRPA[9,10] is generalized to the extended TRPA (ETRPA)[11] by applying the thermo field dynamics[12,13]. In sect. 4, the Hubbard model[14] for electron system in two-dimensional square lattice will be considered, and the instability of the spin density wave (SDW) as the TRPA mode will be interpreted in terms of the Bose condensation[15]. A possible theoretical framework will be proposed to describe the superconductivity in association with the LSF.

2. THE HFB THEORY AS A MEAN FIELD APPROXIMATION AT ZERO TEMPERATURE

Usefulness of the cranked HFB theory has been manifested in

reproducing rotational band structure and providing microscopic basis for the RA mechanism[16]. This approach has several merits, especially (1) the deformation is self-consistently created from the spherical single-particle basis and the Hamiltonian with rotational symmetry, (2) the theory can simulate gradual change of intrinsic structure in association with RA of spins with increasing angular frequency, and (3) every details of single-particle behaviour as well as collective properties can be calculated at the same time.

Here we present two typical examples of our results from the HFB calculations. We assume the phenomenological two-body interaction which includes the quadrupole-pairing interaction in addition to the monopole-pairing interaction plus quadrupole-quadrupole interaction. We have employed fairely wide range of spherical single-particle space covering almost three major shells for both proton and neutron. In the first example, the HFB calculation reproduces simultaneously both backbending and g-factors of even mass Er isotopes from A = 158 to 170. As displayed in Fig. 1 we got amazingly perfect fits to the data[5].

The second example is the reproduction of the superdeformed state in ^{132}Ce[6]. Recent discovery of superdeformed high spin states is interesting from the viewpoint that such states provide excellent tests of the validity of mean field theory in the extremity of large deformation, large rotational frequency and high excitation. We have challenged to this crucial test by trying to reproduce three bands in ^{132}Ce, i. e. the ground band (g. b.), the s. band (s. b.) and the superdeformed band (s.-d. b). Our theoretical level schemes are compared with experiment in Fig. 2. Characteristics of these levels are well explained by the theory. Collective properties such as kinematical moment of inertia, quadrupole moment and deformation parameters along the s.-d. b. are also well reproduced peovided that we introduce a constant effective charge $e_{pol}/e = 0.34$ for the s.-d. b. That the microscopically calculated β is smaller than the experimental one by about 20 % implies a subtle change of charge distribution especially in the nuclear periphery. Therefore, the apparent large value of β assessed from experimental B(E2) value may not necessarily indicate

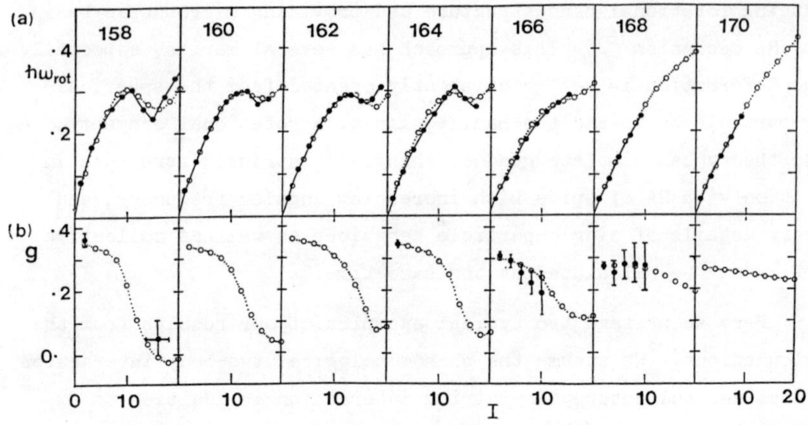

Fig. 1 Theoretical rotational frequency and g-factors versus angular momentum are compared with experiments (solid circles) for even mass Er isotopes.

the increase of prolate deformation in the same rate. This was the first microscopic calculation which was able to explain the superdeformed state based on the fully self-consistent solution to the cranked HFB equation.

3. THE TRPA BUILT ON THE THFB SOLUTION

3.1. The THFB theory and its application

To deal with highly excited state of many particle system like a nucleus, one of the suitable theoretical framework is to extend the HFB theory with statistical recipe. The approximate grand potential is given by

$$F' = \langle H - \lambda\hat{N} - \omega\hat{J}_X \rangle - S'T, \quad (1)$$

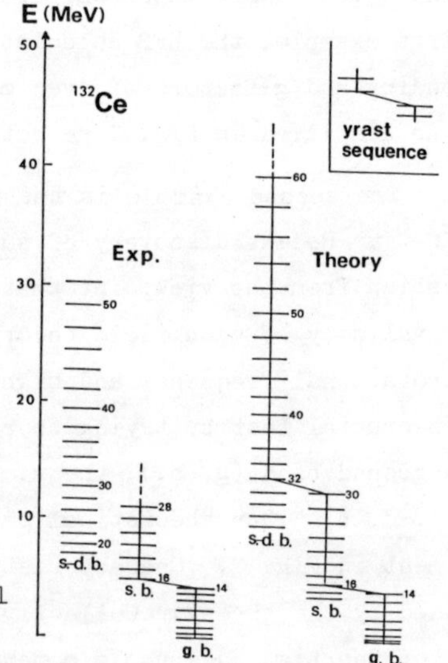

Fig. 2 Comparison of theoretical levels with experiments for ^{132}Ce.

where the approximate entropy is defined by

$$S' = - k\ \mathrm{Tr}(\hat{W}\ \ell n\ \hat{W}). \quad (2)$$

Anticipating the particle-number- and angular momentum- constrained Hamiltonian appearing in the ensemble average bracket in (1) can be brought to diagonal form in the quasiparticle picture, we define the trial density matrix as

$$\hat{W} = \frac{\exp(-\beta \sum_\mu E_\mu \alpha_\mu^\dagger \alpha_\mu)}{\mathrm{Tr}\exp(-\beta \sum_\mu E_\mu \alpha_\mu^\dagger \alpha_\mu)} . \tag{3}$$

The single-particle operator a_k is related to quasiparticle operator α_μ through the generalized Bogoliubov transformation

$$a_k = \sum_\mu (A_{k\mu}\, \alpha_\mu + B_{k\mu}^* \,\alpha_\mu^\dagger). \tag{4}$$

Regarding $A_{k\mu}$, $A_{k\mu}^*$, $B_{k\mu}$, $B_{k\mu}^*$ and E_μ as variational parameters, we derive the THFB equation from the variational requirement

$$\delta\, F' = 0, \tag{5}$$

which is supported by the general inequality for an exact grand potential and an approximate one F',

$$F \leqq F'. \tag{6}$$

The non-linear THFB eigenvalue equation pretends a form similar to zero temperature equation[7],

$$\begin{bmatrix} \xi + \Gamma & \Delta \\ -\Delta^* & -(\xi+\Gamma)^* \end{bmatrix} \begin{bmatrix} A \\ B \end{bmatrix} = \begin{bmatrix} A \\ B \end{bmatrix} E \quad , \tag{7}$$

which is, however, temperature-dependent through the generalized single -particle density

$$\rho_{k\ell} \equiv \langle a_\ell a_k^\dagger \rangle = [B^*(1-f)B^T + AfA^\dagger]_{k\ell} \ , \tag{8a}$$

$$\kappa_{k\ell} \equiv \langle a_\ell a_k \rangle = [B^*(1-f)A^T + AfB^\dagger]_{k\ell} \ , \tag{8b}$$

with

$$f_\mu = 1/[\exp(\beta E_\mu) + 1]. \tag{9}$$

The THFB equation has been solved for deformed nucleus to describe phase transition between superconducting and normal phase under the influence of both the rotational and temperature effect, and also to predict new type of phase with "bidirectional alignment" of spin in the highly excited but low spin region[8].

3.2. The TRPA based on the THFB solution

Recent measurements have been successful to separate the resonance structures from continuum spectrum of high energy γ-rays emitted from the excited compound nucleus[17]. These are the giant dipole resonance (GDR) and the giant quadrupole resonance (GQR) built on the fast rotating hot nucleus. Such collective motions are excited in the nuclear medium whose properties are affected by temperature effect. More specifically, the temperature effect decreases pairing gaps to push the system towards normal phase, while it causes softening of the medium to decrease the vibrational frequency.

Starting from the THFB solution and the same Hamiltonian written in terms of quasiparticles, we can derive the thermal RPA (TRPA) equation again from the variational principle,

$$\delta \langle \exp\{-i(Q^\dagger + Q)\}\, H' \exp\{i(Q^\dagger + Q)\}\rangle = 0, \tag{10}$$

where

$$Q^\dagger = \sum_{\mu<\nu} (X_{\mu\nu}\alpha^\dagger_\mu\alpha^\dagger_\nu - Y_{\mu\nu}\alpha_\nu\alpha_\mu) + \sum_{\mu\nu} Z_{\mu\nu}\alpha^\dagger_\mu\alpha_\nu . \tag{11}$$

Retaining up to the second order in $Q^\dagger$ and Q in the bracket of (10), we get the TRPA equation in the form

$$\mathbf{\Omega}\,\mathbf{M}\,\mathbf{X} = \hbar\omega\,\mathbf{X}, \qquad \mathbf{X} \equiv (X_{\mu\nu}, Y_{\mu\nu}, Z_{\mu\nu}), \tag{12}$$

where $\mathbf{\Omega}$ and $\mathbf{M}$ are the hermitian TRPA matrix and the metric matrix of which concrete forms are given in Ref. 9). In the practice of numerical analysis for the GDR, the imaginary part of the linear response function,

$$\mathbf{R} = \mathbf{M}\left[\hbar\omega - \mathbf{\Omega}\,\mathbf{M}\right]^{-1}, \tag{13}$$

is calculated as a smooth function of ω by adding a proper energy width representing the escape width of the resonance[10].

3.3. The extension of the TRPA

As inferred from the fact that the RPA can be derived from the time-dependent HFB (or HF) formalism, or in the framework of the linear response theory, it is quite interesting to employ the thermo field dynamics (TFD)[12,13] as a real-time formalism for the derivation of the TRPA equation. Applying the TFD, the TRPA operator $Q^\dagger$ is written in larger single-particle space with tilde operator $\tilde{\alpha}_\mu$ in addition to ordinary one α_μ. The increase of the degree of freedom in the variational parameter manifold allows deeper minimum of the grand potential, so that the extended TRPA (ETRPA) equation derived with the present method is able to take more systematic account of thermal correlation than the TRPA. To generalize the first ETRPA to the higher ETRPA is straightforward as for the TRPA. Derivation and detailed account of the ETRPA are presented in Ref. 11).

4. BOSE CONDENSATION OF THE SPIN DENSITY WAVE IN THE HUBBARD MODEL

Our interest here is in the mechanism of realizing the superconductivity with high transition temperature, which has been discovered in the oxide materials like $La_{2-x}Sr_xCuO_4$[18] and $YBa_2Cu_3O_{7-\delta}$[19]. That the isotope effect is hardly observed in these superconductors may imply existence of the other origin of the self-sustaining pairing correlation acting in the electron system. On the other hand, two-dimensional long-ranged antiferromagnetic order has been found in $La_2Cu\,O_4$ at temperatures greater than its Néel temperature.[20] This indicates that spin-spin correlations may play some important roles in the high-T_c superconductivity.

The Hubbard model, which is capable of interpolating two characteristics of magnetism caused by itinerant electrons and localized electrons, is given by

$$H = -t \sum_{\langle k\ell\rangle,\sigma} a^\dagger_{k\sigma} a_{\ell\sigma} + \frac{U}{2} \sum_{k\sigma} a^\dagger_{k\sigma} a^\dagger_{k-\sigma} a_{k-\sigma} a_{k\sigma} \qquad (14a)$$

$$= -t \sum_{\langle k\ell\rangle,\sigma} a^\dagger_{k\sigma} a_{\ell\sigma} + \frac{U}{2} \sum_{k\sigma} a^\dagger_{k\sigma} a_{k\sigma} - \frac{U}{2} \sum_{k} (\sigma^{(+)}_k \sigma^{(-)}_k + \sigma^{(-)}_k \sigma^{(+)}_k). \qquad (14b)$$

In the second expression we observe that, if contributions from the exchange terms occuring in the last term are also included in the THFB equation dealing with first two terms, we can regard the last term giving spin-spin interaction as separable in application of the TRPA. Thus, the TRPA equation for the spin density wave (SDW) of positive helicity, for example,

$$Q^\dagger_+ = \sum_{\mu<\nu} (X^{(+)}_{\mu\nu} \alpha^\dagger_{\mu\uparrow} \alpha^\dagger_{\nu\uparrow} - Y^{(+)}_{\mu\nu} \alpha_{\nu\downarrow} \alpha_{\mu\downarrow}) + \sum_{\mu\nu} Z^{(+)}_{\mu\nu} \alpha^\dagger_{\mu\uparrow} \alpha_{\nu\downarrow} \qquad (15)$$

can be written in the simple form,

$$\sum_\ell (\delta_{k\ell} - U\, M^{(+)}_{k\ell})\, P^{(+)}_\ell = 0, \qquad (16)$$

where

$$M^{(+)}_{k\ell} = \sum_{\mu<\nu} \{ (A^\uparrow_{k\mu} B^\uparrow_{k\nu} - D^\uparrow_{k\mu} A^\uparrow_{k\nu}) \frac{1 - f_{\mu\uparrow} - f_{\nu\uparrow}}{\hbar\omega - E_{\mu\uparrow} - E_{\nu\uparrow}} (A^\uparrow_{\ell\mu} B^\uparrow_{\ell\nu} - B^\uparrow_{\ell\mu} A^\uparrow_{\ell\nu})^*$$

$$- (A^{\downarrow}_{k\mu} B^{\downarrow}_{k\nu} - B^{\downarrow}_{k\mu} A^{\downarrow}_{k\nu})^* \frac{1-f_{\mu\downarrow}-f_{\nu\downarrow}}{\hbar\omega + E_{\mu\downarrow} + E_{\nu\downarrow}} (A^{\downarrow}_{\ell\mu} B^{\downarrow}_{\ell\nu} - B^{\downarrow}_{\ell\mu} A^{\downarrow}_{\ell\nu}) \}$$

$$+ \sum_{\mu\nu} (A^{\uparrow}_{k\mu} A^{\downarrow *}_{k\nu} - B^{\uparrow}_{k\mu} B^{\downarrow *}_{k\nu}) \frac{f_{\nu\downarrow}-f_{\mu\uparrow}}{\hbar\omega - E_{\mu\uparrow} + E_{\nu\downarrow}} (A^{\uparrow *}_{\ell\mu} A^{\downarrow}_{\ell\nu} - B^{\uparrow *}_{\ell\mu} B^{\downarrow}_{\ell\nu}), \tag{17}$$

and

$$P^{(+)}_{\ell} \equiv \langle [\sigma^{(-)}_{\ell}, Q^{\dagger}_{+}] \rangle = \langle [a^{\dagger}_{\ell\downarrow} a_{\ell\uparrow}, Q^{\dagger}_{+}] \rangle. \tag{18}$$

The dimensionality of the TRPA equation (16) for finite two-dimensional square lattice is N_{site}, which is much smaller than the one $(N_{site})^2$ of the ordinary TRPA equation (12).

For simplicity we consider the HF approximation (i. e. $B^{\sigma}_{k\mu} = 0$) for the paramagnetic phase in the finite two-dimensional system, and take Fourier transforms of all the expressions. Then, we can show that the imaginary TRPA solution (i. e. $(\hbar\omega)^2 \leqq 0$) occurs provided that the following instability criterion is satisfied.

$$1 \leqq \frac{U}{(2\pi)^2 N_{site}} \int d^2k \; \frac{f_{\uparrow}(\mathbf{K}+\mathbf{k}) - f_{\downarrow}(\mathbf{k})}{E_{\downarrow}(\mathbf{k}) - E_{\uparrow}(\mathbf{K}+\mathbf{k})} \tag{19}$$

with

$$E_{\sigma}(\mathbf{k}) = \frac{1}{2} U(n-m) - 2t(\cos k_x a + \cos k_y a), \tag{20a}$$

$$n = \langle a^{\dagger}_{\uparrow} a_{\uparrow} \rangle + \langle a^{\dagger}_{\downarrow} a_{\downarrow} \rangle, \tag{20b}$$

$$m = \langle a^{\dagger}_{\uparrow} a_{\uparrow} \rangle - \langle a^{\dagger}_{\downarrow} a_{\downarrow} \rangle. \tag{20c}$$

This condition is satisfied only when U/t is sufficiently large, because the second term in (20a) supplies the factor t in the denominator of (19). Applying the chain approximation to the Bethe-Salpeter equation for the Green function of SDW, we can show that the Boson number diverges as follows.

$$\begin{aligned} \langle Q^{\dagger}_{+} Q_{+} \rangle &= \lim_{t \to +0} \langle T(Q^{\dagger}_{+}(t) Q_{+}(0) \rangle \\ &= \frac{i\hbar}{2\pi} \int_{-\infty}^{\infty} d\omega \; G(\omega) \\ &= \frac{i\hbar}{2\pi} \int_{-\infty}^{\infty} d\omega \; \mathbf{X}^{\dagger} \mathbf{M} [\hbar\omega - \boldsymbol{\Omega} \mathbf{M}]^{-1} \mathbf{X} \longrightarrow \infty \;. \end{aligned} \tag{21}$$

In the last expression we see that two poles on the opposite sides of real axis $\hbar\omega = + i\varepsilon$, come together at the origin, and this pinching of poles results in the Boson condensation of SDW.

Since this instability implies that the THFB solution becomes unphysical, we must look for better approximation to overcome the difficulty. One of such a possibility left us is to extend the variational parameter space in order to attain deeper minimum of the grand potential. For this purpose, introducing the extended Bogoliubov transformation which allows mixing of spin components,

$$a_k^\sigma = \sum_{\mu\sigma'} (A_{k\mu}^{\sigma\sigma'} \alpha^{\sigma'} + B_{k\mu}^{\sigma\sigma'*} \alpha_\mu^{\sigma'\dagger}), \tag{22}$$

and defining

$$\rho_{k\ell}^{\rho\sigma} \equiv \begin{pmatrix} \rho^{\uparrow\uparrow} & \rho^{\uparrow\downarrow} \\ \rho^{\downarrow\uparrow} & \rho^{\downarrow\downarrow} \end{pmatrix}_{k\ell} \qquad \kappa_\ell^{\rho\sigma} \equiv \begin{pmatrix} \kappa^{\uparrow\uparrow} & \kappa^{\uparrow\downarrow} \\ \kappa^{\downarrow\uparrow} & \kappa^{\downarrow\downarrow} \end{pmatrix}_{k\ell}, \tag{23}$$

we derive the following extended THFB (ETHFB) equation

$$(-\lambda - \tfrac{1}{2}g\mu H\sigma + \Gamma_{kk}^{\sigma\sigma})A_{k\mu}^{\sigma\rho} - t \sum_{\ell\in<k\ell>} A_{\ell\mu}^{\sigma\rho} + \Sigma_{kk}^{\sigma-\sigma} A_{k\mu}^{-\sigma\rho} + \Delta_{kk}^{\sigma-\sigma} B_{k\mu}^{-\sigma\rho} = A_{k\mu}^{\sigma\rho} E_\mu^\rho \tag{24a}$$

$$-\Delta_{kk}^{\sigma-\sigma*} A_{k\mu}^{-\sigma\rho} - (-\lambda - \tfrac{1}{2}g\mu H\sigma + \Gamma_{kk}^{\sigma\sigma*})B_{k\mu}^{\sigma\rho} + t \sum_{\ell\in<k\ell>} B_{\ell\mu}^{\sigma\rho} \Sigma_{kk}^{\sigma-\sigma*} B_{k\mu}^{-\sigma\rho} = B_{k\mu}^{\sigma\rho} E_\mu^\rho . \tag{24b}$$

This set of equations must be solved under the constraints for electron density n and spin density m in each site, i. e.

$$\rho_{kk}^{\uparrow\uparrow} + \rho_{kk}^{\downarrow\downarrow} = n, \quad \rho_{kk}^{\uparrow\uparrow} - \rho_{kk}^{\downarrow\downarrow} = m. \tag{25}$$

The chemical potential λ and a combination $\frac{1}{2}g\mu H$ (g is the gyromagnetic ratio of electron in the material, μ magnetic moment of free electron and H magnetic field) are regarded as Lagrange multipliers to be determined by (25). Thus, we have a newly defined order parameter

$$\Sigma_{kk}^{\sigma-\sigma} \equiv -U \rho_{kk}^{\sigma-\sigma} \tag{26}$$

besides the single-particle self-energy and the pairing gap defined by

$$\Gamma_{kk}^{\sigma\sigma} \equiv U \rho_{kk}^{-\sigma-\sigma} \quad , \quad \Delta_{kk}^{\sigma-\sigma} \equiv U \kappa_{kk}^{\sigma-\sigma}. \tag{27}$$

Then, we get

$$\langle H' \rangle = \sum_{k\sigma} \Big[-(\lambda + \tfrac{1}{2}g\mu H\sigma)\rho_{kk}^{\sigma\sigma} - \frac{t}{2} \sum_{\ell\in<k\ell>,\sigma'} (\rho_{k\ell}^{\sigma\sigma'} + \rho_{k\ell}^{\sigma\sigma'*}) + \frac{1}{2}(\Gamma_{kk}^{\sigma\sigma}\rho_{kk}^{\sigma\sigma} + \Delta_{kk}^{\sigma-\sigma}\kappa_{kk}^{\sigma-\sigma*} + \Sigma_{kk}^{\sigma-\sigma}\rho_{kk}^{\sigma-\sigma*}) \Big] \tag{28}$$

in which the last term related to the local spin fluctuation (LSF) contributes to decrease total energy. However, is there in the Hubbard model any source of attractive force strong enough to create Cooper pair of electrons ? Detailed arguments about this problem will be published elsewhere[15].

Since electron spin is reversed by the direct contact of two electrons at a certain site as inferred from the Hamiltonian (14b), propagation of SDW, which is nothing but transfer of magnetization from one site to the other, is associated with the conduction of individual electrons not in pair. This suggests an attractive possibility that new type oj superconductivity may be realized in association with the violation of rotational symmetry in spin space. This is a marked contrast to the violation of number gauge symmetry in usual low-T_c superconductivity. Therefore, macroscopic variable for the new supeconductivity is a phase β of the state, and $\partial E(\beta)/\partial\beta$ corresponds to magnetization transfer through the Spin-Josephson junction. However, it is still an open question if the attractive force is strong enough to create electron Cooper pairs.

Theoretical analysis based on the quantum Monte Carlo simulation[21] suggests that inclusion of oxygen site between neighbouring copper sites is important to create pairing correlation. If this is true, the extended Hubbard model is more suitable to be studied. Our theoretical framework proposed in the present paper is in principle applicable to any two-body interaction including the extended Hubbard model, and to the superconductor associated with LSF.

ACKNOWLEDGEMENTS

The author wishes to express his sincere thanks to Professor H. Miyazawa for his guidance in physics and encouragement. This paper is dedicated to Professor Miyazawa on the occasion of his sixtieth birthday.

REFERENCES

1. B. R. Mottelson and J. G. Valatin, Phys. Rev. Lett. 5, 511 (1960).
2. P. J. Twin et al, Phys. Rev. Lett. 57, 811 (1986).
3. P. J. Nolan et al, J. Phys. G11, L11 (1985);
   A. J. Kirwan et al, Phys. Rev. Lett. 58, 467 (1987).
4. F. S. Stephens and R. S. Simon, Nucl. Phys. A183, 257 (1972).

5. K. Sugawara-Tanabe and K. Tanabe, preprint (1988), to be published.
6. K. Tanabe and K. Sugawara-Tanabe, preprint (1988), to be published.
7. K. Tanabe and K. Sugawara-Tanabe, Phys. Lett. 97B, 337 (1980); K. Tanabe, K. Sugawara-Tanabe and H. J. Mang, Nucl. Phys. A357, 20 (1981).
8. K. Sugawara-Tanabe, K. Tanabe and H. J. Mang, Nucl. Phys. A357, 45 (1981); K. Tanabe and K. Sugawara-Tanabe, Nucl. Phys. A390, 385 (1982).
9. K. Tanabe and K. Sugawara-Tanabe, Phys. Lett. 172B, 129 (1986); Prog. Theor. Phys. 76, 356 (1986).
10. K. Sugawara-Tanabe and K. Tanabe, Prog. Theor. Phys. 76, 1272 (1986); Phys. Lett. 192B, 268 (1987).
11. K. Tanabe, Phys. Rev. C, in press.
12. Y. Takahashi and H. Umezawa, Coll. Phenom. 2, 55 (1975).
13. H. Umezawa, H. Matsumoto and M. Tachiki, Thermo Field Dynamics and Condensed States (North-Holland, Amsterdam, 1982), and references therein.
14. J. Hubbard, Proc. Roy. Soc. London, Ser. A276, 283 (1963); A281, 401 (1964).
15. K. Tanabe, preprint (1988), to be published.
16. H. J. Mang, Phys. Rep. 18, 325 (1975).
17. J. O. Newton et al, Phys. Rev. Lett. 46, 1383 (1981); J. E. Draper et al, Phye. Rev. Lett. 49, 434 (1982); W. Hennerici et al, Nucl. Phys. A396, 329c (1983); J. J. Gaardhøje et al, Phys. Rev. Lett. 53, 148 (1984); C. A. Gossett et al, Phys. Rev. Lett. 54, 1486 (1985).
18. J. G. Bednorz and K. A. Müller, Z. Phys. B64, 189 (1986); S. Uchida et al, Jpn J. Appl. Phys. Part 2, 26, L1 (1987).
19. C. W. Chu et al, Phys. Rev. Lett. 58, 405 (1987); R. J. Cava et al, Phys. Rev. Lett. 58, 408 (1987); M. K. Wu et al, Phys. Rev. Lett. 58, 908 (1987); R. J. Cava et al, Phys. Rev. Lett. 58, 1676 (1987).
20. G. Shirane et al, Phys. Rev. Lett. 59, 1613 (1987).
21. M. Imada, J. Phys. Soc. Jpn 56, 3793 (1987).

Chapter 2 PARTICLE PHENOMENOLOGY

SUM RULES FOR THE $p\bar{p}$ AND pp FORWARD AMPLITUDE TO DETECT ASYMPTOTIC BEHAVIORS*

Keiji Igi

Department of Physics, University of Tokyo
Tokyo, Japan

ABSTRACT

We derive a new sum rule which could detect the odderon term. We also derive a sum rule which would constrain the asymptotic behaviors of the $(p\bar{p} + pp)$ total cross sections.

1. Introduction

A long time ago, Goldberger, Miyazawa and Oehme[1] (GMO) derived a sum rule which gives the s-wave πN scattering lengths in terms of the πN coupling constant and an integral involving the $\pi^{\pm}p$ total cross sections over all energies. This beautiful work[1] is the first application of dispersion relations to the particle physics. Since then, a new page has been opened in the development of the S-matrix theory.

We shall derive a new sum rule which could detect the odderon term in Sec. 2. We also derive a sum rule which constrain the asymptotoc behaviors of the $(p\bar{p} + pp)$ total cross sections in Sec. 3. The final Sec. 4 is devoted to concluding remarks.

2. Sum Rules for Crossing-odd Amplitude

Firstly, we shall investigate the asymptotic behavior of the crossing-odd amplitude. In 1987, the ratio $\rho_{p\bar{p}}$ of real to imaginary part of the $p\bar{p}$ forward scattering amplitude has been measured at $\sqrt{s}$ = 546 GeV in an experiment at the CERN $Sp\bar{p}S$ collider[2]. The value $\rho_{p\bar{p}} = 0.24 \pm 0.04$ found is larger than expected from analyticity according to which $\rho_{p\bar{p}}$ should lie between 0.11 and 0.15[3]. This result has already led to a lot of speculations.

(i) One may be tempted to take this large value of $\rho_{p\bar{p}}$ as evidence for the odderon.[4,5] The price one has to pay for accepting this explanation is that $\sigma_{p\bar{p}}$ and σ_{pp} cross at some energy leading to $\sigma_{pp} > \sigma_{p\bar{p}}$ at very high energies. This consequence is unavoidable as has been derived by Fischer[6] for a large class of odderon-type amplitudes.

(ii) Leader[7] and Martin[3] have pointed out that an unexpected large rise of the crossing-even part of the total cross-section above the energy of the CERN collider may cure the above disease.

(iii) The usual method of analyzing the Coulomb interference data bases on a very simple parametrization of the nuclear amplitude which if far from reality may lead to wrong results for ρ as was argued by Bourrely et al.[8]

A. A sum rule for the odderon.

The motivation of this section is to test the so-called odderon hypothesis for pp and $p\bar{p}$ scattering. Suppose we start from an unsubtracted dispersion relation such as

$$\mathrm{Ref}(\nu) = \frac{P}{\pi} \int_{-N}^{N} d\nu' \, \frac{\mathrm{Imf}(\nu')}{\nu' - \nu} + \mathrm{Re}\left[\frac{1}{\pi i} \int_{c_1} d\nu' \, \frac{f(\nu')}{\nu' - \nu} \right] , \qquad (1)$$

where $f(\nu)$ is a forward scattering amplitude, ν is the laboratory energy and c_1 is a contour along the large circle in the upper-half complex ν plane. Putting ν to be a nucleon mass m, we obtain

$$\mathrm{Ref}(m) = \frac{P}{\pi} \int_{-N}^{N} d\nu \, \frac{\mathrm{Imf}(\nu)}{\nu - m} + \mathrm{ReI}(m), \qquad (2)$$

where $I(m) = \frac{1}{\pi i} \int_{c_1} d\nu \, \frac{f(\nu)}{\nu - m}$,

using the technique due to Lam and Truong.[9]

Let us take as $f(\nu)$ a crossing-odd (spin-averaged) forward amplitude for pp and $p\bar{p}$ scattering :

$$F^{(-)} = \frac{1}{2} \left(F_{p\bar{p}} - F_{pp} \right) \qquad (3)$$

whose numerical evaluation became feasible only recently by virtue of the advent of the LEAR $p\bar{p}$ data on total cross-sections[10] and $\rho_{p\bar{p}}$[11] at low energies and of the $p\bar{p}$ scattering length obtained from measurements of the shift and width of the 1s level of antiprotonic hydrogen.[12]

Assuming $F^{(-)}$ to approach to the maximal odderon term[5] $F_{odd}(\nu) = c\nu \ln^2(-i \frac{\nu}{\nu_0})$ for large ν, we evaluate the contribution of the large semicircle c_1 to obtain for I(m),

$$\frac{c}{\pi i} \int_{C_1} d\nu \ln^2(-i \frac{\nu}{\nu_0}) = \frac{2icN}{\pi} [\ln^2(\frac{N}{\nu_0}) - 2\ln(\frac{N}{\nu_0}) + 2 - \frac{\pi^2}{4}]$$

in the leading order $O(\frac{1}{\nu})$. Therefore ReI(m) receives no contributions in this order. The first non-vanishing contribution arises from the next leading order $O(\frac{m}{\nu^2})$; i.e.

$$I(m) = \frac{cm}{\pi i} \int_{C_1} d\nu \frac{\ln^2(-i \frac{\nu}{\nu_0})}{\nu}$$

$$= cm[\ln^2(\frac{N}{\nu_0}) - \frac{\pi^2}{12}] = \mathrm{Re}I(m) \quad . \tag{4}$$

Therefore, the eqs.(2),(3) and (4) immedialely lead to

$$\mathrm{Re}F^{(-)}(m) = \frac{P}{\pi} \int_{-N}^{N} d\nu \frac{\mathrm{Im}F^{(-)}(\nu)}{\nu - m} + cm[\ln^2(\frac{N}{\nu_0}) - \frac{\pi^2}{12}] \quad . \tag{5}$$

The second term on the right-hand side of eq.(5) was too tiny to detect any odderon term. This corresponds to our previous sum rule.[13]

It is very important to remark the following point. In the original GMO[1], the sum rule for the forward πN amplitude

$$\Delta(\mu) - \Delta(\infty) = \frac{4f^2}{\mu^2} + \frac{1}{2\mu^2} \int_{\mu}^{\infty} \frac{d\nu}{k} [\sigma_{\pi^- p}(\nu) - \sigma_{\pi^+ p}(\nu)] \tag{6}$$

with $\Delta(\mu) = \frac{2}{3}(a_1 - a_3)\frac{1}{\mu}$ has been obtained. The GMO[1] could not prove theoretically $\Delta(\infty)$ to be zero, but argued empirically that $\Delta(\infty)$ is consistent with zero. On the other hand, our sum rule for the πN amplitude which corresponds to eq.(5) is written exactly as

$$\Delta(\mu) - \Delta(N) = \frac{4f^2}{\mu^2} + \frac{1}{2\pi^2}\int_\mu^N \frac{d\nu}{k}\,[\sigma_{\pi^- p}(\nu) - \sigma_{\pi^+ p}(\nu)] \quad , \qquad (7)$$

where $\Delta(N) = 2c[\ln^2(\frac{N}{\nu_0}) - \frac{\pi^2}{12}]$. In general, $\Delta(N)$ contains the information on the high energy amplitude and is shown to be negligible in the present specific model.

B. A new sum rule for the odderon

Applying the above kind of technique to the crossing-even amplitude, the f(P') Regge trajectory had been successfully recovered a long time ago.[14] The reason why the asymptotic behaviors had been predicted from low-energy data is as follows: If one takes a crossing-even amplitude, a contribution to ReI(m) in eq.(2) arises in the leading order.

Therefore we shall derive a new sum rule which could detect the odderon term, by employing a new crossing-even amplitude $f(\nu) = \nu F^{(-)}(\nu)$. Here $F^{(-)}(\nu)$ is defined in eq.(3). We further assume:

$$F^{(-)}(\nu) \xrightarrow[\nu\ \text{large}]{} F_{odd}(\nu) + F_R(\nu). \qquad (8)$$

Here $$F_{odd}(\nu) = c\nu \ln^2(-i\frac{\nu}{\nu_0}) = c\nu[\ln\frac{\nu}{\nu_0} - \frac{1}{2}i\pi]^2 \qquad (9)$$

and the Regge contributions $F_R(\nu)$ are written in the standard form

$$F_R(\nu) = ic_R\nu^{\alpha_R}\exp(-\tfrac{1}{2}i\pi\alpha_R) = \gamma\nu^{\alpha_R}[i+\tan\tfrac{\pi}{2}\alpha_R] . \tag{10}$$

Since $f(\nu)$ is crossing even, the sum rule (2) becomes

$$\mathrm{Re}f(m) = \frac{P}{\pi}\int_0^N d\nu \mathrm{Im}f(\nu)\frac{2\nu}{\nu^2-m^2} + \mathrm{Re}I(m) \tag{11}$$

with $I(m) = \frac{1}{\pi i}\int_{C_1} d\nu \frac{f(\nu)}{\nu-m}$.

Substituting $\nu F_{odd}(\nu) = c\nu^2 \ln^2(-i\frac{\nu}{\nu_0})$ for $f(\nu)$ in the integrand of $I(m)$, we obtain in the leading order $O(\frac{1}{\nu})$,

$$I^{odd}(m) = \frac{1}{\pi i}\int_{C_1} d\nu\, c\nu \ln^2(-i\frac{\nu}{\nu_0})$$

$$= \frac{cN^2}{\pi}\int_0^\pi d\theta e^{2i\theta}[(\ln\frac{N}{\nu_0})^2 + 2i(\theta-\frac{\pi}{2})\ln\frac{N}{\nu_0} - (\theta-\frac{\pi}{2})^2] .$$

Since $\int_0^\pi d\theta e^{2i\theta} = 0$, $\int_0^\pi d\theta e^{2i\theta}(\theta-\frac{\pi}{2}) = -\frac{\pi}{2}i$ and $\int_0^\pi d\theta e^{2i\theta}(\theta-\frac{\pi}{2})^2 = \frac{\pi}{2}$, we have

$$I^{odd}(m) = cN^2[\ln\frac{N}{\nu_0} - \frac{1}{2}] = \mathrm{Re}I^{odd}(m) . \tag{12}$$

Substituting $\nu F_R(\nu)$ for $f(\nu)$ in the integrand of $I(m)$, we obtain

$$I^R(m) = -\frac{2\gamma}{\pi(\alpha_R+1)}N^{\alpha_R+1} = \mathrm{Re}I^R(m) \tag{13}$$

in the leading order. Hence $\mathrm{Re}I(m) = \mathrm{Re}I^{odd}(m) + \mathrm{Re}I^R(m)$.

Therefore, substitution of ReI(m) into eq. (11) immediately leads us to

$$mReF^{(-)}(m) = \frac{P}{\pi}\int_0^N d\nu ImF^{(-)}(\nu)\,\frac{2\nu^2}{\nu^2-m^2}$$

$$+ CN^2[\ln\frac{N}{\nu_0} - \frac{1}{2}] - \frac{2\gamma}{\pi(\alpha_R+1)}N^{\alpha_R+1} \quad . \qquad (14)$$

For the purpose of a numerical evaluation, a more suitable form of eq. (14) is :

$$mReF^{(-)}(m) = (\frac{g_\pi^2}{4\pi})\,\frac{\mu^2}{8m^2}\,\frac{\nu_\pi^2}{\nu_\pi^2-m^2}$$

$$+ \frac{1}{\nu}\int_{\nu(2\mu)}^{m-\varepsilon} d\nu\,\frac{\nu^2 ImF_{p\bar{p}}(\nu)}{\nu^2-m^2} + \frac{2}{\pi}P\int_{m-\varepsilon}^{m+\varepsilon} d\nu\frac{\nu^2}{\nu^2-m^2}ImF^{(-)}(\nu)$$

$$+ \frac{1}{4\pi^2}\int_{k_\varepsilon}^{N} dk\nu[\sigma_{p\bar{p}} - \sigma_{pp}] + CN^2[\ln\frac{N}{\nu_0} - \frac{1}{2}] - \frac{2\gamma}{\pi(\alpha_R+1)}N^{\alpha_R+1}, \qquad (15)$$

wherein $m(\mu)$ is the nucleon (pion) mass, $\nu_\pi = -m + \frac{\mu^2}{2m}$ and $\nu(2\mu) = -m + \frac{2\mu^2}{m}$ is the tip of the unphysical part of the $p\bar{p}$ cut. Here ε is an energy sufficiently small to allow an analytic evaluation of the principal value integral and $k_\varepsilon = \sqrt{(m+\varepsilon)^2-m^2}$. Actually we may choose ε = 1MeV.

3. A Sum Rule for the ($p\bar{p}$ + pp) Forward Amplitude

In this section we shall derive a sum rule which would constrain the asymptotic behaviors of the ($p\bar{p}$ + pp) total cross sections.

Let us define a crossing-even (spin-averaged) forward amplitude for pp and $p\bar{p}$ scattering by

$$F^{(+)} = \frac{1}{2}(F_{p\bar{p}} + F_{pp}) \quad , \tag{16}$$

and substitute this amplitude as $f(\nu)$ in eq. (1). Along the similar line as before, we obtain

$$\mathrm{Re}F^{(+)}(m) = \frac{P}{\pi}\int_0^N d\nu \mathrm{Im}F^{(+)}(\nu)\, \frac{2\nu}{\nu^2-m^2} + \mathrm{Re}\left[\frac{1}{\pi i}\int_{C_1} d\nu\, \frac{F^{(+)}(\nu)}{\nu-m}\right] . \tag{17}$$

We now assume

$$F^{(+)}(\nu)\xrightarrow[\nu\ \mathrm{large}]{} F_p^{(+)}(\nu) + F_f^{(+)}(\nu) + F_1^{(+)}(\nu) + F_2^{(+)}(\nu) \quad , \tag{18}$$

where $F_p^{(+)}(\nu) = ic_p\nu$,

$$F_f^{(+)}(\nu) = -c_f\nu^{\alpha_f}\exp\left(-\frac{1}{2}\, i\pi\alpha_f\right),$$

$$F_1^{(+)}(\nu) = ic_1\nu\left[\,\ell n\frac{\nu}{\nu_1} - \frac{1}{2}\, i\pi\,\right] \quad , \tag{19}$$

and $F_2^{(+)}(\nu) = ic_2\nu\left[\,\ell n\frac{\nu}{\nu_2} - \frac{1}{2}\, i\pi\,\right]^2$.

So, the total cross section is expressed as

$$\sigma^{(+)}(\nu) = \frac{4\pi}{k}\,\mathrm{Im}F^{(+)}$$

$$= 4\pi c_p + 4\pi c_f\left(\sin\frac{\pi}{2}\alpha_f\right)\nu^{\alpha_f-1}$$

$$+ 4\pi c_1\,\ell n\frac{\nu}{\nu_1} + 4\pi c_2\left[\left(\ell n\frac{\nu}{\nu_2}\right)^2 - \frac{\pi^2}{4}\right] \quad . \tag{20}$$

It will be convenient to rewrite eq. (17) in the following form for practical use:

$$\mathrm{Re}F^{(+)}(m) = \frac{f^2}{4\pi}\,\frac{m-\frac{\mu^2}{2m}}{2\mu^2-\frac{\mu^4}{2m^2}}$$

$$+\frac{1}{\pi}\int_{\nu(2\mu)}^{m-\varepsilon} d\nu\,\frac{\nu}{\nu^2-m^2}\,\mathrm{Im}F_{p\bar{p}}(\nu) + \frac{2}{\pi}P\int_{m-\varepsilon}^{m+\varepsilon} d\nu\,\frac{\nu}{\nu^2-m^2}\,\mathrm{Im}F^{(+)}(\nu)$$

$$+\frac{1}{4\pi^2}\int_{k_\varepsilon}^{N} dk[\sigma_{p\bar{p}}+\sigma_{pp}] - \frac{2Nc_p}{\pi} - \frac{2c_f N^{\alpha_f}}{\pi\alpha_f}\sin\frac{\pi}{2}\alpha_f$$

$$-\frac{2c_1 N}{\pi}\left[\ln\frac{N}{\nu_1}-1\right] - \frac{2c_2 N}{\pi}\left[\left(\ln\frac{N}{\nu_2}-1\right)^2+1-\frac{\pi^2}{4}\right]$$

$$\text{with}\quad f^2 = \left(\frac{\mu g_\pi}{2m}\right)^2 . \qquad (21)$$

4. Concluding Remarks

When the dispersion integral diverges, one usually writes a subtracted dispersion relation. In this paper, however, we have written an unsubtracted dispersion relation when the asymptotic behavior of scattering amplitude is known by taking into account the contribution of the large circle. The real part calculated from the unsubtracted dispersion relation receives contribution from all the energy region.

Therefore, we can hope to obtain informations about odderons from the sum rule eq. (15) derived in section 2. B. The sum rule eq.(21) derived in section 3 will constrain the crossing-even asymptotic behaviors from the low- as well as intermediate- energy informations.

ACKNOWLEDGEMENTS

I wish to congratulate Professor Hironari Miyazawa for his 60th birthday. Professor Miyazawa just came back from the states when I became a graduate student of University of Tokyo. I learned a lot of things on the S-matrix theory based on dispersion approach. I am extremely thankful to Professor Miyazawa for all discussions, criticisms and encouragement for over thirty years. I wish him many more active and happy years in the future.

I would also like to thank Professor Peter Kroll, Professor Andre Martin and Dr. Shigeo Yazaki for useful discussions. A part of this article is based on reference 13. I am also thankful to Professor Kyungsik Kang for conversations from which I became aware of the odderon problem.

Finally, I would like to thank the Japan Society for the Promotion of Science for supporting the cooperative research with the Wisconsin group under the Japan-U.S. Cooperative Science Program.

REFERENCES

1. M.L.Goldberger, H.Miyazawa, and R.Oehme, Phys. Rev.99 (1955) 986.
2. D.Bernard et al., CERN Preprint CERN/EP 87-147 (1987).
3. A.Martin - CERN Preprint TH. 4852 (1987).
4. K.Kang and B.Nicolescu, Phys. Rev. D11 (1975) 2461.
5. D.Bernard, P. Gauron, and B. Nicolescu, Phys. Lett. 199B (1987) 125.
6. J.Fischer, Z. Phys. C36 (1987) 273.
7. E.Leader, Phys. Rev. Lett. 59 (1987) 1525.
8. C.Bourrely, J.Soffer and T.T.Wu, Phys. Lett. B196 (1987) 237.
9. W.S.Lam and T.N.Truong, Phys. Lett. 31B (1970) 307.
10. D.Bugg et al., Phys. Lett. 194B (1987) 563.
11. W.Bruckner et al., Phys. lett 158B (1985) 180 and Contributions to the 4th LEAR Workshop, Villars sur Ollon (1987);
    L.Linssen et al., CERN Preprint CERN/EP 87-36 (1987).

12. S.Ahmad et al., Phys. Lett. 157B (1985) 333;
T.P.Gorringe et al., Phys. Lett. 162B (1986) 71.
13. K.Igi and P.Kroll, CERN Preprint CERN-TH. 4891 (1987) with Erratum.
14. K.Igi, Phys. Rev. Lett. 9 (1962) 76; Phys. Rev. 130 (1963) 820.

SLEPTON SEARCH AT MULTI-TEV HADRON COLLIDERS

K. Hidaka and H. Kogai
Department of Physics, Tokyo Gakugei University
Koganei-shi, Tokyo, Japan

ABSTRACT

We compute cross sections for slepton pair production in $\overset{(-)}{p}p$ collisions at TEVATRON and SSC. We find that a detectable number of sleptons should be produced at TEVATRON and SSC if their masses are less than ~100 GeV and ~1.0 TeV, respectively, and that this process deserves detailed Monte Carlo studies to separate signal events from backgrounds.

Recently there has been growing interest in supersymmetric gauge theories which offer a promising solution for the hierarchy probrem of grand unified models. This solution requires that supersymmetric partners of all known particles exist with masses less than about 1 TeV which are thus within reach of the near future colliders. As yet no experimental signature of supersymmetry has been observed.

The production of the supersymmetric partners of the lepton, "the sleptons" ($\tilde{l}$) has been studied by many physicists [1]. The most stringent current limit on the selectron ($\tilde{e}$) mass is $m_{\tilde{e}} > 65$ GeV for the photino ($\tilde{\gamma}$) mass $m_{\tilde{\gamma}} \lesssim 3$ GeV obtained by the experiments in search for $e^- e^+ \to \gamma\tilde{\gamma}\tilde{\gamma}$ events at PEP and PETRA [2]. The UA1 group at CERN $\bar{p}p$ collider obtained a limit $m_{\tilde{e}} \gtrsim 26$ GeV for $m_{\tilde{e}} = m_{\tilde{\nu}}$ from the experiment in search for the slepton pair production through weak boson decay $\bar{p}p \to W^{\pm} X \to \tilde{e}^{\pm}\tilde{\nu} X$ [3]. E. Eichten et al [4] studied the charged slepton pair production $\overset{(-)}{p}p \to \tilde{l}^- \tilde{l}^+ X$ at TEVATRON and SSC. In this article we investigate the slepton pair productions $\overset{(-)}{p}p \to (\tilde{l}^- \tilde{l}^+, \tilde{\nu}\bar{\tilde{\nu}}$ or $\tilde{l}^{\pm}\tilde{\nu}) + X$ at these colliders in a little more detail including QCD radiative corrections and signal-background separation. We show how high masses of the sleptons can be reached at these colliders.

Main contributions to the slepton production in $\overset{(-)}{p}p$ collision are those of the supersymmetric Drell-Yan processes: $\overset{(-)}{p}p \to \tilde{l}\,\bar{\tilde{l}}'\,X$. In the parton model the total cross section for these processes is given by

$$\sigma_{TOT}(\tilde{l}\,\bar{\tilde{l}}') = \sum_{p_1, p_2} \int_0^1 dx_1 \int_0^1 dx_2\, p_1(x_1, Q^2)\, p_2(x_2, Q^2)\, \hat{\sigma}_{TOT}(p_1 p_2 \to \tilde{l}\,\bar{\tilde{l}}') \cdot K(x_1, x_2, Q^2) \qquad (1)$$

, where $p_i(x_i, Q^2)$ are scaling-violating parton densities at $Q^2 = m^2_{\tilde{l}\bar{\tilde{l}}}$, $\hat{\sigma}_{TOT}$ is the total cross section at the parton level in the Born approximation and $K(x_1, x_2, Q^2)$ refers to the so-called K-factor which stems from QCD radiative corrections. As for the parton densities we use those of ref.[4]. By use of the supersymmetric Lagrangian [1] we obtain the subprocess cross section for each lepton flaver with an assumption $m_{\tilde{l}_R} = m_{\tilde{l}_L} = m_{\tilde{l}}$ as follows ($l^- = e^-, \mu^-$ or τ^-):

$$d\hat{\sigma}/d\hat{t}(q\bar{q} \to "\gamma, Z^0" \to \tilde{l}^- \tilde{l}^+) = (\pi\alpha^2/3\hat{s}^2)A\{\hat{s}(\hat{s}-4m^2_{\tilde{l}}) - (2\hat{t}+\hat{s}-2m^2_{\tilde{l}})^2\}$$

$$\hat{\sigma}_{TOT}(q\bar{q} \to "\gamma, Z^0" \to \tilde{l}^- \tilde{l}^+) = (2\pi\alpha^2\hat{s}/9)A\{1-(4m^2_{\tilde{l}}/\hat{s})\}^{3/2} \quad (2)$$

$$d\hat{\sigma}/d\hat{t}(q\bar{q} \to "Z^0" \to \tilde{\nu}_l \bar{\tilde{\nu}}_l) = \{\pi\alpha^2(b_q^2 + a_q^2)/(96\sin^4\theta_W\cos^4\theta_W\hat{s}^2\chi_Z)\} \cdot$$

$$\cdot\{\hat{s}(\hat{s}-4m^2_{\tilde{\nu}}) - (2\hat{t}+\hat{s}-2m^2_{\tilde{\nu}})^2\}$$

$$\hat{\sigma}_{TOT}(q\bar{q} \to "Z^0" \to \tilde{\nu}_l \bar{\tilde{\nu}}_l) = \{\pi\alpha^2(b_q^2 + a_q^2)\hat{s}/(144\sin^4\theta_W\cos^4\theta_W\chi_Z)\} \cdot$$

$$\cdot\{1-(4m^2_{\tilde{\nu}}/\hat{s})\}^{3/2} \quad (3)$$

$$d\hat{\sigma}/d\hat{t}(q_u\bar{q}_d \to "W^+" \to \tilde{l}^+\tilde{\nu}_l) = d\hat{\sigma}/d\hat{t}(q_d\bar{q}_u \to "W^-" \to \tilde{l}^-\bar{\tilde{\nu}}_l) =$$

$$= \{\pi\alpha^2/(48\sin^4\theta_W\chi_W\hat{s}^2)\}[\hat{s}[(1-B^2)\hat{s} - 2\{(1+B)m^2_{\tilde{l}} + (1-B)m^2_{\tilde{\nu}}\}]$$

$$-\{(2\hat{t}+\hat{s}-m^2_{\tilde{l}}-m^2_{\tilde{\nu}})^2 - (m^2_{\tilde{l}}-m^2_{\tilde{\nu}}+B\hat{s})^2\}]$$

$$\hat{\sigma}_{TOT}(q_u\bar{q}_d \to "W^+" \to \tilde{l}^+\tilde{\nu}_l) = \hat{\sigma}_{TOT}(q_d\bar{q}_u \to "W^-" \to \tilde{l}^-\bar{\tilde{\nu}}_l) =$$

$$= \{\pi\alpha^2/(72\sin^4\theta_W\chi_W\hat{s}^2)\}[\{\hat{s}-(m_{\tilde{l}}-m_{\tilde{\nu}})^2\}\{\hat{s}-(m_{\tilde{l}}+m_{\tilde{\nu}})^2\}]^{3/2} \quad (4)$$

with $A = e_q^2/\hat{s}^2 + e_q b_q(1-4\sin^2\theta_W)(\hat{s}-m_Z^2)/(4\sin^2\theta_W\cos^2\theta_W\hat{s}\chi_Z)$

$$+(b_q^2 + a_q^2)\{1+(1-4\sin^2\theta_W)^2\}/(64\sin^4\theta_W\cos^4\theta_W\chi_Z)$$

and $B = (m^2_{\tilde{l}} - m^2_{\tilde{\nu}})/m_W^2$,

where quark masses are neglected, $\hat{s}$ and $\hat{t}$ are the Mandelstam variables ($\hat{t}$ is a squared difference between quark and lepton momenta (or between antiquark and antilepton momenta)), $\chi_Z = (\hat{s}-m_Z^2)^2 + \Gamma_Z^2 m_Z^2$, $\chi_W = (\hat{s}-m_W^2)^2 + \Gamma_W^2 m_W^2$, $m_Z = 93$ GeV, $m_W = 81$ GeV, $\Gamma_Z = 3.0$ GeV, $\Gamma_W = 2.88$ GeV, $\sin^2\theta_W = 0.22$, $\alpha = 1/137$, e_q is the quark charge in units of the proton charge, and $(b_q, a_q) = (1/2-4\sin^2\theta_W/3, 1/2)$ and $(-1/2+2\sin^2\theta_W/3, -1/2)$ respectively for up type and down type quarks. We calculate the cross sections with contributions of 6 quark flavors (u, d, c, s, t and b) including weak quark mixing of Kobayashi-Masukawa. We find that the quark mixing effect and the t quark contribution to the cross sections are negligible. As for the K-factor, according to the analyses of the usual Drell-Yan processes at lower energies [5] we parametrize as

$$K(x_1, x_2, Q^2) = 1 + (\ln(10\,GeV/\Lambda))^2 / \ln(Q^2/\Lambda^2)) \text{ with } \Lambda = 0.2 \text{ GeV}. \quad (5)$$

It is not evident that the K-factor for our supersymmetric Drell-Yan processes is the same as that for the usual Drell-Yan processes. Hence this parametrization should not be taken too seriously. Rather it should be considered as a parametrization in "zero-th order" approximation indicative of representative magnitude of the K-factor. The typical size of this K-factor is 1.5-1.6 and 1.4-1.5 in TEVATRON and SSC energy ranges, respectively. Combining eqs. (1)-(5) with the parton distributions of "Set1" in ref. [4] we calculate the total cross sections of the slepton pair production. The cross sections that follow from "Set2" in ref. [4] are nearly identical.

With the expected luminosities of TEVATRON and SSC ($L=10^{29-30}$ and 10^{33} cm^{-2} sec^{-1}, respectively) one has the expected number of events for the slepton pair production in a run of T = 10^7 sec ($\simeq$ 3 months) or 10^7 x10 sec($\simeq$ 3 months/year x 10years) as shown in Table 1 and Figs.1 and 2. The event rates in $\bar{p}p$ collision are nearly identical to those in pp collision (for the same integrated luminosity) at SSC. In actual experiment one introduces various cuts, such as missing transverse-momentum and acoplanarity angle cuts, to eliminate or suppress background events. Such cuts reduce the number of signal events typycally by about 50-80 %. Assuming that one needs at least 10 (100) events in the TEVATRON (SSC) experiment in order to detect the slepton pair production, one has the "discovery limits" (,i.e.,maximum masses accessible) for the sleptons as shown in Table 2. Strictly speaking these numbers in Table 2 should be taken as "upper limits" of the discovery limits, since 10 events at TEVATRON and 100 events at SSC might be insufficient to detect the slepton pair production due to possible sizable backgrounds which have to be studied by Monte Carlo calculation in detail. Here it is to be noted that by combining $\tilde{l}_e \bar{\tilde{l}}'_e$ and $\tilde{l}_\mu \bar{\tilde{l}}'_\mu$ events with an assumption $m_{\tilde{e}} = m_{\tilde{\mu}}$ and $m_{\tilde{\nu}_e} = m_{\tilde{\nu}_\mu}$ one obtains twice the event rate for the slepton pair production and that by combining $\tilde{l}\,\bar{\tilde{l}}'$ events produced at n interaction-points (n = 2 and 4 at TEVATON and SSC, respectively) one can gain another factor n for the event rates. Here $\tilde{l}_e \bar{\tilde{l}}'_e$ refers to $\tilde{e}^-\tilde{e}^+$, $\tilde{\nu}_e \bar{\tilde{\nu}}_e$ or $\tilde{e}^\pm \tilde{\nu}_e$.

We explain experimentation of the slepton search in detail. The signals and backgrouns for the slepton pair events are as follows:

(i) Signals for the $\tilde{e}^-\tilde{e}^+$ event;

For this event we have signals and backgrounds similar to those for $\tilde{e}^-\tilde{e}^+$ (or $\tilde{e}^\pm\tilde{\gamma}e^\mp$) event in e^-e^+ collision [6]. In case $m_{\tilde{\gamma}} < m_{\tilde{e}}$, the $\tilde{e}^\pm$ decays immediately into $e^\pm$ and $\tilde{\gamma}$ leading to a final state of $e^- e^+ \tilde{\gamma}\tilde{\gamma}$ X. As the $\tilde{\gamma}$ is almost completely inert with respect to interactions with matter and hence escapes detection (in case it is the lightest supersymmetric particle and hence is stable), the signal of this event is an electron pair e^-e^+ acoplanar with respect to the beams with large missing-transverse-momentum ($\not{p}_T$). The UA1 and UA2 data on the transverse-momentum (p_T) distribution of the lepton pair $e^\pm\nu$ from the $W^\pm$ boson decay in the process $pp \to W^\pm X \to e^\pm\nu X$ [3,7] suggest that the event rate of the slepton pair production with no jet activity (where the slepton pair tends to have a very small p_T) is comparable to that with jet activity. Hence we can regard an acoplanar electron pair with large $\not{p}_T$ and no jet activity as the signal of this event. The price to pay for this additional imposition of no jet activity would be the reduction of the number of signal events by about 35% [7]. In case $m_{\tilde{\gamma}} < (m_{\tilde{Z}}, m_{\tilde{W}}$ and $m_{\tilde{\nu}_e}) < m_{\tilde{e}}$, the $\tilde{e}^\pm$ can also decay into $e^\pm \tilde{Z}(\to e^\pm f\bar{f}\tilde{\gamma})$, $\nu_e \tilde{W}^\pm(\to \nu_e f\bar{f}'\tilde{\gamma})$ and $\tilde{\nu}_e$ "$W^\pm$"$(\to \tilde{\nu}_e f\bar{f}')$, where $\tilde{Z}(=\tilde{Z}_1$ (lighter) or $\tilde{Z}_2$ (heavier)) and $\tilde{W}(=\tilde{W}_1$ (lighter) or $\tilde{W}_2$ (heavier)) are the mass eigenstates made mainly of the super-partners of the weak bosons and higgs scalars, "$W^\pm$" is the virtual or real $W^\pm$ boson and $f\bar{f}' = l\bar{l}'$ or $q\bar{q}'$. The branching ratios of these decay modes are determined by the relative sizes of the masses of the involved particles and the gaugino-higgsino mixing entering the relevant coupling constants [8]. The decays of the $\tilde{Z}$ and $\tilde{W}$ into the lightest neutralino (assumed to be the $\tilde{\gamma}$ here) are studied in the literatures [1,9] where the corresponding formulae are given. As a result of these new decay modes there would be additional signals including jets: events with m charged leptons + n jets + $\not{p}_T$ where m=0-6 and n=0-4. Such signal events, however, will be rather difficult to be discriminated from backgrounds. We note here that as suggested in ref.[10] in this case the right- and left-handed scalar electron($\tilde{e}_R$ and $\tilde{e}_L$) decays mainly into $e\tilde{\gamma}$ and $\nu\tilde{W}$ if enough phase spaces are available (i.e., in case $m_{\tilde{\gamma}}$ and $m_{\tilde{W}} << m_{\tilde{e}}$), respectively; The $\tilde{W}$ mode of the $\tilde{e}_R$ decay is forbidden. The sleptons are produced in the ratio $\tilde{e}_L^-\bar{\tilde{e}}_L^- : \tilde{e}_R^-\bar{\tilde{e}}_R^- : \tilde{e}_L^-\bar{\tilde{e}}_R^- + \tilde{e}_R^-\bar{\tilde{e}}_L^- \simeq 1:1:0$ for $m_{\tilde{e}_R} = m_{\tilde{e}_L}$. Hence even in this case one expects that at least about 50 % of the produced $\tilde{e}$ pairs decay into $e^-e^+\tilde{\gamma}\tilde{\gamma}$.

(ii) Backgrounds for the $\tilde{e}^-\tilde{e}^+ \to e^-e^+\tilde{\gamma}\tilde{\gamma}$ event with no jet activity;

There is a background from $\tau^-\tau^+$ production with no jet activity where both τ^- and τ^+ decay electronically ($\tau^\pm \to e^\pm\nu\bar{\nu}$). The UA1 and UA2 data on the p_T distribution of the lepton pair $e^\pm\nu$ from the $W^\pm$ decay [3,7] suggest that the τ pair with no jet activity tends to have a very small p_T and hence that the momenta of the e^- and e^+ (as well as those of the τ^- and τ^+) are almost coplanar with respect to the beam axis ,i.e., back to back in the plane transverse to the beam axis, since the momenta of the $e^-(e^+)$ and $\tau^-(\tau^+)$ are almost parallel. Therefor this source of background should not be serious. There could be "backgrounds" from pair production of L^+L^-, $\tilde{W}^+\tilde{W}^-$ and H^+H^- with no jet activity where both of the two charged particles decay into "$e^\pm$ + missing neutrals(ν and $\tilde{\gamma}$)" at the final stage. Here $L^\pm$ and $H^\pm$ are, respectively, new heavy lepton and charged Higgs scalar. All of these "background" sources, however, could be discriminated from the charged slepton pair production events ($\tilde{l}^-\tilde{l}^+$) on the basis that they also yield acoplanar $e^\mp\mu^\pm$ events with $N(e^-\mu^+) + N(\mu^-e^+) = N(e^-e^+) + N(\mu^-\mu^+)$ in remarkable contrast to the $\tilde{l}^-\tilde{l}^+$ events yielding the acoplanar $e^\mp\mu^\pm$ events with $N(e^-\mu^+)+N(\mu^-e^+)' << N(e^-e^+) + N(\mu^-\mu^+)$. There are also backgrounds from the production of $W^+W^-(\to e^+\nu e^-\bar{\nu})$ and $Z^\circ Z^\circ(\to e^-e^+\nu\bar{\nu})$ with no jet activity. In case $m_{\tilde{e}} \lesssim m_Z$ and m_W, we find that the event rate of $\overset{(-)}{p}p \to (W^+W^-$ or $Z^\circ Z^\circ)X \to e^-e^+\nu\bar{\nu}X$ is negligibly small compared with that of $\overset{(-)}{p}p \to \tilde{e}^-\tilde{e}^+X \to e^-e^+\tilde{\gamma}\tilde{\gamma}X$ mainly due to the very small leptonic branching ratios of the weak bosons: $(B(W\to e\nu))^2 \simeq B(Z^\circ\to e^-e^+)B(Z^\circ\to\nu\bar{\nu}) \simeq 1/100$ (for $N_\nu=3$). In case $m_{\tilde{e}} >> m_Z$ and m_W (where only the SSC experiment is relevant), one can suppress these backgrounds as follows: As for the $Z^\circ Z^\circ$ background event, the decay e^- and e^+ with momenta much larger than m_Z(, say $p_{e\pm} \gtrsim$ 200 GeV) tend to have a small openning angle in contrast to the signal event. As for the W^+W^- background event with no jet activity, the decay e^- and e^+ with momenta much larger than m_W(,say $p_{e\pm} \gtrsim$ 200 GeV) tend to be coplanar by the same reason as for the $\tau^-\tau^+$ background event. Hence by imposing suitable momentum, openning angle and acoplanarity angle cuts on the detected e^- and e^+ one can suppress these backgronds without any significant loss of the signal events. In any case these W^+W^- and $Z^\circ Z^\circ$ backgrounds should not be serious. There could be a background from heavy quark pair production where both quarks decay semileptonically($Q\to qe\nu$). As this event necessarily accompanies jet activity, this background should not be serious. There are also backgrounds from two-photon($q\overset{(-)}{q}{}' \to q\overset{(-)}{q}{}' e^-e^+$ and $q\overset{(-)}{q}{}'\tau^-\tau^+$), $e^-e^+\gamma$, $\tau^-\tau^+\gamma$ events with no jet activity and so on, where the $q\overset{(-)}{q}{}'$ and γ in the final state get into a forward detector gap. However since such events will tend to have p_T balance and/or coplanarity, these sources of background should not be serious.

There are possible supersymmetric processes which also could yield an event of "an acoplanar l^-l^+ + missing neutrals + no jet activity", such as $\overset{(-)}{p}p\to\tilde{Z}\tilde{\gamma}X\to e^-e^+\tilde{\gamma}\tilde{\gamma}X$ and so on. The investigation of the event rate and distributions of the "acoplanar $e^-e^+ + \not{p}_T$ + no jet activity" events observed in actual experiment which satisfy all the cut conditions imposed to suppress all the backgrounds mentioned above would play an important role in discriminating the possible origins of the observed excess of events, such as (A)$\overset{(-)}{p}p\to\tilde{e}^-\tilde{e}^+X\to e^-e^+\tilde{\gamma}\tilde{\gamma}X$, (B)$\overset{(-)}{p}p\to\tilde{Z}\tilde{\gamma}X\to e^-e^+\tilde{\gamma}\tilde{\gamma}X$, (C)$\overset{(-)}{p}p\to\tilde{Z}\tilde{Z}X\to (e^-e^+\tilde{\gamma})(\nu\bar{\nu}\tilde{\gamma})X$, (D)$\overset{(-)}{p}p\to\tilde{\nu}\tilde{\bar{\nu}}X\to(\nu\tilde{Z})(\bar{\nu}\tilde{\gamma})X\to(\nu e^-e^+\tilde{\gamma})(\bar{\nu}\tilde{\gamma})X$, (E)$\overset{(-)}{p}p\to\tilde{\nu}\tilde{\bar{\nu}}X\to(\nu\tilde{Z})(\bar{\nu}\tilde{Z})X\to(\nu e^-e^+\tilde{\gamma})(\bar{\nu}\nu\bar{\nu}\tilde{\gamma})X$ and so on. The analysis of the event rates and distributions of these processes (A)-(E) will be presented elsewhere [11].

(iii) Signals for the $\tilde{e}\tilde{\nu}_e$ event;

In case $m_{\tilde{\gamma}}<(m_{\tilde{Z}}, m_{\tilde{W}}$ and $m_{\tilde{\nu}_e})< m_{\tilde{e}}$, the $\tilde{e}$ decays into $e\tilde{\gamma}$, $e\tilde{Z}(\to ef\bar{f}\tilde{\gamma})$, $\nu_e\tilde{W}(\to \nu_e f\bar{f}'\tilde{\gamma})$ and $\tilde{\nu}_e$"W"$(\to\tilde{\nu}_e f\bar{f}')$. In case $m_{\tilde{\gamma}}<(m_{\tilde{Z}}, m_{\tilde{W}}$ and $m_{\tilde{e}})< m_{\tilde{\nu}_e}$, the $\tilde{\nu}_e$ decays into $\nu_e\tilde{\gamma}$, $\nu_e\tilde{Z}(\to \nu_e f\bar{f}\tilde{\gamma})$, $e\tilde{W}(\to ef\bar{f}'\tilde{\gamma})$ and $\tilde{e}$"W"$(\to\tilde{e}f\bar{f}')$ [12]. The branching ratios of these decay modes are determined by the relative sizes of the masses of the involved particles and the gaugino-higgsino mixing entering the relevant coupling constants. As a result of these decay modes there would be various signals: events with m charged leptons + n jets + $\not{p}_T$ where m and n $\geq$ 0 [12]. The cleanest signal among these is "isolated single $e^{\mp}$ + no jet activity + $\not{p}_T$" which could stem from the $\tilde{e}\tilde{\nu}_e$ production with no jet activity followed by the decays "$(\tilde{e}\to e\tilde{\gamma})$, $(\tilde{e}\to e\tilde{Z}\to e\nu\bar{\nu}\tilde{\gamma})$, $(\tilde{e}\to \nu_e\tilde{W}\to \nu_e e\nu_e\tilde{\gamma})$ or $(\tilde{e}\to\tilde{\nu}_e$"W"$\to\nu_e\tilde{\gamma}e\nu_e)$" and "$(\tilde{\nu}_e\to\nu_e\tilde{\gamma})$ or $(\tilde{\nu}_e\to\nu_e\tilde{Z}\to\nu_e\nu\bar{\nu}\tilde{\gamma})$". We note here that in case $m_{\tilde{W}}<<(m_{\tilde{e}}$ and $m_{\tilde{\nu}_e})$ the $\tilde{e}(=\tilde{e}_L)$ and $\tilde{\nu}_e$ decay mainly into $\nu_e\tilde{W}(\to\nu_e f\bar{f}'\tilde{\gamma})$ and $e\tilde{W}(\to ef\bar{f}'\tilde{\gamma})$, respectively [10], resulting in a relatively small yield rate of the isolated single $e^{\mp}$ events, and that in case $m_{\tilde{\gamma}} < m_{\tilde{\nu}_e} \leq m_{\tilde{e}} <(m_{\tilde{Z}}$ and $m_{\tilde{W}})$ the $\tilde{e}$ and $\tilde{\nu}_e$ decay predominantly into $e\tilde{\gamma}$ and $\nu_e\tilde{\gamma}$, respectively [12], resulting in a relatively large yield rate of the isolated single $e^{\mp}$ events.

(iv) Backgounds for the " $\tilde{e}^{\pm}\tilde{\nu}\to e^{\pm}$ + missing neutrals " event with no jet activity;

Main background for this signal event is that from $\overset{(-)}{p}p\to e\nu X$ with no jet activity. The electron of the signal event would tend to be much softer than that of this background event. There are backgrounds which tend to yield a relatively soft electron, such as those from the $W^{\pm}Z^{\circ}(\to e^{\pm}\nu\nu\bar{\nu})$ and $\tau^{\pm}\nu(\to e^{\pm}\nu\bar{\nu}\nu)$ production with no jet activity and so on. The backgrounds from the two-photon($q\overset{(-)}{q}\to q\overset{(-)}{q}e^-e^+$ and $q\overset{(-)}{q}\tau^-\tau^+$), e^-e^+, $e^-e^+\gamma$, $\tau^-\tau^+$, $\tau^-\tau^+\gamma$ events with no visible jet activity and so on, where all the final state particles expcept one electron are missing due to their escape into the forward detector gaps or so, should not be serious, since such events tend to have very small $\not{p}_T$. Hence an excess over the standerd model expectation in the p_T distribution of the isolated single electrons with no jet activity in lower p_T region would signal the production of the signal events. The estimate of the contributions of the signal and background processes to the isolated single $e^{\mp}$ distribution will be presented elsewhere [11].

There are possible "new physics" processes which also could yield the event of "isolated single $e^{\mp}$ + missing neutrals + no jet activity", such as (A) $\overset{(-)}{p}p\to\tilde{W}\tilde{\gamma}X\to e\nu\tilde{\gamma}\tilde{\gamma}X$, (B) $\overset{(-)}{p}p\to H^{+}Z^{\circ}\to(\tau^{+}(\text{or } L^{\pm})\nu)(\nu\bar{\nu})X\to(e^{\pm}\nu\bar{\nu}\nu)(\nu\bar{\nu})X$, (C) $\overset{(-)}{p}p\to L^{\pm}\nu X\to(e^{\pm}\nu\bar{\nu})\nu X$ and so on. The investigation of the event rate and distribution of the "isolated single $e^{\mp}$ + $\not{p}_T$ + no jet activity" events observed in actual experiment would play an important role in discriminating the possible origines of the observed excess of the events (if any). The analysis of the event rates and distributions of these "new physics" processes (A)-(C) will be presented elsewhere [11].

(v) Signals for the $\tilde{\nu}_e\bar{\tilde{\nu}}_e$ event;

In case $m_{\tilde{\gamma}}<(m_{\tilde{Z}}, m_{\tilde{W}}$ and $m_{\tilde{e}})<m_{\tilde{\nu}_e}$, the $\tilde{\nu}_e$ decays into $\nu_e\tilde{\gamma}$, $\nu_e\tilde{Z}(\to \nu_e f\bar{f}\tilde{\gamma})$, $e\tilde{W}(\to e f\bar{f}'\tilde{\gamma})$ and $\tilde{e}$"W"$(\to \tilde{e} f\bar{f}')$. As a result of these decay modes there would be various striking signals [12]: events with m charged leptons + n jets + $\not{p}_T$, where m and n $\geqq 0$. The most characteristic signal among these would be one-side event (where the visible decay particles are all in one hemisphere of the center of mass system) which could stem from the decays "$\tilde{\nu}_e \to \nu_e\tilde{\gamma}$" and "$\bar{\tilde{\nu}}_e\to$visible particles + missing neutrals(ν and $\tilde{\gamma}$)" in case of a relatively light $\tilde{\nu}_e$.

(vi) Backgrounds for the $\tilde{\nu}_e\bar{\tilde{\nu}}_e$ event;

As for the signal event "acoplanar $l^- l^+$ + $\not{p}_T$ + no jet activity" from the decay "$\tilde{\nu}_e\bar{\tilde{\nu}}_e \to (\nu_e Z)(\bar{\nu}_e\tilde{\gamma}) \to (\nu_e l^- l^+\tilde{\gamma})(\bar{\nu}_e\tilde{\gamma})$" or "$\tilde{\nu}_e\bar{\tilde{\nu}}_e \to (\nu_e \tilde{Z})(\bar{\nu}_e\tilde{Z}) \to (\nu_e l^- l^+\tilde{\gamma})(\bar{\nu}_e\nu\bar{\nu}\tilde{\gamma})$, there are the same backgrounds as those for the $\tilde{e}^-\tilde{e}^+ \to e^- e^+\tilde{\gamma}\tilde{\gamma}$ event with no jet activity. The same argument as for the $\tilde{e}^-\tilde{e}^+$ event leads to a conclusion that these backgrounds should not be serious. As for the signal event "jet(s) + $\not{p}_T$ " from the decays "$\tilde{\nu}_e \to \nu_e\tilde{Z} \to \nu_e q\bar{q}\tilde{\gamma}$" and "$(\bar{\tilde{\nu}}_e \to \bar{\nu}_e\tilde{\gamma}), (\bar{\tilde{\nu}}_e \to \bar{\nu}_e \tilde{Z} \to \bar{\nu}_e\nu\bar{\nu}\tilde{\gamma})$ or $(\bar{\tilde{\nu}}_e \to \bar{\nu}_e\tilde{Z} \to \bar{\nu}_e q\bar{q}\tilde{\gamma})$", there are the same backgrounds as those for the events of " $\tilde{q}\tilde{q}$, $\tilde{q}\tilde{g}$ and $\tilde{g}\tilde{g} \to$ jet(s) + missing neutrals", such as "W"$\to\tau\bar{\nu}\to$jet $\nu\bar{\nu}$, Z° +jet(s) $\to \nu\bar{\nu}$ + jet(s) and so on [13], where $\tilde{g}$ is the gluino. As for the signal event " $e^\mp$ + jet(s) + $\not{p}_T$" from the decays "$(\tilde{\nu}_e \to e^- \tilde{W}^+ \to e^- q\bar{q}'\tilde{\gamma})$ or $(\tilde{\nu}_e \to \tilde{e}^-$"$W^+$" $\to e^-\tilde{\gamma} q\bar{q}')$" and "$(\bar{\tilde{\nu}}_e \to \bar{\nu}_e\tilde{\gamma}), (\bar{\tilde{\nu}}_e \to \bar{\nu}_e\tilde{Z} \to \bar{\nu}_e\nu\bar{\nu}\tilde{\gamma})$ or $(\bar{\tilde{\nu}}_e \to \bar{\nu}_e\tilde{Z} \to \bar{\nu}_e q\bar{q}\tilde{\gamma})$", the main backgrounds are those from $\overset{(-)}{p}p \to W^\pm$jet(s)$X \to e^\pm\nu$ jet(s)X and $\overset{(-)}{p}p \to Q\bar{Q}X \to (qe\nu)(\bar{q}q\bar{q})X$ [3]. The separation of these signal events from the background events in an actual experiment requires detailed Monte Carlo studies (including detector resolution and trigger condition) of the event rates and distributions of these signal and background processes. Such Monte Carlo analyses will be presented elsewhere [11].

It is to be noted here that the effects of the possible extra weak bosons (Z' and W') and their superpartners on the production and decay of the sleptons could be significant.

We conclude that a detectable number of sleptons should be produced at TEVATRON and SSC if their masses are less than ~100 GeV and ~1.0 TeV, respectively, and that the slepton pair production at these hadron colliders deserves detailed Monte Carlo studies.

REFERENCES

[1] H.E. Haber and G.L. Kane, Phys. Rep. 117,75(1985).

[2] M. Davier, Proc. Int. Conf. on High Energy Physics, Berkeley 1986; A.S. Johnson, Boston Univ. preprint BU-HEP / 87-12 (1987) and to be published in Proc. 22nd Moriond Workshop on Electroweak Interactions and Unified Theories, Les Arcs,1987 and references therein.

[3] C. Rubbia, Proc. Int. Symposium on Lepton and Photon Interactions at High Energies, Kyoto,1985; See also R. Barbieri et al., Phys. Lett. 127B,458(1983) and N. Cabibbo et al.,Phys. Lett. 132B,195(1983).

[4] E. Eichten et al., Rev. Mod. Phys. 56,579(1984) and Fermilab preprint Fermilab-Pub-86/75-T(1986).

[5] J.P. Rutherfoord, Proc. Int. Symposium on Lepton and Photon Interactions at High Energies, Kyoto,1985 and references therein.

[6] K. Hidaka, H. Komatsu and P. Ratcliffe, Phys. Lett. 150B, 399(1985) and Tokyo Gakugei Univ. preprint TGU-2,Nov.(1987)(to be published in Nucl. Phys. B).

[7] D. Froidevaux, CERN preprint CERN-EP/87-119(1987) and invited talk at the 1987 Meeting at the American Physical Society-DPF, Salt Lake City, Jan. 1987.

[8] A. Bartl et al., Univ. of Wien preprint HEPHY-PUB 503/87, June 1987.

[9] A. Bartl et al., Nucl. Phys. B278, 1 (1986) and Z. Phys. C30, 441 (1986); N. Oshimo and Y. Kizukuri, Phys. Lett. B186,217 (1987).

[10] H. Baer and E.L. Berger, Phys. Rev. D34,1361(1986) and references therein.

[11] K. Hidaka, in progress.

[12] R.M. Barnett et al., Phys. Lett. 126B, 64(1983) and SLAC preprint SLAC-PUB-3224(1983).

[13] A. Honma, Proc. Int. Conf. on High Energy Physics, Berkeley 1986.

Table 1 Expected number of events ($N=LT\sigma_{TOT}$) for each lepton flavor with various slepton masses and integrated luminosities at TEVATRON and SSC.

TEVATRON ($\sqrt{s}$ = 2 TeV)					
Process	$m_{\tilde{l}}$ (GeV)	$m_{\tilde{\nu}}$ (GeV)	N (events) $LT=10^{36}$ (cm^{-2})	$LT=10^{37}$ (cm^{-2})	$LT=10^{38}$ (cm^{-2})
$\bar{p}\, p \to \tilde{l}^- \tilde{l}^+ X$	50		0.4	3.6	36
	60		0.2	2.1	21
	70		0.1	1.3	13
	80		0.08	0.8	8
	90		0.06	0.6	6
	100		0.04	0.4	4
$\bar{p}\, p \to \tilde{\nu}\bar{\tilde{\nu}} X$		50	0.06	0.6	6
		60	0.04	0.4	4
		70	0.03	0.35	3.5

SSC ($\sqrt{s}$ = 40 TeV)				
Process	$m_{\tilde{l}}$ (GeV)	$m_{\tilde{\nu}}$ (GeV)	N (events) $LT = 10^{40}$ (cm^{-2})	$LT = 10^{41}$ (cm^{-2})
$p\, p \to \tilde{l}^- \tilde{l}^+ X$	200		1480	14800
	400		167	1670
	600		37	375
	800		12	120
	1000		5	48
$p\, p \to \tilde{\nu}\bar{\tilde{\nu}} X$		200	1040	10400
		400	112	1120
		600	25	250
		800	8	78
		1000	3	30

Table 2 Discovery limits (i.e., maximum masses accessible) for the sleptons with various integrated luminosities at TEVATRON and SSC.

TEVATRON ($\sqrt{s}$ = 2 TeV)			
Process	Maximum slepton masses accessible (in GeV units)		
	LT = 10^{36} (cm^{-2})	LT = 10^{37} (cm^{-2})	LT = 10^{38} (cm^{-2})
$\bar{p}p \to \tilde{l}^- \tilde{l}^+ X$	< 50	< 50	75
$\bar{p}p \to \tilde{\nu} \bar{\tilde{\nu}} X$	< 50	< 50	< 50
$\bar{p}p \to \tilde{l}^\pm \tilde{\nu}_l X$	——	The curve with an attached number of 10 in Fig.1	The curve with an attached number of 1 in Fig.1

SSC ($\sqrt{s}$ = 40 TeV)		
Process	Maximum slepton masses accessible (in TeV units)	
	LT = 10^{40} (cm^{-2})	LT = 10^{41} (cm^{-2})
$pp \to \tilde{l}^- \tilde{l}^+ X$	0.47	0.84
$pp \to \tilde{\nu} \bar{\tilde{\nu}} X$	0.42	0.75
$pp \to \tilde{l}^\pm \tilde{\nu}_l X$	The curve with an attached number of 100 in Fig.2	The curve with an attached number of 10 in Fig.2

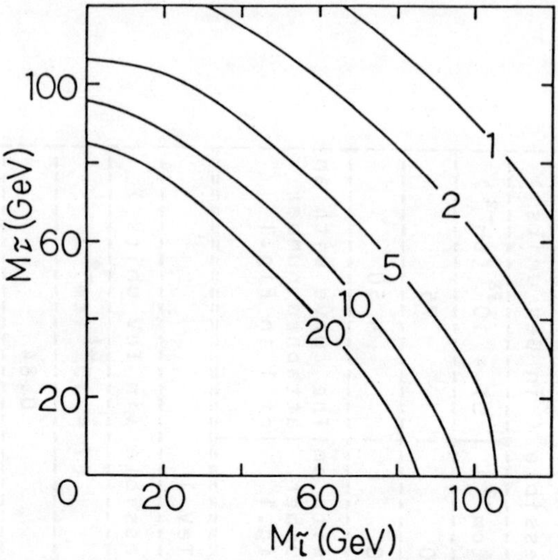

Fig. 1 Expected number of events ($N = LT\sigma_{TOT}$) of the process $\bar{p}p \to \tilde{l}^{\pm}\tilde{\nu}_l X$ for each lepton generation with $LT=10^{37}$ (cm^{-2}) at TEVATRON ($\sqrt{s}$=2 TeV) as a function of $m_{\tilde{l}}$ and $m_{\tilde{\nu}}$. The number attached to each curve refers to the expected number of the events N.

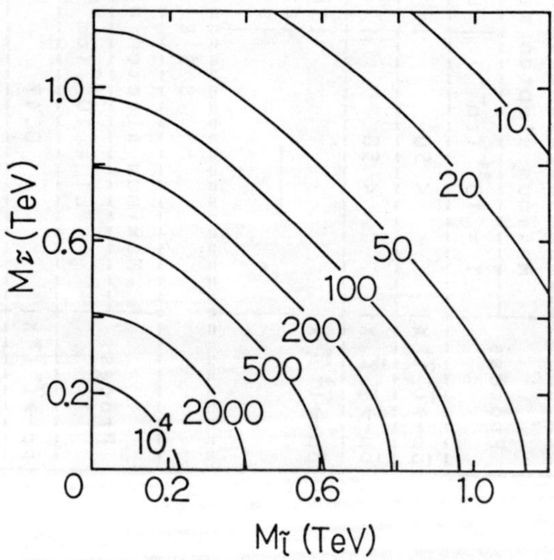

Fig. 2 Expected number of events ($N = LT\sigma_{TOT}$) of the process $pp \to \tilde{l}^{\pm}\tilde{\nu}_l X$ for each lepton generation with $LT=10^{40}$ (cm^{-2}) at SSC ($\sqrt{s}$=40 TeV) as a function of $m_{\tilde{l}}$ and $m_{\tilde{\nu}}$. The number attached to each curve refers to the expected number of the events N.

QUEST FOR THE TOP QUARK

Y. Nagashima
Department of Physics
Osaka University
Toyonaka, Osaka 560, Japan

ABSTRACT

Several methods to search for a new heavy quark are discussed including jet shape analysis for hadronic final states and are applied to the data taken with the VENUS detector at TRISTAN e^+e^- collider.

1. INTRODUCTION

The aim of constructing an e^+e^- collider is, among other things, to search for new fundamental particles, notably for the top quark in the case of TRISTAN. The technique to find it is rather straightforward knowing the fact that it is a charge 2/3 e partner of the known b-quark. In the following I describe several methods to identify the top quark production and some experimental data obtained with the VENUS[1)] detector.

2. METHODS TO IDENTIFY THE TOP QUARK PRODUCTION

2.1. Peak Hunting

A narrow peak due to formation of $t\bar{t}$ bound state is expected to appear just below the open $t\bar{t}$ production threshold. Peaks of radially excited states may also be seen, though it is already marginal at TRISTAN due to the spread of the beam energy. The counting rate is high, about 200/day for the top mass of 25 GeV with the luminosity of $1.5\times10^{31}/cm^2\cdot sec$. Because of the narrowness of the peak, however, the center of mass energy has to coincide exactly with that of the peak to find it. Since it is not economical to scan the wide energy range with such small interval as not to miss the peak, the hunting is usually saved until the beam luminosity becomes high and stable.

2.2. R-ratio

The top quark production at the continuum adds to the value of the ratio R: $\sigma_{tot}/\sigma_{point}$ of the total hadronic crosssection to point-like QED crosssection ($\sigma_{point} = 4\pi\alpha^2/3s$) by an amount of approximately 4/3. Therefore, a rise in the value of R signals the production of the top quark. While a kinematical effect (β^3; $\beta = (1-4m_t^2/s)^{1/2}$) suppresses the production near its threshold, it may be more than compensated by the resonance formation as is the case with $c\bar{c}$ or $b\bar{b}$ production. In a data analysis, it is generally assumed that the production reaches its continuum instantly beyond the threshold.

2.3. Shape Analysis[2)]

A high energy quark is known to produce a jet-like pattern in its hadronic final state. The production of a light quark pair $q\bar{q}$ appears typically as two jet events, while a $q\bar{q}$ pair accompanied by a gluon emission produces a 3-jet event or more generally a planer event. On the other hand, the production of a new heavy quark near its threshold appears as an enhancement of spherically shaped events.

Two parameters used by the VENUS group to sort out the top quark signal are thrust (T) and acoplanarity (A) which are defined by

$$T = \max \frac{\Sigma_i |\mathbf{P}_i \cdot \mathbf{e}_T|}{\Sigma_i \; |\mathbf{P}_i|}$$

$$A = \min 4 \left\{ \frac{\Sigma_i \; |\mathbf{P}_i \cdot \mathbf{e}_A|}{\Sigma_i \; |\mathbf{P}_i|} \right\}^2 ,$$

where $\mathbf{P}_i$ is the momentum of the i-th particle in an event and the maximum (minimum) is found by varying the axis $\mathbf{e}_T$ ($\mathbf{e}_A$). The thrust represents a degree of collimation of final particles into two jets (T = 1 for collinear two-jets, 0.5 for isotropic distributions), whereas the acoplanarity is a measure of deviation from a planer event shape (A = 0 (1) for planer (isotropic) distribution). Associated with the thrust axis $\mathbf{e}_T$ are the major axis $\mathbf{e}_2$ ($\mathbf{e}_2 \perp \mathbf{e}_T$) and the minor axis $\mathbf{e}_3$ ($\mathbf{e}_3 = \mathbf{e}_T \times \mathbf{e}_2$), where $\mathbf{e}_2$ is determined by maximizing the value of

$$F_{major} = \max \frac{\Sigma_i \ |\mathbf{P}_i \cdot \mathbf{e}_2|}{\Sigma_i \ |\mathbf{P}_i|}, \quad \mathbf{e}_2 \perp \mathbf{e}_T$$

For planer events, $\mathbf{e}_T$ and $\mathbf{e}_2$ define an event plane. $\mathbf{e}_A$ is very close to the minor axis $\mathbf{e}_3$ but not identical. Some groups prefer to use a parameter oblateness O instead of the acoplanarity which is defined by

$$O = F_{major} - F_{minor}$$

$$F_{minor} = \frac{\Sigma_i \ |\mathbf{P}_i \cdot \mathbf{e}_3|}{\Sigma_i \ |\mathbf{P}_i|}, \quad \mathbf{e}_3 = \mathbf{e}_T \times \mathbf{e}_2$$

Other frequently used set of parameters are the eigenvalues λ_j ($\lambda_1 \geq \lambda_2 \geq \lambda_3$) of the momentum tensor

$$I_{\alpha\beta} = \Sigma_i \ ((\mathbf{P}_i \cdot \mathbf{P}_i) \ \delta_{\alpha\beta} - P_{i\alpha} P_{i\beta})$$

and corresponding eigen vectors $\mathbf{n}_1$, $\mathbf{n}_2$, $\mathbf{n}_3$ (Fig. 1(a)). One also defines

$$Q_K = 1 - \frac{2\lambda_K}{\lambda_1 + \lambda_2 + \lambda_3} = \frac{\Sigma_i \ (\mathbf{P}_i \cdot \mathbf{n}_K)^2}{\Sigma_i \ (\mathbf{P}_i \cdot \mathbf{P}_i)},$$

$$Q_1 \leq Q_2 \leq Q_3 ,$$
$$Q_1 + Q_2 + Q_3 = 1.$$

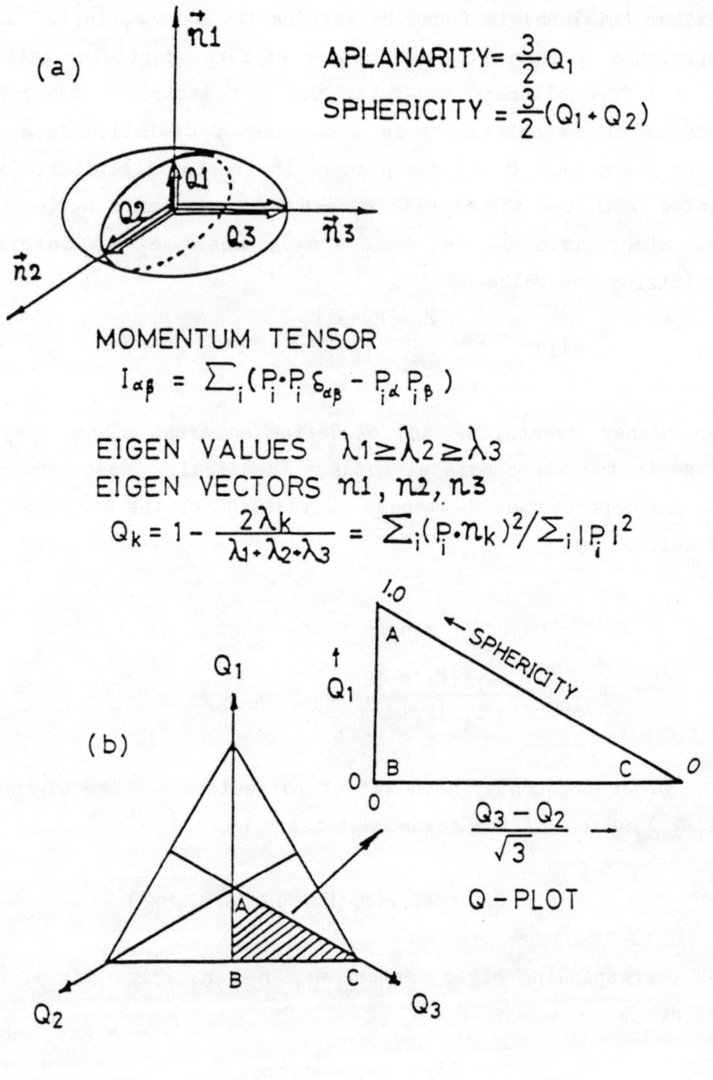

$$I_{\alpha\beta} = \sum_i (P_i \cdot P_i \delta_{\alpha\beta} - P_{i\alpha} P_{i\beta})$$

$$Q_k = 1 - \frac{2\lambda_k}{\lambda_1 + \lambda_2 + \lambda_3} = \sum_i (P_i \cdot n_k)^2 / \sum_i |P_i|^2$$

$$\frac{Q_3 - Q_2}{\sqrt{3}}$$

Fig. 1. Jet analysis of the second order. Collinear 2 jet event corresponds to point C (Q_1=S=0), and planer events are characterized by Q_1=0. Spherical events fall in the neighborhood of point A (S=1, Q_1=1/3).

The sphericity S is defined by

$$S = \min \frac{3\Sigma_i \, P_{iT}^2}{2\Sigma_i \, |P_i|^2} \, .$$

The minimization is done by varying the direction of the axis. This definition is equivalent to

$$S = \frac{3\lambda_3}{\lambda_1+\lambda_2+\lambda_3} = 3/2(Q_1+Q_2)$$
$$0 \leq S \leq 1$$

S = 1 for isotropic distributions, 0 for collinear two jets and $\sim 3/4$ for planer distributions and the eigen vector $\mathbf{n}_3$ corresponding to λ_3 is the sphericity axis or the jet axis. Aplanarity is defined as $A = 3Q_1/2$ $(0 \leq A \leq 0.5)$ and is very similar to the acoplanarity. Because of the constraint $Q_1+Q_2+Q_3=1$, every triplet (Q_1, Q_2, Q_3) of an event corresponds to a point in a shaded triangle in Fig. 1(b). A collection of the points is called a Q-plot.

The advantage of using $(\mathbf{e}_T, \mathbf{e}_2, \mathbf{e}_3)$ and associated parameters determined from the first order moment is that it does not depend on the decay history of particles. The method of momentum tensor does depend on the decay history and also suffers from infrared singularities. However it is fast and has an advantage from experimental point of view.

2.4. Method of Isolated Leptons

As stated in the previous section, a heavy quark production in the threshold neighbor is expected spherical. A certain fraction of the top quark decays semi-leptonically, i.e., $t \to b + \ell + \nu$, producing a jet and a lepton with large transverse momentum relative to the b-jet axis. Thus the existence of an isolated lepton also serves as a signal pointing to possible production of a new particle. The degree of isolation may be expressed by $\cos\delta$, where δ is opening angle between the lepton and the jet axis.

3. VENUS DETECTOR AND PARTICLE SELECTION[3)]

The experiment I describe here has been performed at KEK e^+e^- collider TRISTAN with the VENUS detector. The major parts of the detector which are relevant in the following analysis are a central drift chamber (CDC), two electromagnetic shower calorimeters, and a muon counter (MU). The CDC is placed inside a solenoidal magnetic field of 0.75 tesla parallel to the beam axis, and has an angular coverage of $30^0 < \theta < 150^0$ with momentum resolution of $\sigma_P/P = 0.008 P_T$ (GeV/c). Outside the CDC and the solenoid is the barrel calorimeter which is an array of 5160 lead glass counter (LG). It has angular coverage of $37^0 < \theta < 143^0$ with energy resolution σ_E/E of 3.7 % for electrons of 26 GeV. A pair of liquid argon counters (LA) cover the angular region of $8^0 < \theta < 38^0$ and $142^0 < \theta < 172^0$. However, the LA was not operative at $\sqrt{s}$ = 50 and 52 GeV.

Electrons are identified using the fact that the calorimeters are primarily sensitive to electrons and/or photons and measure only minimum ionizing energy loss of hadrons except for occasional nuclear showers inside the calorimeter. Therefore, the ratio E/P, the energy loss in the calorimeter to the momentum measured by CDC should be close to one for electrons and considerably smaller for hadrons. In the analysis, events with $0.8 < E/P < 1.5$ are treated as electron candidates and the background from hadrons due to showering was subtracted by fitting an exponential shape at low E/P region and extrapolating into the range $E/P > 0.8$. The average efficiency for identifying electrons was 60 %, but it drops to 40 % for $\cos\delta > 0.98$, where separation of the electrons from the crowded hadrons becomes more difficult.

Muons are identified by requiring associated hits in the muon counter with a CDC track, thus ensuring the passage of an iron filter ($\simeq$10 interaction length) without interaction. The backgrounds come from the decay and punch through of pions and kaons. The detection efficiency was estimated using cosmic ray events and was found 92.4 ± 0.7 % and flat above 2.3 GeV. It was also checked against $e^+e^- \to \mu^+\mu^-$ events and agreed within an error.

Major selection criteria of multi-hadron events are i) at least 5 good tracks exist in the events, ii) the total energy in LG is greater than 3 GeV, iii) the total visible energy is greater than $\sqrt{s}/2$. Cuts i), ii) are effective to eliminate beam-gas scattering and two-photon scattering, while cut i) also separates Bhabha scattering and μ or τ pair production from multi-hadrons.

The detection efficiency of the detector was estimated using a Monte Carlo program. Comparisons of observed hadronic events with the expectation generated by the five known quarks as well as by the top quark are made using the LUND string fragmentation model[4]. Various distributions are well reproduced by the Monte Carlo. The integrated luminosity was determined from large angle Bhabha scattering detected by the barrel calorimeter[5].

4. RESULTS

4.1. R-ratio[3]

The value of R was determined from the number of accepted events N_{had}, the number of backgrounds N_{BG}, the integrated luminosity L, the multi hadron detection efficiency ε and the radiative correction factor δ. It is expressed as

$$R = \frac{N_{had} - N_{BG}}{\varepsilon \cdot L \cdot \sigma_{point}\,(1+\delta)}$$

A total of 96 events passed the off-line selection criteria at $\sqrt{s}$ = 50 GeV and 399 events at $\sqrt{s}$ = 52 GeV with the integrated luminosity of 0.69 ± 0.04 pb^{-1} and 2.90 ± 0.08 pb^{-1} respectively. The detection efficiency was 68.3 % at $\sqrt{s}$ = 50 GeV and 67.8 % at $\sqrt{s}$ = 52 GeV. The number of backgrounds for τ pair production (0.8 %) and two photon scattering (0.9 %) were determined by the Monte Carlo. The correction factor δ was calculated[6] to be 0.32 at $\sqrt{s}$ = 50 GeV and 0.31 at $\sqrt{s}$ = 52 GeV. The systematic uncertainties are estimated for 1) δ calculation (2 %), 2) the difference of hadron generator model (1.2 %) as well as the parameter dependence of the model (4.5 %), 3) off-line selection criteria (8 %), and 4) the luminosity (4

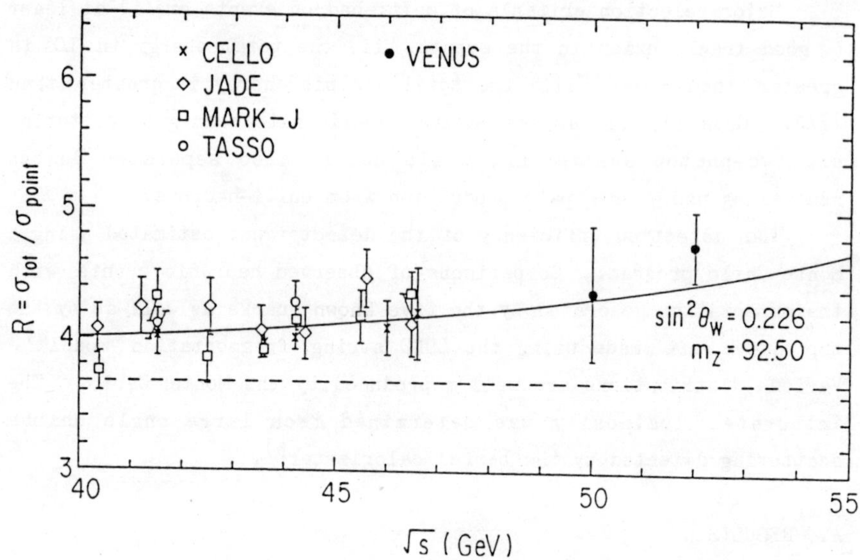

Fig. 2. A plot of R values including PETRA data[9]. The dashed line shows the QPM prediction and the solid line shows the expected R value for the five known quarks including second order QCD and electroweak corrections.

%). Combining all the systematic errors in quadrature the total systematic error of the R value is estimated to be 10 %. The final values of R are

$$R = 4.4 \pm 0.5(\text{stat}) \pm 0.4(\text{sys}) \quad \text{at} \quad \sqrt{s} = 50 \text{ GeV}$$
$$= 4.7 \pm 0.3(\text{stat}) \pm 0.5(\text{sys}) \quad \text{at} \quad \sqrt{s} = 52 \text{ GeV}$$

In Fig. 2, the present values are plotted together with those measured at PETRA energies. They are consistent with the expectation for the five known quarks with QCD as well as electroweak correction. With the present systematic error, we could not exclude the production of the top quark production with a high confidence level. A more stringent test is given by the following shape analysis.

4.2. Shape Analysis[3)]

In order to reduce biases to the shape analysis, two additional selection conditions were applied: (1) $|\cos\theta_T| < 0.7$, where θ_T is the angle between the thrust axis $\mathbf{e}_T$ and the beam axis. This condition reduces the number of events with missing particles in the forward or backward open spaces caused by inoperative LA. (2) $|\cos\theta_n| \geq 0.15$ where θ_n is the angle between the minor axis $\mathbf{e}_3$ (a vector perpendicular to the event plane) and the beam axis. This condition suppresses events accompanying an initial hard photon. We present the observed distribution of the thrust T and the acoplanarity A at $\sqrt{s} = 52$ GeV in Fig. 3a and 3b. The solid histogram shows the Monte Carlo prodiction for the known five quark production. The dashed histogram shows the prediction if the top quark contribution is included assuming the mass of 25.5 GeV and without threshold suppression.

To quantify the above observation, we compare the number of observed events with the expectation at A > 0.2. Wc used the acoplanarity distribution because, compared to the thrust, it is more sensitive to events with a spherical shape. We observed 7 events while 33 top events were expected from the Monte Carlo. Without subtracting possible contribution from the known five quarks, the upper limit of the crosssection of a new charge 2/3 e quark is estimated to be

$$\sigma_{t\bar{t}} < 17 \text{ pb} \quad 95\ \%\ \text{CL}.$$

to be compared with 34 pb at continuum.

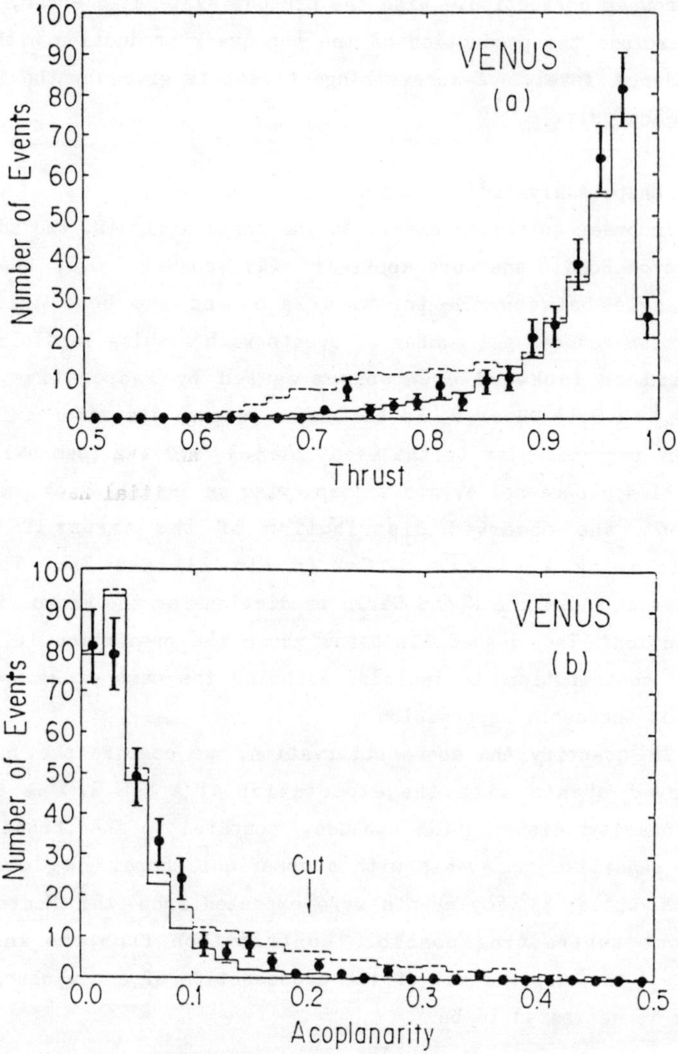

Fig. 3 (a) The observed thrust distribution (filled circles), and (b) the observed acoplanarity distribution. The solid histograms show the Monte Carlo prediction for the five known quarks and the dashed histograms show the prediction if the top production is included, assuming top mass of 25.5 GeV.

Table 1. Summary of the inclusive lepton events. The first column is the observed number of leptons in the various cut conditions. The second column is the expected contributions from the Monte Carlo with the known five quarks. The numbers in the parenthesis are estimated contribution from the backgrounds.

	#observed leptons	#expected from the Monte Carlo
μ inclusive events		
Total	30	29.7 ± 1.8 (11.4 ± 1.4)
T < 0.8	2	1.5 ± 0.4 (0.5 ± 0.2)
T < 0.8 & |cosδ| < 0.7	1	0.1 ± 0.1 (0.0)
e inclusive events		
Total	41	39.7 ± 5.5 (20.0 ± 5.3)
T < 0.8	1	1.9 ± 0.4 (1.1 ± 0.3)
T < 0.8 & |cosδ| < 0.7	1	0.6 ± 0.3 (0.3 ± 0.1)

4.3. Isolated Leptons[7)]

Recently Mark J and Jade group reported an excess of multi hadron events with isolated muons in the low thrust region at the energy $46.3 < \sqrt{s} < 48$ GeV[8)]. Because the top quark production had already been excluded in this energy region, the observed excess of muons was interpreted as a possible existence of the 4th generation of a charge -1/3 e (b' quark). However, the anomaly was neither confirmed by other groups nor observed in electron events. If it signals onset of a new phenomenon with the threshold at $\sqrt{s} = 46$ GeV, the experiments at TRISTAN at higher energy should detect such events unambiguously.

Our data which contains isolated lepton samples consists of 76 events at $\sqrt{s}$ = 50 GeV with the integrated luminosity L of 0.61 pb^{-1}, 397 events at $\sqrt{s}$ = 52 GeV for L = 2.9 pb^{-1} and 274 events at $\sqrt{s}$ = 55 GeV with L = 2.5 pb^{-1}. The LA was operative at $\sqrt{s}$ = 55 GeV. Table 1 summarizes the number of observed inclusive leptons together with the estimated background contributions. Also shown are the expected numbers of prompt leptons from the multi hadron production with the five known quarks. As seen from the table, the results obtained from the Monte Carlo calculation agree quite well with the observed inclusive leptons. We may conclude we did not observe an excess of large angle ($\cos\delta < 0.7$) isolated leptons in the low thrust ($T < 0.8$) region. The result does not change if we assume 1/s or constant behavior for the production crosssection of the new phenomenon.

5. SUMMARY

I discussed several methods for detecting production of a new heavy quark. Applying those methods to the data taken by the VENUS detector at TRISTAN, we observed no evidence of the top quark production up to $\sqrt{s}$ = 52 GeV. We did not confirm the anomaly observed by Mark J and JADE group at $\sqrt{s}$ = 46 GeV.

References

1) VENUS Collaboration, J. Phys. Soc. Japan 56 (1987) 3763.
2) S. Brandt, Si-85-09, talk presented at the International Workshop on Advanced Methods in Nuclear Scattering Data, Berlin, June 18-20, 1985.
3) VENUS Collaboration, Phys. Lett. 198 (1987) 570.
4) B. Anderson, G. Gustafson, G. Ingelman and T. Sjöstrand, Phys. Rep. 97 (1983) 33.
5) VENUS Collaboration, J. Phys. Soc. Japan 56 (1987) 3767.
6) F. Berends, R. Kleiss, and S. Jadach, Nucl. Phys. B202 (1982) 63, Computer Phys. Comm. 29 (1983) 185.
7) VENUS Collaboration, to be published in Phys. Lett.
8) Mark J Collaboration, B. Adeva et al., Phys. Rev. D34 (1986)

681. JADE Collaboration, W. Bartel et al., Z. Phys. C36 (1987) 15.

9) CELLO Collaboration, H.-J. Behrend et al., Phys. Lett. B183 (1987) 400.

LIGHT-LIKE DISTRIBUTION OF CONSTITUENTS INSIDE HADRONS

Kisei Kinoshita
Physics Department, Faculty of Education, Kagoshima University
Kagoshima 890, Japan

ABSTRACT

We discuss the momentum distribution of constituents inside hadrons in different models, utilizing the light-like description of composite systems.

1. INTRODUCTION

The deep inelastic lepton-hadron scattering, together with massive lepton-pair production, revealed single quark-parton distributions inside nucleons and mesons. However, this type of information is not enough for the understanding of dynamical structure of hadrons, which are many particle systems of quarks and gluons in confined domains, possibly clustered into a few numbers of dressed quarks.

The standard approach to attack this problem has been the investigation of static properties such as the mass levels and decay properties. On the other hand, it is expected to extract informations on the internal structure of hadrons from the fragmentation pattern of hadrons in high energy collisions, as the longitudinal distributions of produced particles are characterized by the flow of valence quarks from incident hadrons.[1] Furthermore, it is possible to connect the picture of hadrons at rest and in flight by utilizing the boost invariant description of composite systems in terms of light-like variables.[2,3] In this essay, we discuss this issue and related problems.

2. LIGHT-LIKE VARIABLES

Let us consider a hadron of mass M in flight with momentum p, and its dissociation into a particle of mass m with momentum k and others, as shown in Fig. 1.

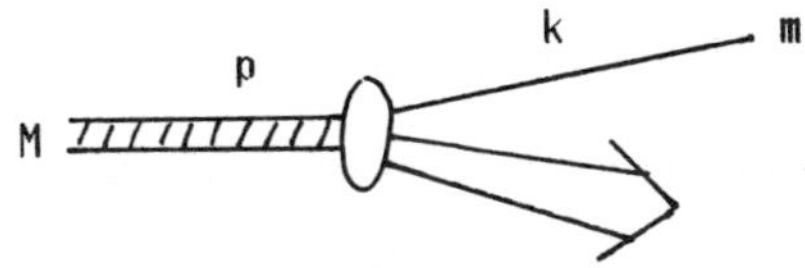

Fig.1.

Taking the longitudinal axis along p, we may adopt the light-like parametrization of four-momenta as

$$k = [k_+, k_T, k_-] = [k_0+k_L, k_T, k_0 - k_L],$$

$$p = [p_+, 0, p_-], \quad p_- = M^2/p_+ .$$

We define the light-like fraction as

$$z = k_+ / p_+ .$$

These expressions can be utilized irrespective of whether k is real or virtual.

If k is on-shell, we have

$$k = [m_T \exp(y), k_T, m_T \exp(-y)],$$

where y is the rapidity, and m_T is the transverse mass

$$m_T = (m^2 + k_T^2)^{1/2} .$$

Then, the light-like fraction z is invariant under longitudinal boost, which makes only the additive change of the rapidities of p and k. The Lorentz invariant phase space is proportional to

$$d^3k/E_k = dz/z \, d^2k_T .$$

3. ON-SHELL DISSOCIATION

Let us consider the dissociation $p \rightarrow k_1 + k_2 + .. + k_n$, satisfying the energy momentum conservation, where p and k_i with masses m_i are on-shell. Kinematically, this is the case for the decay of a resonance or a fire-ball, and approximately the one for the dissociation of loosely bound state as we come later on. From the conservation of energy and longitudinal momentum, we have

$$1 = z_1 + z_2 + .. + z_n , \qquad 0 < z_i < 1.$$

The ranges of z can be easily obtained in the rest frame of p where

$$z_i = (E_k^* + k_L^*)_i/M .$$

Therefore, we can immediately learn a few lessons :

(i) The maximum(minimum) of z_1 is the configuration that k_{1L} takes the maximum(minimum) value balancing with all the others in the opposite direction, as shown in Fig. 2.

Fig. 2.

(ii) If the dissociation is close to the threshold, the distribution is sharply peaked at the mass ratio $z_i \sim m_i/M$.

(iii) If the dissociated state is highly relativistic so that the masses are negligible compared with the size of the three-momenta available, the limit of z reaches close to 0 and 1.

(iv) The average $<z_i>$ turns out to be small for large number of final state particles, or for the coexistence with other massive particles.

Such properties as above can be confirmed by the phase-space computation of the longitudinal distribution $x_L = k_L / p_L$ for the decay of high momentum resonance.[4] The merit to use the light-like fraction is that it is boost invariant and positive definite satisfying $0 < z < 1$.

The boost invariance is a kinematical argument and valid for particles on the mass shell. One may apply to the structure of loosely bound state after absorbing a small amount of excitation energy. Another application may be the on-shell parton model explicitly assuming a small parton mass.[5]

4. TWO-BODY BOUND STATE

Let us study the effect of constituent mass on the light-like distribution in the virtual dissociation of a hadron, $p \to k + k'$. In the covariant approach to probe k, we may treat k' as on-shell states in the forward absorptive amplitude $p + (-k) \to p + (-k)$ as shown in Fig. 3.[6)]

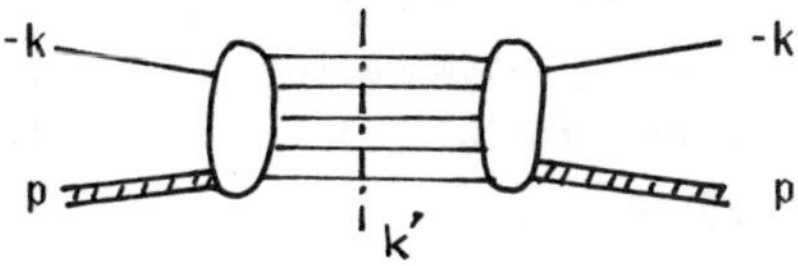

Fig. 3.

Since k is now off-shell, we put

$$k = [\, zp_+ \,,\, k_T \,,\, (k^2 + k_T{}^2)/zp_+ \,] \quad .$$

Denoting $s' = k'^2$, we have

$$s' = (p-k)^2 = k^2 - (k^2 + k_T{}^2)/z + (1-z)M^2 \; .$$

In the case of the two-body bound state with $s' = m'^2$, we have

$$k^2 - m^2 = \{\, k_T{}^2 + z\, m'^2 + (1-z)\, m^2 - z(1-z)\, M^2 \,\} \,/(z-1) \; .$$

If the dissociation vertex is regarded as a constant g in the spinless model as shown in Fig. 3, we obtain the following distribution in dz and dk_T (see the next section and ref. 3)):

$$F(z,k_T) = [2(2\pi)^3]^{-1}\, g^2\, z(1-z)/[k_T{}^2 + M_{eff}{}^2(z)]^2 \; ,$$

where

$$M_{eff}{}^2(z) = \; z\, m'^2 + (1-z)\, m^2 - z(1-z)\, M^2,$$

which is possitive for $0 < z < 1$, as $m + m' > M$. The constraint $0 < z < 1$ follows naturally from the analyticity of the forward absorptive amplitude. Integrating over dk_T , we have the light-like distribution of k in p :

$$F(z) = (g/4\pi)^2 \;\; z(1-z)/\, M_{eff}{}^2(z) \; .$$

We see that the mass dependence of F(z) is in accord with our intuition: it is symmetric around $z = 1/2$ for $m = m'$, while it is strongly peaked at $z \sim 1$ for $m >> m'$, or oppositely at $z \sim 0$ for $m << m'$. Since $M_{eff}{}^2(z)$ is rewritten as

$$M_{eff}{}^2(z) = [\, z\, m' - (1-z)\, m \,]^2 + z(1-z)\, [(m+m')^2 - M^2] \; ,$$

the distribution becomes very narrow as $m+m' \to M$, with $\langle z \rangle \sim m/M$. In Fig. 4, we show the quark distributions $F_{q/PS}(z)$ in pseudoscalar mesons with different flavors, assuming the constituent quark mass for u (=d), s, c and b as 0.35, 0.5, 2.0, and 5.2 GeV/c^2, respectively, where the coupling constant g is adjusted to satisfy the normalization condition

$$\int_0^1 F_{q/PS}\, dz = 1.$$

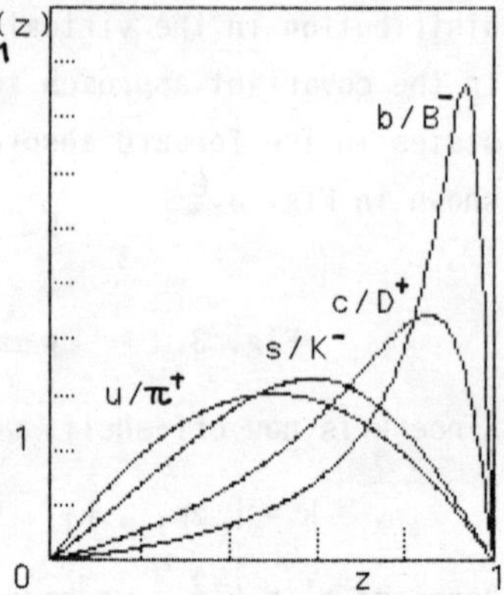

Fig. 4. The distribution of q/PS.

We note here that the mass dependence of F(z) is quite similar to the situation in the covariant model for the quark fragmentation, heavy quark → heavy meson + light quark, as discussed in 7).

5. REGGE BEHAVIOR IN THE COVARIANT MODEL

Now we consider more general situations in the spinless model for $F(z,k_T)$:3)

$$F(z,k_T) = [2(2\pi)^4]^{-1}\ z/(1-z)\ \int ds'\ \mathrm{Im}\ T(s',k^2)/[k^2 - m^2]^2 ,$$

where Im T is the forward absorptive amplitude for p +(-k) → p + (-k). The pole at $s' = m'^2$ gives the two-body contribution discussed in sect. 4, and continuum gives higher-body contributions. As for the latter, let us assume the following form:

$$\mathrm{Im}\ T(s',k^2) = C\ s'^{\alpha}\ /[k'^2 - M'^2]^{2N} \quad \text{for } s' > s_t ,$$

where C is a constant constrained by the normalization, M' is a phenomenological constant, and α is the Regge intercept for p + (-k) → p + (-k). Then, we obtain the continuum contribution to F(z) as follows :

$$F_{cont}(z,k_T) = C\ [2(2\pi)^4]^{-1}\ z^{-\alpha}\ (1-z)^{2N+1}\ \times$$

$$\int_{zs_t}^{\infty} ds''[\ s''^{\alpha}\ \{k_T^2 + M_{eff}^2(z,s'')\}^{-2}\{k_T^2 + M'_{eff}{}^2(z,s'')\}^{-2N}\ ],$$

where s" = zs' and

$$M_{eff}^2(z,s") \;=\; s" + (1-z)\,m^2 - z(1-z)\,M^2 \;,$$

$$M'_{eff}{}^2(z,s") = s" + (1-z)\,M'^2 - z(1-z)\,M^2 \;.$$

Then we obtain the behavior

$$F_{cont}(z,k_T) \sim C'\, z^{-\alpha}\, (1-z)^{2N+1} / \{k_T{}^2 + M^*_{eff}{}^2\}^{2N+2} \;,$$

where $M^*_{eff}{}^2$ is chosen to approximate the exact equation.

Thus, we may understand the power behavior of the quark distribution for 1-z and the Regge behavior at z $\sim$ 0 from the properties of the quark-hadron amplitude. The value of N is constrained by the correspondence to the dimensional counting, as discussed in ref. 3).

6. CONSTITUENT DISTRIBUTIONS IN THE QUARK-DIQUARK CASCADE MODEL

In the quark-diquark cascade model developed by Noda, Tashiro, the present author and others,[8-9)] we may treat rich variety of inclusive distributions in terms of the quark-diquark cascade, which starts from the constituent quark, diquark or their anti-particles inside a hadron. There, it is assumed that a meson is regarded as $M \sim q + \bar{q}$, and the baryon as $B \sim q + qq$, in contrast to the additive quark picture, $B \sim q + q + q$. The colors of qq and $\bar{q}$ are anti-triplet, opposite to q. The constituents q, $\bar{q}$ and qq may be clusters of many partons.

As the structure of hadrons in the soft interaction, the following forms of joint distribution functions in dz_i is assumed, where z_i denotes the lightlike fraction of constituent i's momentum with respect to the parent hadron along the direction of its motion:

$$H_M(z_q,z_{\bar{q}}) = z_q{}^{R-1}\, z_{\bar{q}}{}^{R-1}\, \delta(1-z_q-z_{\bar{q}}) \;,$$

$$H_B(z_q,z_{qq}) = z_q{}^{R-1}\, z_{qq}{}^{R'-1}\, \delta(1-z_q-z_{qq}) \;.$$

Then, one constituent distributions inside a hadron are

$$H_{q/M}(z) = H_{\bar{q}/M}(z) = z^{R-1}(1-z)^{R-1} \,/\, B(R,R),$$

$$H_{q/B}(z) = z^{R-1}\ (1-z)^{R'-1}\ /\ B(R,R'),$$

$$H_{qq/B}(z) = z^{R'-1}\ (1-z)^{R-1}\ /\ B(R,R').$$

The parameters R and R' are related to the effective meson trajectories in the Regge theory with the following correspondence:

$$H_{q/M}(z=m^2/s) \sim s^{1-R} \sim s^{\alpha_M(0)},$$

$$H_{qq/B}(z=m^2/s) \sim s^{1-R'} \sim s^{\alpha_{M4}(0)},$$

where $\alpha_i(0)$ are Regge intercepts and M4 denotes the exotic meson state $qq\bar{q}\bar{q}$. Thus, we get

$$R = 1 - \alpha_M(0), \quad R' = 1 - \alpha_{M4}(0).$$

These are also related to the baryon Regge intercept $\alpha_B(0)$ as

$$R + R' = 2 - 2\,\alpha_B(0),$$

by considering the energy dependence of the backward MB scattering. If we adopt $0 < \alpha_M(0) < 0.5$, $-1 < \alpha_{M4}(0) < -0.5$ and $-0.5 < \alpha_B(0) < 0$, we have

$$0.5 < R < 1, \quad 1.5 < R' < 2.$$

To be realistic, we should take into account the symmetry properties of diquark states and the flavor dependence due to the difference of effective quark mass. On the basis of phenomenological studies, we have found the following results.

(i) Concerning non-strange quarks and diquarks, we have

$$R = 0.5, \quad R'_{[qq]} = 1.75, \quad R'_{\{qq\}} = 2.25,$$

where [qq] and {qq} denote antisymmetric and symmetric diquark states respectively for the flavor indices. The difference between [qq] and {qq} suggests that antisymmetric states, with higher intercepts, are more tightly bound than symmetric states.

(ii) The strange quark parameters are[9]

$$R_s = 1.0,$$

$$R_{[qs]} = 1.5\ R_{[qq]},\quad R_{\{qs\}} = 1.5\ R_{\{qq\}}\ ,$$

$$R_{\{ss\}} = 2.0\ R_{\{qq\}}\ ,$$

where q denotes non-strange quark. These parameters for strange diquarks are obtained from new data on hyperon beam experiments [10].

As typical examples, we show in Fig. 5 the distributions u/π^+, s/K^-, $[ud]/p$ and $\{ss\}/\Xi^-$. In these distributions, the behavior at $z \sim 1$ is singular as $(1-z)^{-0.5}$, since the remaining constituent is the light quark. On the other hand, the behavior at $z \sim 0$ depends strongly on each species. As a whole, there is a general trend that massive constituents take over large light-like fractions.

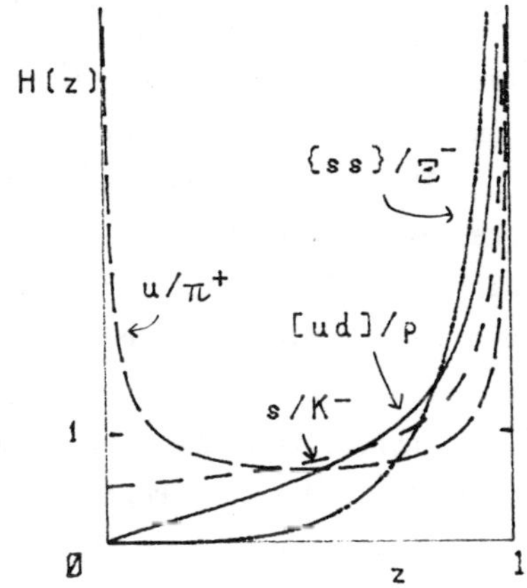

Fig. 5. The constituent distributions obtained in ref. 9).

7. CONCLUDING REMARKS

We have discussed a few models of constituent distributions inside hadrons, and found a general trend that the light-like fraction of a constituent is closely related to the size of its mass compared with other components. On the other hand, shapes of distributions are quite different in different approaches. Now, let us discuss the interrelation among them, concentrating our attention to the u quark distribution in π^+.

In the case of spinless model with constant dissociation vertex discussed in sect. 4, we obtain the behavior $F(z)_{u/\pi^+} \sim z(1-z)$ peaked at the average value $z = 1/2$, since the z dependence of $M_{eff}{}^2(z)$ is mild. This is in contrast to the behavior $H_{q/M}(z) \sim z^{-0.5}(1-z)^{-0.5}$ peaked at $z \sim 0$ and 1 in the quark-diquark cascade model discussed in sect. 6.

A link between these may be obtained from the model discussed in sect. 5. Namely, the exponent of z controling the behavior at $z \sim 0$ is 1 as far as m' of the remaining system is finite, while it changes to $-\alpha$ if the mass spectrum of the remaining system is characterized by $(s')^{\alpha}$ as in sect. 5. Therefore, if we look at the light-like fraction $z' = 1-z$ of the remaining system, we have the corresponding behavior at $z' \sim 1$ as $(1-z')^{-\alpha}$. The continuum of s' can be understood that the system s' is not the single quark-parton state but some dressed quark state with multi-parton configuration.

The physical content of the multi-parton configuration is presently not clear yet. It may be multiperipheral arm reaching to wee parton domain to cause the Pomeron exchange,[11] or the quark with surrounding field of gluons to be understood on the basis of QCD-bag dynamics.

Anyhow, from the above arguments concerning the boundary behavior characterized by the Regge intercept, we may understand the constituent distribution adopted in the quark-diquark cascade model for soft collision as follows: the behavior at $z \sim 0$ is controlled by the high mass continuum of the remaining state, while at $z \sim 1$ by the way of heavily dressing up itself, as illustrated in Fig. 6.

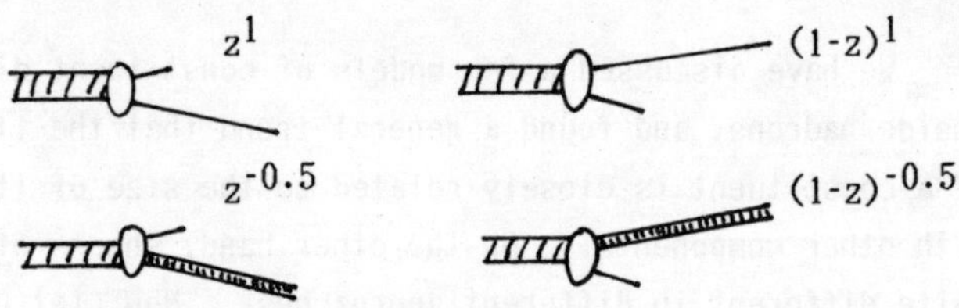

Fig. 6. The correspondence of the behavior at $z \sim 0$ and 1 versus thin and dressed quarks.

ACKNOWLEDGEMENTS

It is my great pleasure to dedicate this paper to Professor H. Miyazawa, who led me to start the works on particle physics in very inspiring atmosphere for young researchers, and opened his eyes to the importance of the interplay among theories and experiments.

REFERENCES

1) For early works,
Satz, H., Phys. Rev. Lett. **19**, 1453 (1967); Budapest Conf.(1977).
Ezawa, Z. F., Maharana, J. and Miyazawa, H., Prog. Theor. Phys. **51**, 877 (1974): Maharana, J.,ibid., **51**, 1624 (1974).
Kinoshita, K. and Noda, H., Prog. Theor. Phys. 52, 1622 (1974).
For a review of additive quark model,see
Anisovich, V. V. et al., Quark Model and High Energy Collisions, World Scientific, 1985.

2) First suggestion on light-like coordinates by
Dirac, P. A. M., Rev. Mod. Phys. **21**, 392 (1949).
See also
Yan, T. M. et al., Phys. Rev. **D7**, 1133, 1147, 1760, 1780 (1973).
Weinberg, S., Phys. Rev. **150**, 1313 (1966).

3) Kinoshita, K., Kinoshita, Y. and Myozyo, Y., Prog. Thoeor. Phys. 55, 211 (1976).

4) Noda, H. and Tashiro, T., Prog. Theor. Phys. **62**, 1172 (1979).

5) Akama, K., Prog. Theor. Phys. **51**, 1879 (1974).

6) For an early review,
Landshoff, P.V. and Polkinghorne, J.C., Phys. Rep. **5** , 1 (1972).

7) Kinoshita, K., Prog. Theor. Phys. **75**, 84 (1986).

8) Kinoshita, K., Noda, H., Tashiro, T. and Mizouchi, M., Zeit. Phys. **C4**, 103 (1980), for the basic scheme.
Kinoshita, K., Noda, H. and Tashiro, T., Prog. Theor. Phys. **68**, 1699, 2086, (1982); ibid. **74**, 1084 (1985).
Tashiro T., Noda, H., Kinoshita K. and Iso, C., Z. Phys. **C35**, 21 (1987).

9) Tashiro T., Noda, H. and Kinoshita K., Z. Phys. **C39**(1988),in press.

10) Biagi, S. F. et. al., Z. Phys. **C34**, 187 (1987).

11) Kogut,J. and Susskind, L., Phys. Rep. **8**, 76 (1973).
Koplik, J. and Mueller, A. H., Phys. Rev. **D12**, 3638 (1975).

QCD Parton Shower in e^+e^- Annihilation

Kiyoshi KATO
Kogakuin University, Department of Physics
Shinjuku, Tokyo 160

and

Tomo MUNEHISA
Yamanashi University, Faculty of Engineering
Takeda, Kofu 400

Abstract

The distribution of jets in the e^+e^- annihilation is studied using a cluster analysis. We compare a new parton shower model for jets which includes the full next-to-leading logarithmic order contribution with the leading logarithmic order parton shower and the conventional matrix element method.

1.Introduction

The event generators, the Monte-Carlo models for jets based on the perturbative QCD, are widely used for the study of e^+e^- annihilation. Since we cannot calculate the all diagrams of QCD, we use a part of perturbative series. There are two classes of models, the matrix element method[1] and the parton shower[2,3]. In the former, momenta of partons (quarks and gluons) are generated according to the matrix elements which are calculated up to a fixed order of α_s. Due to the technical limitation, only calculated are those for up to 4 jets. In this method, the distribution of partons is precise up to the calculated order. Since these matrix elements have mass singularity we must introduce a cutoff to define finite cross sections. On the other hand, we sum up the leading contribution of all order diagrams in the parton shower method.

To illustrate these methods we consider the three-jet cross secton, $e^+e^- \rightarrow$

$q\bar{q}G$,

$$\frac{d\sigma}{dz_1 dz_2} = \sigma_0 \frac{\alpha_s}{2\pi} C_F \frac{z_1^2 + z_2^2}{(1-z_1)(1-z_2)} \tag{1}$$

where z_1 and z_2 are the energy fraction of quark and antiquark, σ_0 is the lowest order cross section, $C_F = 4/3$ and α_s is the running coupling constant of QCD. As is seen in Eq.(1), the cross section is singular where $z_2 = 1$ ($z_1 = 1$), i.e., the quark (antiquark) and the gluon is collinear and where $z_1 = z_2 = 1$, i.e., the gluon is soft. To remove the singularity, the introduction of a cutoff is required. In this paper, we use the invariant mass cutoff y as $(p_i + p_j)^2 > ys$ for any pair of partons where s is the squared center-of-mass energy. Thus we obtain a cutoff dependent, finite three-jet fraction,

$$R(y) = \frac{1}{1+\alpha_s/\pi} \frac{\alpha_s}{2\pi} C_F [2\log^2 y + 3\log y - \frac{\pi^2}{3} + \frac{5}{2} + O(y)]. \tag{2}$$

Since the value of α_s is not very small, we must choose a rather large value of y if we want to keep the perturbation valid. For instance, R is almost 1 when $\sqrt{s} = 100$[GeV] and $\sqrt{ys} = 7$[GeV]. This means that we cannot resolve a jet of mass$\approx$ 7[GeV], so that the distribution of events generated by this method has a 'discontinuity' at a small jet mass.

To describe the parton shower, it is convenient to use the infinite momentum frame, i.e, we boost the system along the opposite direction of the momentum of antiquark. Using variables in this frame, the Eq.(1) is rewritten as follows:

$$\frac{d\sigma}{dx dt} = \sigma_0 \frac{\alpha_s}{2\pi} C_F \{\frac{1}{t} P(x) + 2(x - \frac{1}{1-x}) + t(1 - x + \frac{1}{1-x})\} \tag{3}$$

where $P(x) = (1+x^2)/(1-x)$, x is the light-cone fraction of momentum of quark and t =(invariant mass of quark and gluon)$^2/s$. $t = 0$ corresponds to the collinear singularity and the first term in the curly bracket in Eq.(3) is dominant in the collinear region. In the leading-logarithmic approximation, we only keep this dominant term. When we calculate diagrams using a physical gauge like axial gauge, the leading term does not appear in interference terms, so that it is possible to treat the parton shower as a stochastic branching process. The probability that a quark with normalized virtuality t does not branch for $t \to t - dt$ is

$$1 - \frac{\alpha_s}{2\pi} C_F \frac{dt}{t} \int P(x) dx. \tag{4}$$

Integrating this, we obtain the non-branch probability in the leading logarithmic order,

$$\Pi(t_1, t_0) = \exp[-\int_{t_0}^{t_1} \frac{dt}{t} \frac{\alpha_s(ts)}{2\pi} C_F \int_{\epsilon}^{1-\epsilon} dx P(x)]. \tag{5}$$

$\Pi(t_1, t_0)$ is the probability that a quark of virtuality t_1 does not branch and turns to a quark of lower virtuality t_0. Here, $\epsilon = Q_0^2/ts$ is the cutoff for the infrared singularity where Q_0^2 is the cutoff virtuality. By this, we obtain a continuous distribution of jet's mass down to a cutoff mass ($\sim Q_0$). Summarizing the above arguments, the matrix element method is correct for the hard three jets but it gives discontinuous distribution for the mass of jets while the parton shower gives continuous distribution for jet's mass but it gives results which differ from the exact one for the finite, non-leading contribution.

The last defect of the parton shower is removed by introducing next-to-leading order corrections into the parton shower[3]. We review this method in the next section. In section 3, we present the distribution of jets using our event generator. First we discuss the three jets to check our model. We also study the four-jet sector where the finite term differs from the exact one even in our improved method. Most of the analyses in this paper are done for a system of generated partons. When we refer to the effect of hadronization, we use EPOCS[4] to convert partons into hadrons in its local mode which is similar to the LUND string fragmentation[5].

2. Our model

The parton shower including the full next-to-leading order is discussed in Ref.[3]. The corrections are classified into five categories; the corrected two-body branching, the three-body branching, the corrected $P-$functions, the precise treatment of phase space and the hard scattering cross section for three jets which corresponds to the coefficient function. The guiding principles of our model are as follows:

(*1*) The perturbative treatment should be kept valid, i.e., the higher order is smaller than the lower order.

(*2*) Distributions of generated partons are consistent with the prediction of the conventional analysis for inclusive quantities.

(*3*) The branching functions are positive.

The parton shower in leading logarithmic order gives incorrect distribution for hard three jets. To remove this defect we generate three-jet events according to Eq.(1) (or equivalent one, Eq.(3)). This formula is also used in some models(e.g., the last one in Ref.[2]). However, the introduction of Eq.(1) is only a part of next-to-leading corrections and if the remaining parts are formulated according to the leading order, the model is incorrect in the sense that the model does not stand on any theoretically well-defined scheme. We introduce a cutoff δ to separate the two and three jets. We calculate the three-jet fraction integrating the Eq.(3) for $t > \delta$. By this fraction we decide to generate a cascade which starts either as two jets or as three jets. When an event starts as two jets, its maximum virtuality at the first branching is limited by δs and when an event starts as three jets, it is generated by Eq.(3) for $t > \delta$. The above procedure is to connect Eqs.(3) and (5) at $t = \delta$ and physical quantities are almost insensitive to the value of δ. In Fig.1 we show $\Pi_B = \frac{d}{dt}\Pi(1,t)$ and $\frac{1}{\sigma_0}\frac{d\sigma}{dt} = \int dz \frac{1}{\sigma_0}\frac{d\sigma}{dzdt}$ and $\delta \approx t$ where two curves cross. From this figure we can see that Eq.(3) is not good for small t while Eq.(5) is an overestimation for large t. The use of Eq.(3) is to define a new coefficient function. It is well known that only the combination, $P^{(1)}(x) - \beta_0 C^{(1)}(x)/2$ where $P^{(1)}$ and $C^{(1)}$ are the P-function and the coefficient function in the next-to-leading order, respectively, is independent of renormalization scheme.[6] This causes to add $P^{(1)}(x)$ a $\delta-$dependent term,

$$\frac{\beta_0}{2}[\int_\delta^1 (inside\ of\ \{\ \}\ in\ Eq.(3)\ )dt - C^{(1)}(x)] \tag{6}$$

in our model. Here $C^{(1)}(x)$ is the conventional coefficient function.[7]

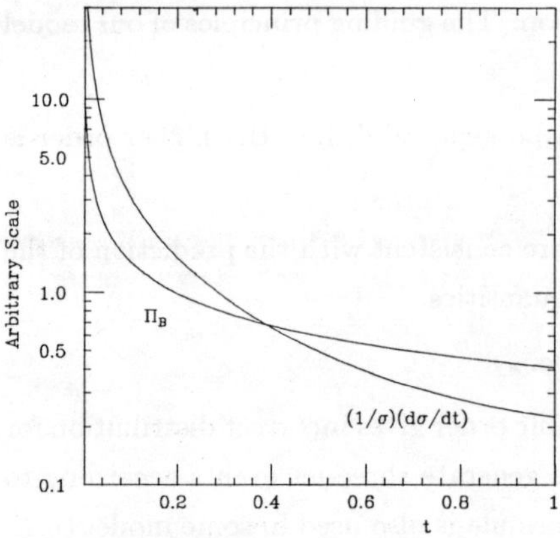

Fig.1 Distribution of t in the leading-logarithmic approximation for $\sqrt{s} =$ 50GeV and $\Lambda = 0.2$ GeV.

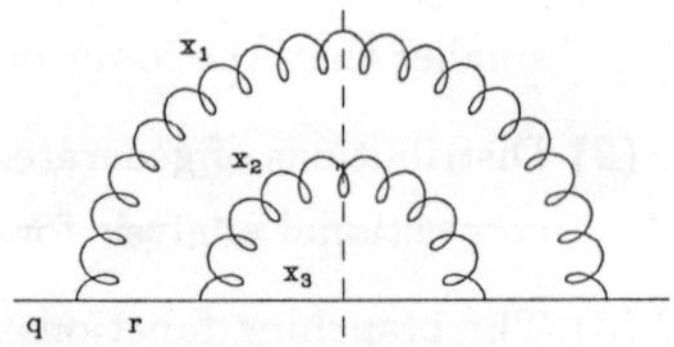

Fig.2 An example of diagram to contribute $q \to q + G + G$. A solid line is a qurak and a curly line is a gluon.

To see the importance of the kinematics, we consider the diagram shown in Fig.2. This diagram contributes both to a sequence of two tow-body branches, $q \to q + G$, and to a three-body branch, $q \to q + G + G$. Omitting factors irrelevant to the present discussion, this diagram gives

$$\frac{dq^2}{q^2} \int_{Q_0^2}^{(x_2+x_3)q^2} \frac{dr^2}{r^2} P(x_2 + x_3) \frac{1}{x_2 + x_3} P(\frac{x_3}{x_2 + x_3}) dx_2 dx_3. \tag{7}$$

In the above equation, the upper limit of r^2-integration is determined by the momentum conservation. If we replace this upper limit by q^2 and subtract it from Eq.(7), we obtain

$$-\log{(x_2 + x_3)} P(x_2 + x_3) \frac{1}{x_2 + x_3} P(\frac{x_3}{x_2 + x_3}) \tag{8}$$

which contributes to the next-to-leading order as a part of three-body decay functions. In the generation of partons, we of course keep the momentum conservation. However, in the leading logarithmic order there is no difference in *the formulation* whether we keep the momentum conservation or not. Another kinematical constraint comes from the angular ordering introduced by the careful study of infra-red structure in QCD.[8] The angular ordering is only shown for a limiting case like $x_1 \gg x_2 \gg x_3$. In the simulation, we must deal with a case like $x_1 > x_2 > x_3$. This allows us to apply a new restriction to phase space if it is identical to the angular ordering in the limiting case. These delicate points on kinematics are seldom recognized in the inclusive analysis but they are very important for the exclusive case, i.e., the Monte Carlo simulation of parton shower. Functions for three-body decays and the correction to two-body decays are given in Ref.[9] by the calculation of a number of two loop diagrams. Some of these functions are negative. Using the freedom in kinematics, we redefine these functions in Ref.[3] under principles *(1)* and *(3)*. The argument of α_s is ambiguous in the leading logarithmic order. Considering in the next-to-leading order, it should be p_T^2 rather than Q^2 to cancel the singularity in the second order P-function.

We have written our Monte Carlo code taking account of the features discussed above and have checked that the principle *(2)* really holds.[3] This means that we can determine the QCD Λ using our model, which is impossible in the leading logarithmic model.

Since we use the light-like axial gauge in our formulation, the antiquark does not radiate. This causes small but non-negligible asymmetric distribution between quark and antiquark. To avoid it, we have devised the 'two-jet scheme' whose details will be described in another paper.[10] The essential point of this scheme is that the content in the curly bracket of Eq.(3) is

$$\frac{2}{t(1-x)} - \frac{2-(1-x)}{t} - \frac{2-t}{1-x} + 2x + t(1-x) \tag{9}$$

which is symmetric for $1-x$ and t in the singular region, $t \approx 0$ or $x \approx 1$. In this region the energy fraction in the center-of-mass system is

$$z_1 = x + (1-x)t \approx x \text{ and } z_2 = 1-t \tag{10}$$

and the strong coupling constant is

$$\alpha_s(p_T^2) \approx \alpha_s((1-x)tQ^2). \tag{11}$$

Using this symmetry, $t \leftrightarrow 1-x$ and $z_1 \leftrightarrow z_2$, we can formulate our model in a symmetric manner.

3.Results of our model

First, we describe the cluster method used here.[11] For generated events we calculate invariant masses for all combinations of two partons. For the pair of the minimum mass, the two partons are merged as a single parton whose momentum is the sum of those of the pair. This procedure is repeated until the square of minimum mass exceeds a fixed value, ys. Remaining partons are considered as jets.

Partons are generated under the conditions; $\sqrt{s} = 50\text{GeV}$, $Q_0 = 1\text{GeV}$, $\delta = 0.05$ and $\Lambda = 0.2\text{GeV}$. In Fig.3(a) we show the y-dependence of the multiplicity of clusters. The effect of hadronization is very large in small y region. The distribution generated by the parton shower in the leading logarithmic order is given in Fig.3(b).* In this case, as is discussed in Section 1, the faction of three jets is larger than that in the next-to-leading order. From this analysis, we choose $y = 0.04$ for the following analysis. With this value, reasonable fractions of clean three jets and four jets are separated well and also the effect of hadronization is not large.

We calculate the thrust distribution for the events of three clusters as

$$\frac{1}{N_0}\frac{dN}{dT} \tag{12}$$

where the normalization, N_0, is the number of events of two and three clusters. In Fig.4 we compare it with the that calculated by the $O(\alpha_s)$ matrix element[12]

$$\frac{1}{\sigma}\frac{d\sigma}{dT} = \frac{1}{(1+\alpha_s/\pi)}\frac{\alpha_s}{2\pi}C_F\left[\frac{2(3T^2-3T+2)}{(1-T)T}\log\frac{2T-1}{1-T} + \frac{3(3T^2-8T+4)}{1-T}\right]. \tag{13}$$

If clusters are massless, there is no events for $1 - T < y$. In Fig.2 we have, however, events for $1 - T < y$ due to massive clusters. The agreement of both

* In our Monte Carlo code, we can switch to the leading logarithmic mode by an input parameter. Here we use $\Lambda = 0.8\text{GeV}$ which gives almost the same magnitude of scaling violation as the case of $\Lambda = 0.2\text{GeV}$ in the next-to-leading order.[3]

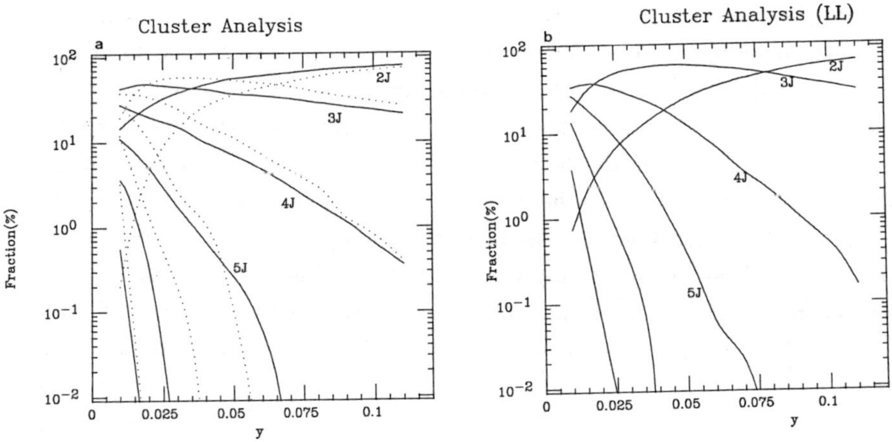

Fig.3 Multiplicity distibution of clusters in (a)next-to-leading logarithmic model and in (b)leading logarithmic model. In the former, the dotted line represents the results after the hadronization.

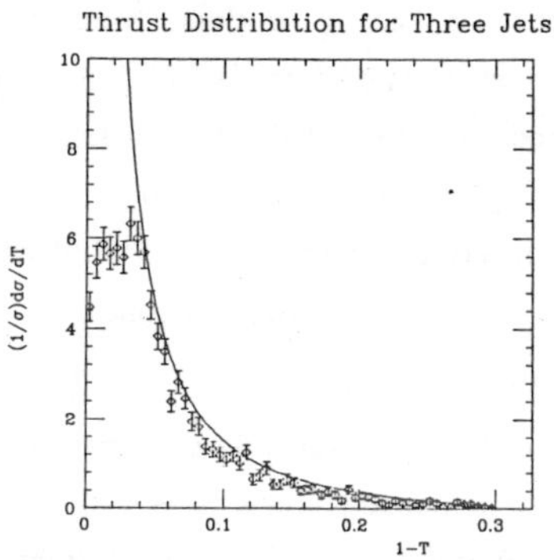

Fig.4 Thrust distribution for three jets. The solid line represents Eq.(13) and the points result from 10,000 events generated by our parton shower model.

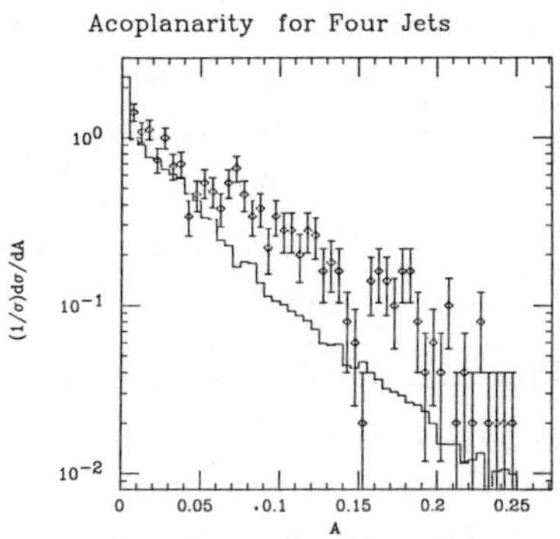

Fig.5 Acoplanarity distribution for four jets. The histgram represents the calculation by matrix elements and the points are the results of our model.

distribution seems to be good and the small discrepancy may be originated from kinematical effects.

Next we calculate the acoplanarity for events of four clusters and compare it that calculated by the $O(\alpha_s^2)$ four-jet matrix element in EPOCS. The normalization of both distributions is chosen as in the thrust analysis. As is shown in Fig.5, the parton shower gives larger four-jet fraction. In the small acoplanarity region, i.e., planar configulation, the agreement of two distributions is as good as in the thrust distribution. However the deviation is not negligible in the large acoplanarity region. We also calculate the angle distribution between jets in the four-jet samples and found that not only the normalization but also the correlation of jet direction is slightly different. These facts may suggest that we need more correction for non-planar events even in the parton shower of next-to-leading order.

4.Conclusion and discussion

In this paper, we have checked that our improved generator for parton shower is consistent with the three-jet events. After the clustering of partons we can distinguish the difference of leading and the next-to-leading models. Since the formulas given in section 1 are for the massless parton, the distribution is a little modified due to the finite virtuality and the effect of kinematics. In the four-jet events, there is a small discrepancy between distributions of parton shower and those of matrix elements. The difference partly comes from the fact that we neglect the terms beyond the next-to-leading logarithmic approximation.

In literatures[11] the matrix element method is not favorable since it is insufficient to explain the multiplicity of jets. This method also has the discontinuity problem. However, this method should not be discarded since it is still useful to estimate some channels, e.g., four-parton production which is not exactly calculated in the parton shower.

The authors thank to the colleagues of the theoretical working group for TRISTAN. They also thank to prof.T.Kamae for stimulating discussions.

References

[1] A.Ali, E.Pietarinen, G.Kramer and J.Willrodt, Phys.Lett. **93B** (1980), 155;

P.Hoyer, P.Osland, H.G.Sander, T.F.Walsh and P.M.Zerwas, Nucl.Phys. **B161** (1979), 349.

[2] R.Odorico, Nucl.Phys. **B172** (1980), 157;
G.C.Fox and S.Wolfram, Nucl.Phys. **B168** (1980), 285;
T.D.Gottschalk, Nucl.Phys. **B214** (1983), 201; Nucl.Phys. **B227** (1983), 413;
G.Marchesini and B.R.Webber, Nucl.Phys. **B238** (1984),1;
B.R.Webber, Nucl.Phys. **B238** (1984),492;
T.Sjöstrand, Comput.Phys.Commun. **39** (1986), 347;
T.Sjöstrand and M.Bengtsson, Comput.Phys.Commun. **43** (1987), 492.

[3] K.Kato and T.Munehisa, Mod.Phys.Lett. **A1** (1986),345; Phys.Rev. **D36** (1987), 61.

[4] K.Kato and T.Munehisa, *KEK Report* 84-18 (1984); 87-5 (1987).

[5] B.Andersson and G.Gustafson, G.Ingelman and T.Sjöstrand, Phys.Rep. **97** (1983), 31 *and references therein.*

[6] W.A.Bardeen, A.J.Buras, D.W.Duke and T.Muta, Phys.Rev. **D18** (1978), 3998.

[7] G.Curci, W.Furmanski and R.Petronzio, Nucl.Phys. **B175** (1980), 27.

[8] A.Bassetto, M.Ciafaloni and G.Marchesini, Phys.Rep. **100** (1983), 201.

[9] J.Kalinowski, K.Konishi and T.R.Taylor, Nucl.Phys. **B181** (1981), 221;
J.Kalinowski, K.Konishi, P.N.Scharbach and T.R.Taylor, Nucl.Phys. **B181** (1981), 253;
J.F.Gunion and J.Kalinowski, Phys.Rev. **D29** (1984), 1545.

[10] K.Kato and T.Munehisa, *in preparation.*

[11] W.Bartel *et al.*,(JADE Collaboration), Z.Physik **C33** (1986), 23.

[12] A.De Rújula, J.Ellis, E.G.Floratos and M.K.Gaillard, Nucl.Phys. **B138** (1978), 387.

Annihilation of e^+e^- into $\gamma\gamma\gamma$ through Z

Masataka Igarashi
Tokai University, Department of Physics
Hiratuka, Kanagawa 259-12, JAPAN
and
Nobuya Nakazawa
Kogakuin University, Department of Physics
Shinjuku, Tokyo 160, JAPAN

Abstract

The three gamma decay of Z-boson in electron positron annihilation process is considered by calculating the fourth rank vacuum polarization tensor. The process is potentially expected to be the background to the reactions $e^+e^- \to \gamma\gamma$ and $e^+e^- \to \nu\bar{\nu}\gamma$. The rate is so small that we can neglect this process as the backgrounds even when CM-energy is equal to Z-boson mass.

1.Introduction

The determination of the number of almost massless neutrinos is very important both in particle physics and in cosmology. The operation of SLC and

LEP in near future is expected to give a means of determining that number by allowing the direct measurement of the decay width of Z-boson and/or the cross section of $e^+e^- \rightarrow \nu\bar{\nu}\gamma$[1]. We here consider the so-called neutrino counting reaction $e^+e^- \rightarrow \nu\bar{\nu}\gamma$ in the CM energy around or slightly above M_z. This reaction is brought about by the s-channel Z-boson and, for the case of electron neutrino,also by the t-channel W-boson. Since the latter contribution is known to be negligibly small in the energy range of our interest, we limit ourselves to the reaction mediated Z-boson only.

In order to deduce the number of the massless neutrinos, one must have the precise knowledge of the cross section which cannot get rid of higher order corrections. On this point of view, we have calculated the QED radiative corrections to this process and showed that the corrections are rather large as is usual for resonances[2]. The virtual corrections due to pure weak one loop diagrams, such as Z and W boson exchange correction to Z-electron vertex, are expected to be small compared with QED ones. The possible exception may be the case that the virtual Z-boson is close to the mass-shell[3]. The calculatoions of these corrections are now in progress.

Another important study which must be done before the experiment is the estimate of the backgrounds[4]. The main background reactions to this process are the radiative Bhabha scattering [5] and $e^+e^- \rightarrow \gamma\gamma\gamma$ [6]. According to ref.[4],these backgrounds can be reduced considerably under suitable experimental set-ups and it is really possible to determine the number through the reaction $e^+e^- \rightarrow \nu\bar{\nu}\gamma$.

The possible background $e^+e^- \rightarrow \gamma\gamma\gamma$ is also induced by a higher order diagram shown in Fig.1. It is interesting to note that,when QED is extended to the standard SU(2)xU(1) theory, the Z-boson contribution for this reaction sets in only at the one loop level[7]. This process is a potential background to the neutrino counting reaction as well as to $e^+e^- \rightarrow \gamma\gamma$. In this paper, we calculate the process Fig.1 by using the expression of the fourth rank vacuum polarization tensor given by V.Costantini et.al.[8]. The kinematics and the cross section are given in the next section and the numerical results are shown in section 3. Discussions are given in the final section.

2.Kinematics and cross section

The lowest order diagram of e^+e^- annihilation into $\gamma\gamma\gamma$ through virtual Z contains the fourth rank vacuum polarization tensor $G_{\mu\nu\lambda\sigma}(1,2,3,4)$ corresponding to Fig.2. The definition of the polarization tensor is as follows.

$$G_{\mu\nu\lambda\sigma}(1,2,3,4) = \frac{1}{i\pi^2}\int d^4q$$

$$\otimes Tr(\gamma_\mu \frac{1}{(\not{q}-m)}\gamma_\nu \frac{1}{(\not{q}-\not{k}_2-m)}\gamma_\lambda \frac{1}{(\not{q}+\not{k}_1+\not{k}_4-m)}\gamma_\sigma \frac{1}{(\not{q}+\not{k}_1-m)})$$

$$+ \quad \{\text{ two more permutations in photon indices }\} \tag{1}$$

Photons with index 2,3 and 4 are on shell and index 1 represents the vector part of the virtual Z-bson. Axial vector part of Z-boson fermion coupling does not contribute brecause of Bose statistics of photons. The fourth rank vacuum polarization tensor is related to the lowest-order nonlinear interacton between four electromagnetic fields and important for the processes, such as photon photon scattering, Delbrück scattering , photon splitting, etc. In our case three external photons are on-shell and thus the situation is similar to the photon splitting in an external field.

We use the expressions given by V.Costantini et al. in the framework of helicity amplitude[8]. The above tensor is then denoted as $G_\mu{}^{(\lambda_2\lambda_3\lambda_4)}(1,2,3,4)$ with $\lambda_2, \lambda_3, \lambda_4$ being the helicity of three real photons,respectively.

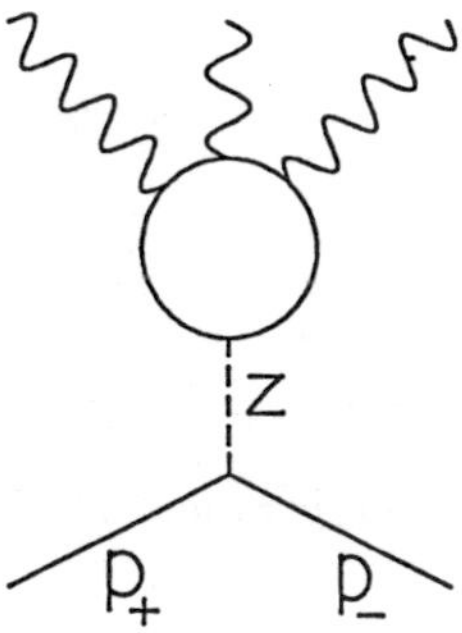

Fig.1. Feynman diagram of e^+e^- annihilation into $\gamma\gamma\gamma$ through virtual Z-boson

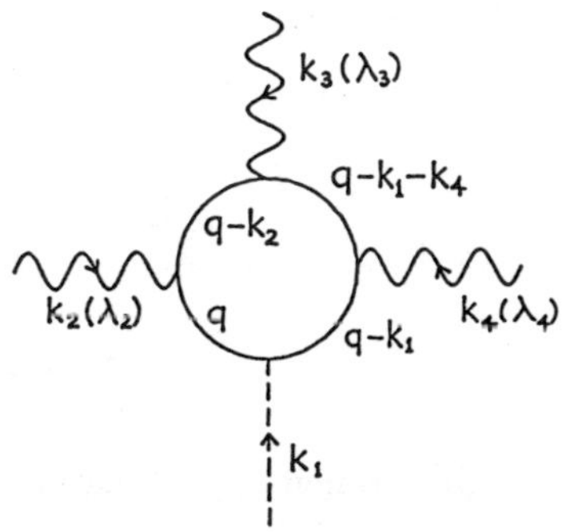

Fig.2. One of the 6 diagrams contributing to the fourth rank vaccum polarization;the polarization tensor in Eq.(1) is defined by only the half of them.

We have the following formulas.

$$G_\mu{}^{(+,+,+)}(1,2,3,4) = \frac{i}{4\sqrt{2\Delta}}(2W^2(1-y)E^{(1)}(x,y,z)(k_{2\mu} - \frac{y}{x}k_{1\mu})$$

$$-2W^2(1-z)E^{(1)}(x,z,y)(k_{3\mu} - \frac{z}{x}k_{1\mu}) + 4E^{(2)}(x,y,z)\Psi_\mu), \tag{2}$$

$$G_\mu{}^{(+,-,+)}(1,2,3,4) = G_\mu{}^{(+,+,+)}(1,2,3,4)[(x \leftrightarrow z),(k_2 \leftrightarrow k_4)], \tag{3}$$

$$G_\mu{}^{(+,+,-)}(1,2,3,4) = G_\mu{}^{(+,+,+)}(1,2,3,4)[(x \leftrightarrow y),(k_2 \leftrightarrow k_3)], \tag{4}$$

$$G_\mu{}^{(-,+,+)}(1,2,3,4) = \frac{i}{4\sqrt{2\Delta}}(-8)\Psi_\mu, \tag{5}$$

where $W^2 = {k_1}^2 = (p_+ + p_-)^2$ is the CM-energy and $x = 2\omega_2/W, y = 2\omega_3/W, z = 2\omega_4/W$ are the energy fraction of on-shell photons,respectively $(x + y + z = 2)$;

$$\Delta = W^6(1-x)(1-y)(1-z)/4, \tag{6}$$

$$\Psi_\mu = -i\varepsilon_{\mu\nu\rho\sigma} k_2^\nu k_3^\rho k_4^\sigma . \tag{7}$$

In the light fermion approximation, the functions $E^{(1)}(x, y, z)$ and $E^{(2)}(x, y, z)$ are given by

$$E^{(1)}(x,y,z) = 2\frac{(1-z)}{y} + (3 - \frac{1}{y} + 2\frac{(1-y)}{1-x} - 2\frac{(1-x)}{y^2}) log(1-y)$$

$$+(\frac{1-z}{z} + 2\frac{1-z}{1-x}) log(1-z) + (\frac{y-z}{1-x} + 2\frac{(1-y)(1-z)}{(1-x)^2}) G(y,z), \tag{8}$$

$$E^{(2)}(x,y,z) = (1 - \frac{1}{y} + 2\frac{1-y}{1-x}) log(1-y) + (1 - \frac{1}{z} + 2\frac{1-z}{1-x}) log(1-z)$$

$$+(\frac{x}{1-x} + 2\frac{(1-y)(1-z)}{(1-x)^2}) G(y,z), \tag{9}$$

$$G(x,y) = log(1-x) log(1-y) - \frac{\pi^2}{6} + Sp(x) + Sp(y) \tag{10}$$

and the function $Sp(x)$ is the dilogarithm defined by

$$Sp(x) = -\int_0^x \frac{log(1-t)}{t} dt. \tag{11}$$

These expressions are consistent with those used by M.L.Laursen et al.[9] to estimate the rate of $\gamma\gamma\gamma$ decay of on-shell Z-boson. It should be noted that the ${G_\mu}'s$ are free from the mass singularity.

Next step is to derive the cross section by contracting the tensor indices with the tensor $L_{\mu\nu}$of electron-positron system.

$$L_{\mu\nu} = (-k_{1\mu}k_{1\nu} + Q_\mu Q_\nu + W^2 g^{\mu\nu}) \tag{12}$$

where $Q_\mu = (p_+ - p_-)_\mu$. The first part of tensor $L_{\mu\nu}$ does not contribute since G_μ is conserved. For the on-shell free Z-boson, only the last term remains.

We consider the cross section in the coordinate system where the positron momentum is chosen as z-axis; the scattering angle of detected photon is defined as θ_x and the polar and azimuthal angle of the second photon and the third photon as θ_y, ϕ_y, θ_z and ϕ_z ,respectively. The final expression of the cross section except the statistical factor is given by

$$\frac{d^2\sigma}{dx dcos\theta_x} = \frac{\alpha^5}{4(4\pi)^3}(3\sum_i f_v^i(q^i)^3 + \sum_l f_v^l(q^l)^3)^2(f_v^{e2} + f_a^{e2})$$

$$\otimes \int dcos\theta_y \int d\phi_y \frac{W^2}{(W^2 - M_z{}^2)^2 + M_z{}^2\Gamma_z{}^2} \frac{x(1-x)}{(1-ax)^2} Samp \tag{13}$$

where,$a = (1-(sin\theta_x sin\theta_y cos\phi_y + cos\theta_x cos\theta_y))/2$ and $M_z(\Gamma_z)$ is the mass(width) of Z-boson,respectively; in the unit of electric charge, $f_v(f_a)$ are the vector(axial-vector) coupling constant of lepton(l)/quark(i) with Z-boson,respectively, and q is coupling constant with photon. The square of amplitude $Samp$ is given as follows.

$$Samp = Samp1 + Samp2, \tag{14}$$

$$Samp1 = \{16\frac{y(1-y)}{x(1-x)}E^{(1)}(x,y,z)^2 + 16\frac{z(1-z)}{x(1-x)}E^{(1)}(x,z,y)^2$$

$$-32\frac{(1-y)(1-z)}{x(1-x)}E^{(1)}(x,y,z)E^{(1)}(x,z,y) + 16E^{(2)}(x,y,z)^2 + \frac{64}{3}\}$$

$$+\{x \leftrightarrow y\} + \{x \leftrightarrow z\}, \tag{15}$$

$$Samp2 = \{\frac{-4}{(1-x)(1-y)(1-z)}(y(1-y)(cos\theta_y - cos\theta_x)E^{(1)}(x,y,z)$$

$$-z(1-z)(cos\theta_z - cos\theta_x)E^{(1)}(x,z,y))^2 + |Q\cdot\Psi|^2(E^{(2)}(x,y,z)^2 + \frac{1}{3})\}$$

$$+\{x \leftrightarrow y\} + \{x \leftrightarrow z\}, \tag{16}$$

$$|Q\cdot\Psi| = |Q^\mu \Psi_\mu| = xy\vec{n}\cdot\vec{n}_2\times\vec{n}_3, \tag{17}$$

where $\vec{n}, \vec{n}_2$ and $\vec{n}_3$ are unit vector of positron,photon 2 and photon 3 momentum,respectively. The $Samp1$ corresponds to on-shell Z-boson decay and the $Samp2$ is the new term arisen from the contraction with the $Q_\mu Q_\nu$ term.

3.Numerical results

The integration of Eq.(13) over the phase space is performed by using an adoptive Monte-Carlo integration program BASES[9] based on VEGAS[10]. The functions $E^{(i)}$'s do not have mass singularity. The infrared singularity ($x, y, z \to 0$)and the collinear singularity($(1-x), (1-y), (1-z) \to 0$) are logarithmic and integrable. We introduce a cutoff parameter ϵ to the variable x, y and z for convenience. The resultant value is independent of ϵ.

First we consider the case that two photons are lost in the beam pipes,namely

$$|cos\theta_y| \quad \text{and} \quad |cos\theta_z| \quad \geq cos\xi \tag{17}$$

with $cos\xi = 0.96$. This constitutes the background of the reaction $e^+e^- \to \nu\bar{\nu}\gamma$. When integrated over x ($x = 0$ to 1),

$$\sigma(e^+e^- \to Z \to \gamma + \text{missing}) \simeq 2.2\times 10^{-9}pb \tag{18}$$

In the second case the photons 2 and 3 are detected and they are almost antipararell with each other;namely, the acollinearity angle is less than η,say 10 degrees.

$$(1-2a) < -cos\eta. \tag{19}$$

These events simulate $e^+e^- \to \gamma\gamma$. If these photons have energy larger than $W/3$,

$$\sigma(e^+e^- \to Z \to \gamma\gamma + \text{missing}) \simeq 6.0 \times 10^{-8} pb \tag{20}$$

The numerical values are obtained by taking $W = M_z$,$M_w = 83$ GeV ,$M_z =$ 93.2 GeV, $\Gamma_z = 2.5$ Gev and including four light quarks and three charged leptons as internal fermion loops . The results are quite small and negligible compared to the Born cross sections. This is also valid for the differential cross sections.

4.Discussion and conclusions

At $W = M_z$, we have evaluated the annihilation cross section of e^+e^- Annihilation into $\gamma\gamma\gamma$ through Z-boson by using the fourth rank vacuum polarization tensor under the condition that it may become the background to $e^+e^- \to \nu\bar{\nu}\gamma$ and $\gamma\gamma$. The cross section is free from the mass singularity as a result of the gauge invariance[9]. The contribution from $Q_\mu Q_\nu$ term in Eq.(12) due to the off-shell effect of Z-meson is negative definite and reduces the cross section further. By these two reasons ,the results are negligibly small in comparison with the Born cross section. It is thus not necessary to worry about this process and its interference with the direct annihilation process as the backgrounds. For the sake of reference,we cite the total cross section at $W = M_z$.

$$\sigma(e^+e^- \to Z \to \gamma\gamma\gamma) = 2.5 \times 10^{-7} pb \tag{21}$$

The most cumbersome background reaction to $e^+e^- \to \nu\bar{\nu}\gamma$ is the radiative Bhabha scattering and it is important to calculate the radiative corrections to this process in order to determine the number of massless neutrinos precisely by the reaction $e^+e^- \to \nu\bar{\nu}\gamma$.

Acknowledgement

We thank to Dr.K.Hikasa for calling our attention to the ref.[9]. This work was performed in the course of theoretical studies for the TRISTAN e^+e^- project at the National Laboratory for High Energy Physics.

References

[1] E.Ma and J.Okada,Phys.Rev.Lett. **41** (1978) 287;*ibid.* **41**(1978)1759.

[2] M.Igarashi and N.Nakazawa,Nucl.Phys. **B288** (1987),301.

[3] M.Igarashi, N.Nakazawa,T.Shimada and Y.Shimizu, Nucl.Phys. **B263** (1986),347.

[4] G.Barbiellini,B.Richter and J.L.Siegrist, Phys.Lett. **106B** (1981),414; S.Whitaker, Proceedings of the 23 International Conference on High Energy Physics (1986),602.

[5] J.Fujimoto,M.Igarashi and Y.Shimizu,Progr.Theor.Phys. **77**(1987),118.

[6] K.Tobimatu and Y.Shimizu,Progr.Theor.Phys. **74**(1985),567; *ibid.***76**(1986),334.

[7] J.Bartels,A.Fridman,A.Schwarz and T.T.Wu, Z.Physik **C23**(1984),295.

[8] V.Costantini,B.De Tollis and G.Pistoni, Nuovo.Cimento. **2A** (1971), 733.

[9] M.L.Laursen,K.O.Mikaelian and A.Samuel, Phys.Rev. **D23** (1981),2795.

[10] S.Kawabata Comput.Phys.Commun. **41** (1986),127.

[11] G.P.Lepage J.Comp.Phys.**27** (1978),192.

Chapter 3 MODELS OF ELEMENTARY PARTICLES

A MODEL OF LEPTONS

Hiroshi Suura
School of Physics and Astronomy, University of Minnesota,
Minneapolis, Minnesota 55455 USA

ABSTRACT

Born-Oppenheimer approximation is applied to a lepton wave equation in a multi-dimensional space and the Mead-Truhlar effect is investigated, under an assumption that rotational symmetry in iso-space is broken explicitly at the level of the c-number theory.

On this memorable occasion for my friend Professor Miyazawa, who has never been shy in proposing bold ideas, I may be allowed to indulge in a small fantasy of my own. It is a model of leptons, and instead of assuming a composite structure I will consider a wave equation in a multi-dimensional space-time. Fermions in a compactified seven or six dimensional space have been studied in the context of Kaluza-Klein theory before.[1)] Although very elegant, these models do not seem appropriate to formulate a kind of problem I would like to address in this note. My main interest here is two-fold. One is the application of the Born-Oppenheimer approximation to the extra-dimensional motion, analogous to the treatment of the electronic motion in molecular dynamics, with a particular attention to the Mead-Truhlar effect[2)] (also known as Berry's phase)[3)]. The other focus of my interest is the nature of symmetry-breaking that splits the electron-neutrino degeneracy. Thus, although it may be a step backward I consider a flat seven dimensional space-time, with the extra Cartesian three dimensions (ξ-space) defining the isospace. I may either confine leptons near the origin of the ξ-space by a strong confining potential or restrict them to the surface of a sphere. I will adopt the former description in the following. Then, my model Hamiltonian will be

$$H = -i\rho_1 \sigma \cdot \nabla_x + H_\xi$$

$$H_\xi = \rho_2 \, (-i\tau \cdot \nabla_\xi - g\tau \cdot A) + \rho_3 M + U \tag{1}$$

where $\{\rho_i\}$, $\{\sigma_i\}$ and $\{\tau_i\}$ are three sets of Pauli spins, each with an obvious physical meaning. In the absence of the directional potential A to be specified below, and with the potential U rotationally invariant in ξ space, the total isospin

$$\vec{I} = \vec{L}_\xi + \frac{1}{2}\vec{\tau}$$

is conserved. I will further assume that

$$[\rho_1, U]_+ = [\rho_1, H_\xi]_+ = 0 \ . \tag{2}$$

The spectrum of $H_0 = H_\xi(A = 0)$ is given by

$$H_0 \phi^{(\pm)}_{n',I,I_z} = \pm E_{n',I} \phi^{(\pm)}_{n',I,I_z} , \tag{3}$$

with

$$\phi^{(-)}_{n',I,I_z} = \rho_1 \phi^{(\rho)}_{n',I,I_z} , \tag{4}$$

Here n′ is a radial quantum number, and the doubly degenerate ground state ($n' = 0$, $I = \frac{1}{2}$, $I_z = \pm \frac{1}{2}$) corresponding to the electron-neutrino doublet is assumed to lie at, or very close to, $E = 0$. (I am considering $\phi^{(+)}$ and $\phi^{(-)}$ together.) All the excited states ($n' \geq 1$ or $I \geq \frac{3}{2}$) are assumed to have extremely high energy eigenvalues. The characteristic frequencies associated with excitations in ξ-space are much higher than those associated with motions in the ordinary space. This is in direct analogy with molecular motions where the electronic excitation frequencies are much higher than those of vibrational and rotational modes. Thus, the Born-Oppenheimer approximation[4)] as applied to molecular problems can be used here also, and I may neglect the coupling of the ground state to all the excited states.

In the standard model of the electro-weak interaction of Glashow, Salam and Weinberg the electron-neutrino degeneracy is removed through the quantized Higgs field. In this note, however, I have made a basic assumption that the degeneracy is lifted already at the level of the c-number theory by a directional force in ξ-space ($-g\rho_2\tau\cdot A$ term in H_ξ), which defines a charge axis $\hat{n}$. As a simplest choice for this perturbation I take

$$\vec{A} = \frac{B}{2}\,\hat{n} \times \vec{\xi} \ ,$$

a uniform 'magnetic' field in the direction $\hat{n}$. A rationale for this assumption could be that the lepton masses are now of dynamical nature and thus are derivable in principle. Besides, it will be seen in the following that without the symmetry breaking term in H_ξ no physical consequence of any interest would follow from (1) without knowing an explicit form of the potential U. Now in the vacuum, the potential U is independent of four dimensional space-time coordinate $(\vec{x},t)$. It is strongly confining and yet must produce, together with the directional potential, the two low lying states

$$H_\xi\phi_\nu = 0 \ ,$$

$$H_\xi\phi_e^{(\pm)} = m_e\,\phi_e^{(\pm)} \ . \tag{5}$$

Furthermore, the neutrino state must satisfy the chirality condition,

$$\rho_1 \phi_\nu = - \phi_\nu \ . \tag{6}$$

Whether or not a suitable U and a certain limiting process can be found such as to produce (5) and (6) avoiding the fine tuning problem is an open question. A singular limiting process must be involved because otherwise existence of the right handed massless neutrino could not be avoided. In the following I tacitly assume that equations (5) and (6) are both satisfied and investigate its consequences. In doing so I have further to specify the nature of the breaking of the rotation symmetry in ξ-space. Since the breaking is explicit, the direction of the charge axis $\hat{n}$ must have a definite physical meaning. In the vacuum, however, there exists no particular direction as long as the universe is isotropic as is commonly believed. Therefore, I am led to propose that the charge axis $\hat{n}$ at a space point $\vec{x}$ coincides with a hitherto undetected 'cosmic axis' $\hat{R}$, and then proceed to show that physical processes resulting from existence of such a direction have rates too small to be detected. I may take as $\hat{R}$

$$\hat{R} = (\vec{x}-\vec{x}_c)/|\vec{x}-\vec{x}_c| \tag{7}$$

where $\vec{x}_c$ is a far away point in space. Now that H_ξ depends on space coordinate $\vec{x}$ parametrically through $\hat{n} = \hat{R}$, wave functions $\phi_e^{(\pm)}$ and ϕ_ν do also, though m_e does not. The lepton wave equation in x-space will now involve non-abelian gauge fields reflecting the x-dependence of ϕ's, a Mead-Truhlar effect formulated in this way by Moody, Shapere and Wilczek[5] in the case of molecular problems. In case local singularities exist, U will have x-dependent terms leading to x-dependence of ϕ's and lepton masses. With a small modification to allow x-dependence of masses the following formulation will accomodate such local interactions.

I now explore consequences of the assumed spectrum in ξ-space. The total wave function $\Psi(x,t;\xi)$ in the seven dimensional space can be expanded in the following way in the Born-Oppenheimer approximation stated before. For simplicity I consider only static cases, so that the spectrum of H_ξ is independent of time.

$$\Psi = \Phi_e^{(+)}(x,t)\phi_e^{(+)}(\xi;x) + \Phi_e^{(-)}(x,t)\phi_e^{(-)}(\xi;x) + \sqrt{2}\Phi_\nu(x,t)\phi_\nu(\xi;x) \qquad (8)$$

Since ϕ's are Dirac spinors on which $\{\rho_i\}$ and $\{\tau_i\}$ operate, Φ's must be two-component Pauli spinor on which $\{\sigma_i\}$ operate. Introducing Ψ into

$$i\,\frac{\partial\Psi}{\partial t} = (-\,i\rho_1\sigma\cdot\nabla + H_\xi)\Psi$$

and using (5) and (6) I obtain

$$i\dot{\Phi}_e^{(\pm)} = -i\sigma\cdot\nabla\,\Phi_e^{(+)} \pm m_e\Phi_e^{(\pm)} - g\sigma\cdot\bar{b}_{ee}\,\Phi_e^{(\pm)}$$

$$-\,g\sigma\cdot\bar{b}_{ee}\,\Phi_\rho^{(+)} \pm g\sigma\cdot b_{e\nu}\,\Phi_\nu \qquad (9a)$$

and

$$i\,\dot{\Phi}_\nu = i\sigma\cdot\nabla\,\Phi_\nu + g\sigma\cdot b_{\nu\nu}\,\Phi_\nu + \frac{g}{2}\,\sigma\cdot b_{\nu e}\,(\Phi_e^{(+)} - \Phi_e^{(-)})\ , \qquad (9b)$$

where the gauge fields $\bar{b}_{ee}$ etc. are defined by

$$gb_{ee} = i(\phi_e^{(+)}, \nabla_x \phi_e^{(+)}) = i(\phi_e^{(-)}, \nabla_x \phi_e^{(-)})$$

$$g\bar{b}_{ee} = i(\phi_e^{(+)}, \nabla_x \phi_e^{(-)}) = i(\phi_e^{(-)}, \nabla_x \phi_e^{(+)})$$

$$gb_{e\nu} = i\sqrt{2}\, (\phi_e^{(+)}, \nabla_x \phi_\nu) = -i\sqrt{2}\, (\phi_e^{(-)}, \nabla_x \phi_\nu)$$

$$gb_{\nu e} = i\sqrt{2}\, (\phi_\nu, \nabla_x \phi_e^{(+)}) \tag{10}$$

Introducing four component spinors

$$\psi_e = \begin{bmatrix} \Phi_e^{(+)} \\ \Phi_e^{(-)} \end{bmatrix}, \quad \psi_\nu = \begin{bmatrix} \Phi_\nu \\ -\Phi_\nu \end{bmatrix},$$

and new 4 x 4 matrices α, β, γ_5 that operate on ψ_e and ψ_ν, equations (9a) and (9b) can be cast into Dirac equations

$$i\dot{\psi}_e = (-i\alpha \cdot \nabla_x + \beta, m_e)\, \psi_e - \alpha \cdot b_{ee}^R\, \psi_e^R$$

$$- \alpha \cdot b_{ee}^L\, \psi_e^L - \alpha \cdot b_{e\nu} \psi_\nu , \tag{11a}$$

$$i\dot{\psi}_\nu = -i\alpha \cdot \nabla_x\, \psi_\nu - \alpha \cdot b_{\nu\nu}\, \psi_\nu - \alpha \cdot b_{\nu e}\, \psi_e , \tag{11b}$$

where

$$\psi_e^{L(R)} = \frac{1}{2}(1 + \gamma_5)\, \psi_e ,$$

and

$$b_{ee}^{L(R)} = b_{ee} + \bar{b}_{ee} = i(\phi_e^{(+)}, (1 + \rho_1)\, \nabla_x\, \phi_e^{(+)}) . \tag{12}$$

The gauge fields b's cannot be transformed away because of the electron mass term. If so, effects of these static fields should not be detectable. For instance, the last term of equation (11a) describes charge decay $e \to \nu$ under the static field $be\nu$, whose magnitude must be examined. Taking the gradient ∇_x of equation (5), I derive

$$g\, \dot{b}_{\nu e} = \frac{1}{m_e}\, i\, (\phi_\nu,\ \nabla_x\, H_\xi\, \phi_e^{(+)}) \ . \tag{13}$$

The only x-dependent term in H_ξ is the directional perturbation

$$H' = -g\rho_2\ \tau\cdot A = -\frac{1}{2} gB\ \ \rho_2\ \hat{R}\cdot\xi \times \tau \ . \tag{14}$$

which gives

$$\nabla_x\ H_\xi = -\frac{1}{2} gB\ \rho_2\ \{\ \frac{1}{R}\ \xi \times \tau - \frac{\hat{R}}{R}\ \hat{R}\cdot\xi \times \tau\ \}\ . \tag{15}$$

To first order in perturbation H', the eigenfunction $\phi_e^{(+)}$ and ϕ_ν may be replaced in terms of the eigenfunctions of H_0, eq. (3), like

$$\phi_e^{(\pm)} = \phi_{-1/2}^{(\pm)} \ ,$$

and

$$\phi_\nu = \frac{1}{\sqrt{2}}\ [\phi_{1/2}^{(+)} - \phi_{1/2}^{(-)}\ ] \ , \tag{16}$$

where an abbreviation

$$\phi_m^{(\pm)} = \phi_{n'=0, I=\frac{1}{2}, m}^{(\pm)}$$

has been used. Matrix elements of H' and $\nabla_x H'$ reduce, by Wigner-Eckart theorem, to

$$(\phi_{m'}^{(k')}\ ,\ \rho_2 \xi \times \tau\ \phi_m^{(k)}) = D_{kk'} \langle I=\tfrac{1}{2}, m' | \vec{I} | I=\tfrac{1}{2}, m \rangle\ , \qquad (17)$$

where $k,k' = \pm$. It is easy to see that D_{+-} and D_{-+} vanish if the radial functions involved in the ground state are real. I find then from (14) and (18) choosing $\hat{R}$ as the quantization axis,

$$m_e \sim \langle H' \rangle_e = -\frac{gB}{2} (\phi_{-1/2}^{(+)}, \hat{R}\cdot\xi \times \tau\ \phi_{-1/2}^{(+)})$$

$$= \frac{gB}{4} D_{++}\ . \qquad (18)$$

From (13, (15), (16) and (17) on the other hand,

$$g\ b_{\nu e} = -\frac{i\ gB}{2m_e R} \left(\frac{\phi_{1/2}^{(+)} - \phi_{1/2}^{(-)}}{\sqrt{2}}\ ,\ \xi \times \tau\ \phi_{-1/2}^{(+)} \right)$$

$$= -\frac{i\ gB}{4m_e R} D_{++} \frac{\hat{n}_x - i\hat{n}_y}{\sqrt{2}}$$

$$= -\frac{i}{R} \frac{\hat{n}_x - i\hat{n}_y}{\sqrt{2}} \qquad (19)$$

The life time for an atomic electron is estimated to be

$$T = \frac{1}{16} m_e^6\ a^5 R^2$$

where a is an average Bohr radius. For $a \sim 10^{-9}$ cm and $R \sim 10^5$ light years I obtain $\tau \sim 10^{43}$ year, too long to be seen.

Finally, equations (11a,b) can be cast into the standard $SU(2)_L \times U(1)$ form by introducing a doublet $L = \begin{bmatrix} \psi_\nu \\ \psi_e^L \end{bmatrix}$ and a singlet $R = \psi_e^R$, and then making a gauge transformation

$$R \to e^{i\Lambda} R \text{ and } L \to e^{i\Lambda} L\ ,$$

where

$$\nabla\Lambda = g(b_{\nu\nu} + b^L_{ee}) - g\, b^R_{ee}.$$

Such a Λ exists because it can be shown that $\nabla \times b^R_{ee} = \nabla \times (b^L_{ee} + b_{\nu\nu}) = 0$. The interaction lagrangian now reads

$$L_{int} = 2\, g\, R^{+}\alpha\cdot CR + gL^{+}\alpha\cdot CL$$

$$+ gL^{+}\alpha\cdot \underset{\sim}{b}L - (m_e L^{+}\eta\beta R + \text{e.e.})$$

where

$$gC = gb^R_{ee} - \frac{1}{2}g(b_{\nu\nu} + b^L_{ee}),$$

$$g\underset{\sim}{b} = g \begin{bmatrix} \frac{1}{2}(b_{\nu\nu} - b^L_{ee}) & b_{\nu e}, \\ b_{e\nu} & -\frac{1}{2}(b_{\nu\nu} - b^L_{ee}) \end{bmatrix},$$

and $\eta = [{}^0_1]$. It is readily be seen that

$$\nabla \times \underset{\sim}{b} = i\underset{\sim}{b} \times \underset{\sim}{b}\ ,$$

meaning the vanishing of the magnetic field. However, the field C and $\underset{\sim}{b}$ cannot be gauged away because of the existence of the mass term. It is quite amusing that the weak hypercharge ratio of 2 to 1 for R and L followed from the formulation, rather than being put in by hand. It is also interesting that the spectrum in ξ-space determines uniquely the type of interaction in x space corresponding to any interaction mediated through ξ-space.

REFERENCES

1. For instance, Hara, O. and Goto, T., Prog. Theor. Phys. 33, 907 (1965); Randjbar-Daemi, S., Salam, A. and Strathdee, J., Nucl. Phys. B214, 491 (1983).

2. Mead, C. A. and Truhlar, D. G., J. Chem. Phys. 70, 2284 (1979).

3. Berry, M. V., Proc. Roy. Soc. Lond. A392, 45 (1984).

4. For instance see Schiff, L. I., Quantum Mechanics, pp. 447, McGraw-Hill.

5. Moody, J., Shapere, A. and Wilczek, F., Phys. Rev. Lett. 56, 893 (1986).

YUKAWA COUPLINGS AND THE NATURE OF ZERO MODES IN THE SKYRME MODEL

Ken Kawarabayashi

Institute of Physics
University of Tokyo (Komaba)
Tokyo

ABSTRACT

Several issues related, directly or indirectly, to the Yukawa coupling in the Skyrme model are discussed. We try to shed a new light on the physical nature of the zero modes associated with translation (rotation) invariance of the model.

1. It is well known that to leading order in 1/N expanded QCD, pseudoscalar mesons are uniquely described at low energies by the non-linear sigma model of spontaneously broken chiral symmetry. The simplest choice of the lagrangian is given by

$$\mathcal{L}_0 = \frac{F_\pi^2}{16} \mathrm{Tr}(\partial_\mu U \partial^\mu U^\dagger) \tag{1}$$

where U represents a unitary $N_f \times N_f$ matrix field and F_π ($\simeq$ 186 MeV) is the pion decay constant of the order of $\sqrt{N}$. Assuming two flavors for simplicity, U is parametrized, in the vacuum sector, by the pion field $\pi^a(x)$ (a=1,2,3) as

$$U = \exp[i \frac{2}{F_\pi} \vec{\tau} \cdot \vec{\pi}(x)] \tag{2}$$

It has been shown[1] that with a slight modification of (1), the model has solitons as topologically non-trivial solutions. Moreover, those solitons have precisely the right quantum numbers for baryons in the large N QCD, provided one takes into account the effects coming from the Wess-Zumino-Witten term.[2] Thus baryons can be consistently described as solitons - Skyrmions - in the non-linear sigma model.

The static properties of the nucleon have been computed by a number of authors[3] in the semi-classical approximation. The results obtained are generally in fair agreement with experiments, a fact of which seems to strongly support the idea of "baryons as solitons" in these models. The pion nucleon interactions were next examined by expanding the small (quantum) fluctuations around the Skyrmion. It is found that because of stability of the Skyrmion, there remains no term in the interaction lagrangian linear with respect to the pion (fluctuation) field. This has led to some controversial arguments in the literature[4] about the pion nucleon coupling in the Skyrme model.

The pion nucleon scattering has also been studied elaborately but here apparent discrepancies were found in lower partial waves.[5,6] The most severe of these is the failure of the model to reproduce the resonant behavior associated with the Δ resonance.

Usually this is attributed, not to the failure of the model itself, but to a great sensitivity of small perturbations arising from higher order 1/N corrections to these amplitudes. Indeed, it is shown that the translational and the rotational zero-modes couple only to lower partial waves in the semi-classical approximation. However, there does not seem to exist a recipe to cure these deficiencies in a consistent manner.

The purpose of this note is to discuss several issues related to the Yukawa coupling in the Skyrme model and to shed a new light on the physical nature of the zero modes appeared in the meson-soliton scattering amplitude.

2. THE PION-NUCLEON COUPLINGS CONSTANT

In the Skyrme model, one usually starts from the hedge-hog configuration ansatz for the classical solitons, i.e.,

$$U_0 = \exp[i\,\vec{\tau}\cdot\hat{x}\,F(r)] \tag{3}$$

with the boundary conditions $F(0) = \pi$ and $F(\infty) = 0$. Here $\hat{x}$ is defined by $\mathbf{x}/r$.

To study pion nucleon couplings in this model, one breaks up pion field, in the one soliton sector, into two pieces; a classical part and a small quantum fluctuation, latter of which is identified with physical pions. For $N_f = 2$, the classical pion field can be read off from (3)

$$\begin{aligned}\pi^a_{cl} &\equiv -i\,\frac{F_\pi}{8}\,\mathrm{Tr}\,(\tau^a A(t)(U_0 - U_0^\dagger)A(t)^{-1}) \\ &= \frac{1}{4}F_\pi\,\mathrm{Tr}\,(\tau^a A\,\tau^j A^{-1})\,\hat{r}_j \sin F(r)\end{aligned} \tag{4}$$

where $\hat{r} = \mathbf{r}/r$ and $\mathbf{r} = \mathbf{x}-\mathbf{X}(t)$.

Here we have explicitely introduced collective quantum variables to maintain translational as well as rotational invariances. It should be noted that the classical pion field given by (4) is of the order of $\sqrt{N_c}$

In terms of π^a_{cl}, we decompose

$$\pi^a(\mathbf{x},t) = \pi^a_{cl}(\mathbf{x}-\mathbf{X}(t);A(t)) + \pi^{a\prime}(\mathbf{x},t) \tag{5}$$

with $\pi^{a\prime}$ being the fluctuation field.

The Yukawa coupling is then defined by[7,8]

$$\langle N'(\mathbf{p}')|\,(\Box + m_\pi^2)\,\pi^a(\mathbf{x},0)\,|N(\mathbf{p})\rangle = \frac{i\,g_{\pi NN}}{2m_N}\langle N'|\,\tau^a\,\boldsymbol{\sigma}\cdot(\mathbf{p}'-\mathbf{p})|N\rangle\, e^{-i(\mathbf{p}'-\mathbf{p})\cdot\mathbf{x}} \tag{6}$$

where $\mathbf{P}'$ and $\mathbf{P}$ are the momenta of the Skyrmion.

Noticing that to the leading order of $1/N$, π^a_{cl} completely dominates the matrix element (6), we evaluate the left-hand side of with the aid of (1).

Making use of the following relation[3]

$$\langle N'|\mathrm{Tr}(\tau^a A \tau^j A^{-1})|N\rangle = -\frac{2}{3}\langle N'|\tau^a \sigma^j|N\rangle$$

we get

$$\frac{g_{\pi NN}}{2m_N} = \frac{F_\pi}{6}\lim_{q^2\to -m_\pi^2}(q^2+m_\pi^2)\Gamma(q^2) \tag{7}$$

with

$$q_j\Gamma(q^2) = -i\int d^3x\, e^{iq\cdot x}\,\hat{r}_j \sin F(r) \tag{8}$$

It is not difficult to show that the coupling constant defined above is equivalent to that given from the asymptotic behavior of the Skyrmion solution.

It is of crucial importance, as is stressed in reference (8), that the coupling constant obtained in this way is of the order of $\sqrt{N}$ and thus the Skyrme model reveals itself as the strong coupling theory, when it is viewed as the effective theory of pions and nucleons.

3. GOLDBERGER-TREIMAN RELATION

The G-T relation should be satisfied in any chiral model involving pions and nucleons. To derive it in the case of finite pion mass in the Skyrme model, we note that the divergence of the axial vector current is given by

$$\partial^\mu A_\mu^a = -i\,\frac{F_\pi^2}{16}\,m_\pi^2\,\mathrm{Tr}(\tau^a(U-U^\dagger)) \tag{9}$$

where we have introduced the pion mass term as an explicit breaking of the chiral SU(2) symmetry.

Comparing the right-hand side of (9) with (4) and (5), again to the leading order on 1/N, we find†

$$\partial^\mu A_\mu^a = \frac{1}{2} F_\pi m_\pi^2 \pi_\alpha^a (x - x ; A) \qquad (10)$$

which is the "PCAC" relation in the this model. The G-T relation can be derived by evaluating the matrix element of (10) taken between one nucleon states with momenta $\mathbf{P}'$ and $\mathbf{P}$, respectively. In fact, starting from

$$\langle N'(\mathbf{p}')| \partial^\mu A_\mu^a | N(\mathbf{p})\rangle = \frac{1}{2} F_\pi m_\pi^2 \langle N'(\mathbf{p}')| \pi_\alpha^a | N(\mathbf{p})\rangle \qquad (11)$$

we first evaluate the left-hand side, assuming that the pion pole dominates the (induced) form factor and find, in the non-relativistic limit,

$$\text{l.h.s of (11)} = i \left(g_A(t) - F_\pi \cdot \frac{g_{\pi NN}}{2m_N} \left(\frac{t}{t + m_\pi^2} \right) \right) \langle N' | \frac{\tau^a}{2} \sigma \cdot q | N \rangle \qquad (12)$$

with $t \equiv q^2$.

The right-hand side of (11), on the other hand, can be directly evaluated from (5) and (6), to find

$$\text{r.h.s of (11)} = i F_\pi \frac{g_{\pi NN}}{2m_N} \left(\frac{m_\pi^2}{t + m_\pi^2} \right) \langle N' | \frac{\tau^a}{2} \sigma \cdot q | N \rangle \qquad (13)$$

Comparing (12) with (13), we get for $t \simeq 0$

$$g_A(0) = F_\pi \cdot \frac{g_{\pi NN}}{2m_N} \qquad (14)$$

which is the G-T relation.

---

† We have neglected higher order term arising from $\mathring{A}(t)$.

4. ADLER-WEISBERGER SUM RULE

The A-W sum rule is an another consequnce of the chiral symmetry. As such, it should be obeyed in the Skyrme model also.

The sum rule, in its most conventional form,[9] can be expressed as

$$1-\frac{1}{g_A^2}=\left(\frac{2m_N}{g_{\pi NN}}\right)^2\left(\frac{1}{K(0)}\right)^2\frac{1}{2\pi}\int_0^\infty\frac{d\nu}{\nu}\left(\sigma_{\pi^+p}(\nu)-\sigma_{\pi^-p}(\nu)\right) \quad (15)$$

where $\sigma_{\pi^\pm p}(\nu)$ denote the total crosssections for scattering of zero mass $\pi^\pm$ by protons, at the incident pion energy ν in the lab. system, and K(t) is the pion form factor normalized so that $K(-m_\pi^2) = 1$.

The first question one might ask is the question how is the sum rule acually saturated in a manner consistent with the model prediction for the axial coupling g_A. For, naively we expect that the integral over the crosssections of the right-hand side of (15) should come out to be positive, since the Δ resonance will certainly dominate over the integral and will give positive contributions. The sum rule (15) then seems to imply that $g_A^2 > 1$. However, the axial coupling actually evaluated in the Skyrme model turns out to be less than one, typically of the order of ~0.6.[3]

To see how the sum rule works out consistently, we examine it more closely.[10] The A-W sum rule can be viewed as a direct consequence of the current algebra formula for S-wave π-N scattering lengths combined with the dispersion relation sum rule, first derived by Goldberger, Miyazawa and Oehme[11] in 1955.

The latter sum rule amounts to

$$\frac{2\pi}{3}(a_{1/2}-a_{3/2})\frac{1}{m_\pi}=\left(\frac{g_{\pi NN}}{2m_N}\right)^2+\frac{1}{2\pi}\int_{m_\pi}^\infty\frac{d\nu}{k}\left(\sigma_{\pi^-p}(\nu)-\sigma_{\pi^+p}(\nu)\right) \quad (16)$$

where $a_{1/2}$ ($a_{3/2}$) denotes the S-wave scattering length for π-N scattering with the total isospin T = 1/2 (T = 3/2).

The current algebra, on the other hand, give predictions[12] for the S-wave scattering lengths, in terms of F_π, as

$$a_{1/2} - a_{3/2} = \frac{3}{2\pi}\left(1+\frac{m_\pi}{m_N}\right)^{-1}\frac{m_\pi}{F_\pi^2} \tag{17}$$

These two results [(16) and (17)] can be combined with the G-T relation (14) and lead to the A-W sum rule, provided one neglects the difference coming from off the mass-shell extrapolation of the order of m_π^2/m_N^2.

It is straightforward to verify, on the other hand, that the current algebra formula (17) is indeed satisfied in the Skyrme model. We only note here that it is of the order of 1/N.

It remains then to check the G-M-O sum rule. For this purpose, we first evaluate a particular contribution for the integral over the crosssections coming from Δ resonance, assuming that the Δ is located nearby at threshold such that $m_\Delta - m_N \simeq O(1/N)$ and its width is narrow. We make use of the coupling relation $g_{\pi N\Delta} = \frac{3}{2} g_{\pi NN}$ and find that to the leading order, its contribution turns out to be of the order of N and precisely cancels the first term of the right-hand side of the G-M-O sum rule. The next leading contribution, being superficially of the order 1, comes from the remaining integral over the continuum. However, it also vanishes since, up to this order, $\pi^- p$ and $\pi^+ p$ total crosssections are equal. The equality of these crosssections itself is a known consequence from the static strong coupling theory and reflects a close relationship between Skyrmions and baryons in the large N QCD.

To summarize, the saturation of the A-W sum rule is automatic in the Skyrme model up to order 1. To extract meaningful informations, one must go one step further and take into account higher order corrections.

5. ZERO MODES

As is well known, the presence of the zero frequencly modes in the meson soliton scattering is a general phenomenon, reflecting the invariance of the theory under continuous transformation, such as translations and rotations.

To manifestly maintain those invariances in perturbations, we introduce, in the standard method of quantization, collective coordinates and eliminate zero mode states from the physical spectra by setting appropriate subsidiary conditions.[7,13,14]

In the ϕ^4 theory in two space-time dimensions, it is known that the t matrix for the meson-soliton scattering contains pole terms. In fact, one can show that the t matrix satisfies the dispersion relation

$$t(\omega) = -\frac{48mk^2}{\omega^2} - \frac{24mk^2}{\omega_1^2-\omega^2} - \frac{2k^2}{\pi}\int_{2m}^{\infty}\frac{\omega' d\omega' \, \mathrm{Im}\, t(\omega')}{(\omega'^2-4m^2)(\omega'^2-\omega^2)} \tag{18}$$

where $\omega^2 = k^2 + 4m^2$ and $\omega_1 = \sqrt{3}m$.

Recently we have shown[15] that the first two terms in (18) cocincide, in the strong coupling limit ($1/g \to \infty$), with the Born terms (direct plus crossed diagrams) arising from the classical part of the Green's function. In this limit, the meson-soliton coupling constant f is defined, in completely parallel to (6) and (7), by†

$$\begin{aligned} \langle p'|(-\tfrac{d^2}{dx^2}+4m^2)\phi(x,0)|p\rangle &= i(p'-p)f((p'-p)^2) \\ &= ((p'-p)^2+4m^2)\int dx_0\, e^{i(p'-p)x_0}\,\phi_{cl}(x-x_0) \end{aligned} \tag{19}$$

where $\phi_c(x)$ is the kink solution.

Furthermore in an attempt to understand why the classical Born contribution is present in the scattering amplitude obtained from the

---

† The transition coupling for the intermediate excitation state ($\omega_1 = \sqrt{3}m$) can be defined in a similar way.

quantum fluctuation, we found that the first term ($\equiv t^B$) in (18) also coincides with the contribution from the zero-frequency mode ($\equiv t^{(0)}$). It can be shown that this coincidence of t^B with $t^{(0)}$ is not accidental and that the distinction between the two merely reflects different choices for the gauge fixing conditions in the constraint gauge system.[13,14]

Our finding in the ϕ^4 theory indicates that the zero modes in the Skyrme model should be interpreted as the Born amplitudes coming from N and Δ states in the large N limit. The Yukawa coupling of pions and nucleons, in this limit, is completely predicted by the classical Skyrmion.

ACKNOWLEDGEMENTS

This note is dedicated to Professor Hironari Miyazawa on the occasion of this sixtieth birthday. I owe Professor Miyazawa, since my graduate student years, a great deal for my understanding on pion physics from threshold up to infinity. I would like to take this opportunity to mention that among his many important works on dispersion relations, the equality of particle and anti-particle total crosssections at high energies, known as the Pomeranchuk theorem, has been independently pointed out by Professor Miyazawa in 1957.[16] He did not publish this important observation, however.

The materials of this artice are taken from a joint work with Kohichi Ohta (reference 15). This work is supported in part by the Grant-in-Aid for Cooperative Research (#6330210) and for Scientific Research (#63540207).

REFERENCES

1. T.H.R. Skyrme, Nucl. Phys. 31 (1962) 556.
2. E. Witten, Nucl. Phys. B160 (1979) 57; B223 (1983) 422, 433.
3. G. Adkins, C. Nappi and E. Witten, Nucl. Phys. B228 (1983) 552.
4. M. Uehara, Prog. Theor. Phys. 75 (1986) 212; 78 (1987) 984; C. Hajduk and B. Schwesinger, Nucl. Phys. A453 (1986) 620; S. Saito, Proceedings of 1987 International Workshop on Low Energy Effective Theory QCD, Nagoya (1987).
5. A. Hayashi, G. Echart, G. Holzwarth and H. Wallister, Phys. Lett. B147 (1984) 5; G. Echart, A. Hayashi and G. Holzwarth, Nucl. Phys. A448 (1986) 732; G. Echart and B. Schwesinger, Nucl. Phys. A458 (1986) 620.
6. M.P. Mattis and M. Karliner, Phys. Rev. D31 (1985) 2833; M.P. Mattis and M. Peshkin, Phys. Rev. D32 (1985) 58; M. Karliner and M.P. Mattis, Phys. Rev. D34 (1986) 1991; M. Karliner, Phys. Rev. Lett. 57 (1986) 523.
7. J.L. Gervais, A. Jevicki and B. Sakita, Phys. Rev. D12 (1975) 1038.
8. J.L. Gervais and B. Sakita, Phys. Rev. D30 (1984) 1795.
9. S.L. Adler, Phys. Rev. Lett. 14 (1965) 1051; W.I. Weisberger, Phys. Rev. Lett. 14 (1965) 1047.
10. See also, M. Uehara, Preprint SAGA-HE-25 (1988).
11. M.L. Goldberger, H. Miyazawa and R. Oehme, Phys. Rev. 96 (1955) 986.
12. S. Weinberg, Phys. Rev. Lett. 17 (1966) 616.
13. A. Hosoya and K. Kikkawa, Nucl. Phys. B101 (1975) 271.
14. J.L. Gervais, A. Jevicki and B. Sakita, Phys. Reports 23C (1976) 281.
15. K. Kawarabayashi and K. Ohta, to appear in Phys. Lett. B.
16. H. Miyazawa, Talk at the Symposium on anti-nucleons held at Kyoto, 1957; Soryūshi-Ron Kenkyū (Mimeographed Circular in Japanese) 17 (1958) 207.

COMPOSITE WEAK BOSONS

Mahiko Suzuki

Department of Physics and Lawrence Berkeley Laboratory
University of California, Berkeley, California 94720
U.S.A.

ABSTRACT

Dynamical mechanism of composite W and Z is studied in a 1/N field theory model with four-fermion interactions in which global weak SU(2) symmetry is broken explicitly by electromagnetic interaction. Issues involved in such a model are discussed in detail. Deviation from gauge coupling due to compositeness and higher order loop corrections are examined to show that this class of models are consistent not only theoretically but also experimentally.

1. INTRODUCTION

Gauge bosons have a nice geometrical interpretation for their existence. In the standard picture, weak bosons are gauge bosons which acquire mass by spontaneous symmetry breakdown. Experimentally, the most persuasive evidence for this picture is the relation between the weak boson masses and the weak mixing angle. As was emphasized by Hung and Sakurai[1)] and by Bjorken[2)], however, such a relation can be naturally realized at a phenomenological level even if W and Z are not gauge bosons. A conclusive test must probe the couplings among W, Z and photon in future experiment. From a theoretical viewpoint, renormalizability of the spontaneously broken theory guarantees approximate validity of tree-diagram predictions except for Higgs bosons masses. Elementary Higgs bosons are, until now, an artifice which mars the otherwise beautiful standard theory. By contrast, if W and Z are composite and much lighter than their constituents, they behave just like gauge bosons and Higgs bosons are not needed. There is a long history in construction of gauge-like bosons as composites of fermion-antifermion. I would like to revisit this problem with focus on the following aspects:

(a) Relation between gauge symmetry and lightness of composites.

(b) How general the dynamical mechanism of binding can be.

(c) Loop corrections to the order of W and Z loops.

(d) How to include quarks and leptons.

2. COMPOSITE GAUGE-LIKE BOSONS

The mechanism to generate nonabelian gauge-like bosons from four-fermion interactions of preons was first demonstrated by Eguchi and Sugawara.[3] It goes as follows: Start with an SU(n) globally symmetric interaction of N families of preons in the fundamental representation,

$$L(\psi) = \sum_{j=1}^{N} \bar{\psi}_j (i\not\partial - M)\psi_j - \frac{2}{v^2} \sum_{j,k=1}^{N} (\bar{\psi}_j \gamma_\mu \tfrac{1}{2}\lambda_a \psi_j)(\bar{\psi}_k \gamma^\mu \tfrac{1}{2}\lambda_a \psi_k). \qquad (2.1)$$

Introduce a set of auxiliary vector fields A^μ_a to rewrite $L(\psi)$ into[4]

$$L(\psi,A) = \sum_j \bar{\psi}_j (i\not\partial - M)\psi_j - \frac{2m_0}{v} \sum_j (\bar{\psi}_j \gamma_\mu \tfrac{1}{2}\lambda_a \psi_j) A^\mu_a + \tfrac{1}{2} m_0^2 A^\mu_a A_{a\mu}. \qquad (2.2)$$

By computing radiative corrections to the leading order in 1/N for $L(\psi,A)$ and making wave-function renormalization, one finds

$$L_{eff} = \sum_j \bar{\psi}_j (i\not\partial - M)\psi_j - g \sum_j (\bar{\psi}_j \gamma_\mu \tfrac{1}{2}\lambda_a \psi_j) A^\mu_a + \tfrac{1}{2}(m^2 + \delta m^2) A^\mu_a A_{a\mu}$$

$$- \tfrac{1}{4} G^{\mu\nu}_a G_{a\mu\nu} - i(\lambda/M^2) f_{abc} G^{\mu\nu}_a G^\kappa_{b\nu} G_{c\mu\kappa} + \cdots\cdots , \qquad (2.3)$$

where $G^{\mu\nu}_a = \partial^\mu A^\nu_a - \partial^\nu A^\mu_a - g f_{abc} A^\mu_b A^\nu_c$ and all couplings are evaluated at zero four-momenta. Note that couplings of dimension higher than four are suppressed by powers of the preon mass M. It is remarkable that not only the three-vector and four-vector couplings are given by the same coupling g, but also is the Yukawa coupling to the preons given by g. Realization of gauge symmetry is in fact not surprising if one notices that $L(\psi,A)$ possesses SU(n) gauge symmetry up to the mass term of A^μ_a.

Since the pioneering work of Bjorken[5] on abelian gauge symmetry, it has been postulated that the mass term $m^2 + \delta m^2$ can be set equal to zero to turn the vector particles into genuine gauge bosons. This postulate poses some serious problem.[6)] The bare mass m_0 must not be zero in order for the Lagrangians of (2.1) and (2.2) to be equivalent. On the other hand, the current $\bar{\psi}_j\gamma_\mu\frac{1}{2}\lambda_a\psi_j$ is conserved, so δm^2 must be zero when evaluated at zero four-momentum. Since m^2 is equal to m_0^2 up to the wave function renormalization, it is impossible to require that A_a^μ is massless, namely $m^2+\delta m^2=0$. Diagramatically, the procedure from (2.1) to (2.3) is equivalent to summing an infinite series of vector bubbles as shown in Fig.1,

$$\sim (\bar{u}\gamma_\mu\tfrac{1}{2}\lambda_a u)(\tfrac{1}{4}v^2 - N\Pi(q^2))^{-1}(\bar{v}\gamma^\mu\tfrac{1}{2}\lambda_a u), \tag{2.4}$$

where $(g^{\mu\nu} - q^\mu q^\nu)\Pi(q^2)$ is the vacuum polarization. The mass of A_a^μ

Fig.1. Formation of vector composites.

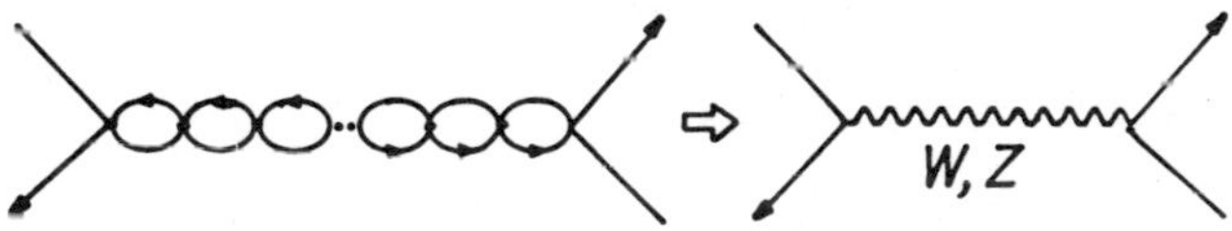

is given by $m_V^2 = (\frac{1}{4}v^2 - N\Pi(0))/N\Pi'(0)$, where $\Pi'(q^2)$ is the first derivative of $\Pi(q^2)$. Since $\Pi(0) = 0$ by global current conservation, one can attain $m_V^2=0$ only by taking the singular limit of $v = 0$ for a given large finite N. If one breaks global SU(n) in (2.3), $\Pi(0)$ would be no longer equal to zero. But one must make a precise tuning of parameters to achieve $m_V^2=0$. Besides being artificial, such a tuning involves cancelling of terms of different orders in 1/N in the present parametrization of N. Therefore, if we insist on that v^2 is O(1), the only possibility is to accept $\Pi(0)=0$ and give up $m_V=0$. As the effective force N/v^2 of the four-fermion interaction grows stronger, the composite vector boson mass becomes smaller.

This remark on the composite mass is relevant to Higgs boson mass as well. If one introduces an attractive four-fermion force of scalar-scalar type, one can generate spin-0 composites. However, the natural mass scale of such scalar composites is $O(\Lambda^2)$ where Λ is the cutoff energy of the theory, identified with the scale of the preon mass M. The scalar force must be fine tuned in order to keep the composite scalar boson mass much lighter than M. Furthermore, such a fine tuning involves cancellation of terms of different orders in 1/N. Therefore, we are unable to obtain composite Higgs bosons in a consistent manner, as was emphasized recently by Cohen, Georgi and Simmons.[7] As we will see later, there is no need for Higgs bosons in composite W and Z models since an explicit global symmetry breaking by electromagnetic interaction generates the correct relation between the weak boson masses and the mixing angle.

3. HOW GENERAL IS THE MECHANISM ?

The interaction in (2.1) is carefully set up to guide us to the gauge coupling for the vector composites. The crucial ingredient is conservation of the current $\partial^\mu(\Sigma\ \overline{\psi}_j\gamma_\mu\frac{1}{2}\lambda_a\psi_j) = 0$. This is in fact much stronger a requirement than just SU(n) global symmetry. If one replaces the interaction in (2.1) by

$$L_{int}(\psi) = -\frac{2}{v^2}J^\mu_a J_{a\mu},$$
$$J_{a\mu} = \sum_j (\overline{\psi}_j\gamma_\mu\tfrac{1}{2}\lambda_a\psi_j + \frac{\kappa}{2M}\partial^\nu(\overline{\psi}_j\sigma_{\mu\nu}\tfrac{1}{2}\lambda_a\psi_j)), \tag{3.1}$$

this simple dynamical generation of the composite gauge-like bosons fails immediately; the Yukawa coupling with preons, the three-boson coupling and the four-boson coupling are given by $g(1+O(\kappa^2\Lambda^2/M^2))$, $g(1+O(\kappa))$, and g^2, respectively. The replacement with L_{int} of (3.1) does not violate global SU(n) symmetry. However, once the nonabelian magnetic moment interaction is added, the composite vector boson mass no longer remains light. The origin of this puzzle can be solved when the Lagrangian is written with the auxiliary field A^μ_a, as in (2.2):

$$L = \sum_j \overline{\psi}_j(i\not{\partial} - M)\psi_j - (2m_0/v)\sum_j(\overline{\psi}_j\gamma_\mu\tfrac{1}{2}\lambda_a\psi_j + \frac{\kappa}{2M}\partial^\nu(\overline{\psi}_j\sigma_{\mu\nu}\tfrac{1}{2}\lambda_a\psi_j))A^\mu_a$$
$$+\tfrac{1}{2}m_0^2 A^\mu_a A_{a\mu}. \tag{3.2}$$

This Lagrangian does not possess local SU(n) symmetry any longer even after the mass term is removed. The derivative ∂^ν in the magnetic interaction destroys the gauge symmetry. A Lagrangian with local SU(n) symmetry would be obtained from (3.2) by replacing ∂^ν by the covariant derivative D^ν. Then we can not write any simple Lagrangian like the one in (2.1) in terms of preon fields alone.

In the late 1960's, Kroll, Lee and Zumino[8] showed in the context of the field current identity that if massive vector fields A_a^μ transform like the adjoint representation of a Lie group and satisfy

$$\partial_\mu A_a^\mu = 0, \tag{3.3}$$

and if the three-body interaction of dimension four obeys global symmetry of the group, the four-body coupling is exactly equal to the one required by local symmetry. According to this theorem, the auxiliary field in (3.2),

$$A_a^\mu = \frac{2}{m_0 v} \sum_j (\bar{\psi}_j \gamma^\mu \tfrac{1}{2}\lambda_a \psi_j + \frac{\kappa}{2M}\partial^\nu (\bar{\psi}_j \sigma_{\mu\nu} \tfrac{1}{2}\lambda_a \psi_j)), \tag{3.4}$$

must satisfy (3.3) in order to generate massive gauge bosons. However, the right-hand side of (3.4) is not a conserved current; the Noether current obtained from the Lagrangian (3.1) is

$$\begin{aligned} J_a^\mu = & \sum_j \bar{\psi}_j \gamma^\mu \tfrac{1}{2}\lambda_a \psi_j \\ & + \frac{2\kappa}{v^2 M} f_{abc} \sum_{j,k} [\bar{\psi}_j \gamma_\rho \tfrac{1}{2}\lambda_b \psi_j + \frac{\kappa}{2M}\partial^\nu (\bar{\psi}_j \sigma_{\rho\nu} \tfrac{1}{2}\lambda_b \psi_j)] (\bar{\psi}_k \sigma^\rho_\mu \tfrac{1}{2}\lambda_c \psi_k) \end{aligned} \tag{3.5}$$

Therefore, we do not obtain gauge couplings. We cannot build a composite gauge boson model by replacing $\bar{\psi}\gamma^\mu \frac{1}{2}\lambda_a \psi$ by the nonminimal current, J_a^μ of (3.5); once we inroduce $J_a^\mu J_{a\mu}$ term with J_a^μ of (3.5) as an interaction, the Noether current computed from such a Lagrangian would contain six-fermion terms and J_a^μ of (3.5) would be no longer a conserved current. We have to keep going to find finally a conserved current which contains infinite-body-fermion terms. Therefore, for any simple calculation to be carried out in such a composite model, the minimal fermionic current is essential.

This embarrassing observation applies to composite vector bosons made of bosonic preons also. For the bosonic preons, the minimal current contains a derivative

$$\phi^\dagger \tfrac{1}{2}\lambda_a \overleftrightarrow{\partial}_\mu \phi . \tag{3.6}$$

With $\overleftrightarrow{\partial}_\nu$, we can not build a simple model Lagrangian of composite gauge-like bosons. The iterative procedure of adding interactions of dimension up to infinity can be worked out in case of SU(2). It gives us a compact nonpolynomial Lagrangian

$$L = \sum_j (|\partial^\mu \phi_j|^2 - M^2 |\phi_j|^2) - \frac{2}{v^2} J^\mu_a J_{\mu a}, \qquad (3.7)$$

where

$$J_{\mu a} = i \sum_j \phi_j^\dagger \tfrac{1}{2}\lambda_a \overleftrightarrow{\partial}_\mu \phi_j / (1 - \sum_j |\phi_j|^2/v^2). \qquad (3.8)$$

Eqs.(3.7) and (3.8) are reminiscent of the CP^{N-1} model. Here, none of the ϕ fields develop a vacuum expectation value. Unlike the CP^{N-1} model, our model does not impose a subsidiary condition on $\Sigma_j|\phi_j|^2$.

Does this exercise imply that the dynamically generated gauge interaction is a freak ? I do not think so. In the four-fermion model of (2.1), lightness of the vector composites is ensured by global current conservation. On the other hand, in the alternative models like (3.5), there is no reason for vector composites to appear as light particles. Their natural mass scale is the preon mass itself unless one makes a fine tuning of parameters. Though it may sound expedient, we may put forth the conjecture that when vector composites are formed as light particles by dynamical necessity rather than by accident or by fine tuning, such vector particles behave like gauge bosons. Until it is proved in a model in which lightness of vector composites is realized by a mechanism different from the four-fermion model (2.1), the conjecture is not much more than a wishful thinking. In the sense of the Kroll-Lee-Zumino theorem[8], we have only to show that an appropriately defined composite wave function Ψ_μ obeys the transversality condition

$$\frac{\partial}{\partial X_\mu} \Psi_\mu(X, x_i) = 0 \quad (x_i = \text{relative coordinates}) \qquad (3.9)$$

not only on mass shell of composites but also off mass shell. We have to be careful about the ambiguity of the wave functions off shell.

4. COMPOSITE MODEL OF W AND Z WITH EXPLICIT SYMMETRY BREAKING

When we build a concrete model of composite W and Z from our study in the preceding Sections, we must keep in mind the followings:

(a) To build a simple model, we assume that preons are fermions without derivative interaction. If one chooses bosons as preons, onw would have to start with a nonpolynomial effective Lagrangian as was seen in Section 3. Then we would lose calculability of model.

(b) To suppress nongauge couplings of dimension six and higher for vector composites, the preon mass M must not be smaller than the electroweak scale. When energy approaches M, nongauge couplings of higher dimensions become important. The larger M is, the more accurate the dynamical gauge symmetry is for vector composites.

(c) Since the scalar composites of mass much lighter than M is unnatural or inconsistent with 1/N expansion, the mass splitting between W and Z must be realized by explicit breaking. We can introduce an elementary photon and incorporate the γ-Z mixing mechanism proposed by Hung and Sakurai ten years ago.[1)]

With these reminders, we are ready to write a concrete model.[6)] It consists of N fermion doublets as preons and an elementary photon;

$$L = \sum_j \overline{\psi}_j (i\partial\!\!\!/ - M)\psi_j - \frac{2}{v^2}\sum_{j,k} (\overline{\psi}_j\gamma_\mu \tfrac{1}{2}\tau_a\psi_j)(\overline{\psi}_k\gamma^\mu\tfrac{1}{2}\tau_a\psi_k)$$
$$- \tfrac{1}{4}F^{\mu\nu}F_{\mu\nu} - e\sum_j (\overline{\psi}_j\gamma_\mu(Y + \tfrac{1}{2}\tau_3)\psi_j)A^\mu , \qquad (4.1)$$

The SU(2) gauge coupling and the W mass are obtained by calculating the infinite series of bubbles;

$$g^2 = [\,N\Pi'(0)]^{-1} , \qquad (4.2)$$

$$m_W^2 = \tfrac{1}{4}v^2[\,N\Pi'(0)]^{-1}. \qquad (4.3)$$

The neutral component W^0 of the triplet vector composites mixes with photon through a preon loop since electromagnetic interaction is SU(2) breaking. The mixing is in the form of wave-function mixing

$$\tfrac{1}{2}(e/g)F_{\mu\nu}(\gamma)F^{\mu\nu}(W^0). \qquad (4.4)$$

Since g^2 is O(1/N), we regard the electromagnetic coupling squared as also of O(1/N) for the purpose of power counting of 1/N. Then the weak mixing angle is a well-defined finite number in the large N limit. The two mass eigenstates of the neutral vector bosons are

$$\begin{aligned} \gamma &= A_\mu, \\ Z &= [W_\mu^0 - (e/g)A_\mu]/[1 - (e/g)^2]^{\frac{1}{2}} \end{aligned} \qquad (4.5)$$

with eigenvalues $m_\gamma = 0$ and

$$m_Z^2 = m_W^2/[1 - (e/g)^2]. \qquad (4.6)$$

The neutral weak current of Z is given by

$$J^{\mu}_{NC} = \{g/[1-(e/g)^2]^{\frac{1}{2}}\}[J^{\mu}_3 - (e/g)^2 J^{\mu}_{em}]. \qquad (4.7)$$

By defining the weak mixing angle θ_W by $\sin\theta_W = e/g$, Eqs.(4.5)-(4.7) are identical with those of the standard electroweak theory. In this model, unlike other dynamical models, there is no need to struggle for making $\sin\theta_W$ large enough to be compatible with experiment.

5. DEVIATION FROM GAUGE COUPLING DUE TO COMPOSITENESS

To the lowest order of electromagnetic coupling e, all interactions of dimension four are equal to the nonabelian gauge couplings at zero external momenta. Couplings of dimension four defined on the W and Z mass shells deviate from the gauge couplings by inverse powers of the preon mass M. These deviations are effective interactions of dimension six and higher. Such deviations have been computed explicitly for the electromagnetic vertex of W. The result is as follows:[6)]

$$\begin{aligned} L_{int} = & \, ief_{\gamma}(W^{\dagger}_{\mu\nu}W^{\mu} - W^{\dagger}_{\mu}W^{\mu}_{\nu})A^{\nu} + ie\kappa_{\gamma}W^{\dagger}_{\mu}W_{\nu}A^{\mu\nu} + ie\lambda_{\gamma}W^{\dagger}_{\rho\mu}W^{\mu}_{\nu}A^{\nu\rho} \\ & + ie\mu^{(1)}_{\gamma}(\partial^{\lambda}W^{\dagger}_{\mu}\partial^{\mu}W_{\nu}\partial^{\nu}A_{\lambda} - \partial^{\nu}W^{\dagger}_{\mu}\partial^{\lambda}W_{\nu}\partial^{\mu}A_{\lambda}) \\ & + ie\mu^{(2)}_{\gamma}[(W^{\dagger}_{\mu}\overleftrightarrow{\partial}_{\lambda}W_{\nu})(g^{\mu\nu}\Box - \partial^{\mu}\partial^{\nu})A^{\lambda} + \text{cyclic permutations}], \end{aligned} \qquad (5.1)$$

where f_{γ}, κ_{γ}, λ_{γ}, $\mu^{(1)}_{\gamma}$ and $\mu^{(2)}_{\gamma}$ are constants, $W_{\mu\nu} = \partial_{\mu}W_{\nu} - \partial_{\nu}W_{\mu}$ and so forth, and cyclic permutations among $W^{\dagger}_{\mu}$, W_{ν} and A_{λ} are meant in the third line. The constants are given to the first order in q_i^2/M^2 (q_i = W and γ momenta) by

$$\begin{aligned} f_{\gamma} &= 1 + [(1/5)/\ln(\overline{\Lambda}/M)^2](q_1^2+q_2^2+q_3^2)/M^2, \\ \kappa_{\gamma} &= 1 + [(1/5)/\ln(\overline{\Lambda}/M)^2](q_1^2+q_2^2+q_3^2)/M^2, \\ \lambda_{\gamma} &= -[(9/20)/\ln(\overline{\Lambda}/M)^2](1/M)^2, \\ \mu^{(1)}_{\gamma} &= -[(1/5)/\ln(\overline{\Lambda}/M)^2](1/M)^2, \\ \mu^{(2)}_{\gamma} &= [(1/5)/\ln(\overline{\Lambda}/M)^2](1/M)^2, \end{aligned} \qquad (5.2)$$

where the first terms in f_{γ} and κ_{γ} represent the gauge symmetric terms. The momentum dependent corrections grow with energy to enhance compositeness effect in high energy processes. The sign of the momentum dependent terms in f_γ and κ_γ are such that it is consistent with form factor damping in the space-like direction. When SU(2) breaking beyond the lowest order W^0-γ mixing is included, the coefficients of $f_\gamma \sim \mu_\gamma^{(2)}$ generally acquire additional correction terms of $O(e^2)$.

The characteristics observed in the WWγ vertex given above emerge in all other multi-boson vertices. In case of four-point functions, couplings of dimension four receive SU(2) symmetric corrections of $O(p^2/M^2)$ where p^2 stands for properly symmetrized Lorentz scalar variables made of four external momenta. They are form factor effects. Couplings of interaction operators with dimension D (>4) are of the order of $(1/M)^{D-4}$ with SU(2) symmetric coefficients up to the W^0-γ mixing. Symmetry breaking effects are down by a factor of $O(\alpha/4\pi)$.

6. ABSENCE OF DANGEROUS LOOP CORRECTIONS

The important issue is how large deviation from the standard theory is when one computes loop diagrams of physical W and Z. Since the W and Z propagators obtained from the bubble summation of Fig.1 are those of the unitary gauge, there is a legitimate reason to worry about the potentially dangerous contributions from the longitudinal polarizations of W and Z. On the other hand, the effective Lagrangian of our model is identical to the minimal standard theory with the physical Higgs boson mass let to infinity or set equal to the preon mass M. By this observation, we may feel optimistic to believe that W and Z loop corrections are no worse than those of the standard theory with the Higgs boson mass equal to O(M). Demonstrating this expectation is not so trivial because we must deal with quantities in the next leading order in 1/N. We look into this problem in this Section.

The operators which receive potentially the largest correction are those of the lowest dimension, namely the W and Z mass terms. If we compute the difference $\delta m_Z^2 \cos^2\theta_W - \delta m_W^2$ from the diagrams of one W, Z or γ loop, we find that the dangerous terms of $O(1/m_{W,Z}^4)$ and $O(1/m_{W,Z}^2)$ are cancelled between δm_W^2 and $\delta m_Z^2 \cos^2\theta_W$ by global SU(2) symmetry and the right mixing relation $m_Z^2 \cos^2\theta_W = m_W^2$. The surviving contribution turns out to be to the lowest order in e^2

$$\delta m_W^2 - \delta m_Z^2 \cos^2\theta_W \sim (\alpha/4\pi)(\alpha_g/4\pi)\, M^2 , \tag{6.1}$$

where $\alpha_g = g^2/4\pi$ and the cutoff is chosen to be equal to M. This correction stays small enough if M is not much larger than O(1 TeV). However, this is not sufficient. The mass shifts themselves must remain small in order for our composite model to be viable. We will see that the potentially dangerous terms proportional to the inverse powers of $m_{W,Z}$ are cancelled out in δm_W^2 and in δm_Z^2 separately to the lowest order in electromagnetic coupling and that the explicit symmetry breaking due to electromagnetic interaction generates nonnegligible contributions when the preon mass scale is too high.

In the present parametrization of the four-fermion coupling, which is different from Ref.6 and follows Ref.7, the longitudinal polarization term $k^\mu k^\nu/m_{W,Z}^2$ of the vector composite propagators is of O(N). If this term survives in W and Z diagrams, our model would no longer be consistent with 1/N expansion at higher loop levels and immediately lose its beauty. As we wish, however, such dangerous loop diagram corrections are completely absent thanks to global SU(2) current conservation in the lowest order of electromagnetic interaction beyond the W^0-γ mixing. Once higher order electromagnetic effects are included, the contributions of the longitudinal polarization stay with coefficients of $O(\alpha/4\pi)$.

Let us examine a class of diagrams in which a weak boson propagator appears as shown in Fig.2.

Fig.2. Diagrams of physical W or Z loop.

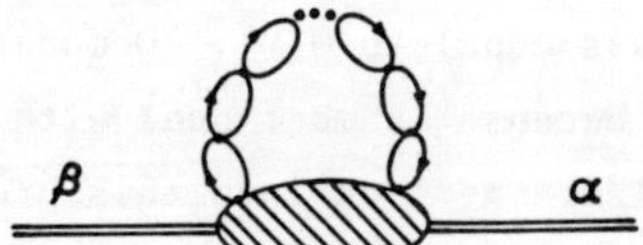

The matrix element can be written as

$$M = i\int \langle\beta|T^*(J_\mu^i(x)\; J_\nu^j(y))|\alpha\rangle\; \delta^{ij}\; \Delta^{\mu\nu}(x-y)\; d^4x\; d^4y, \tag{6.2}$$

where

$$\Delta^{\mu\nu}(x-y) = \int\{(-g^{\mu\nu} + \frac{k^\mu k^\nu}{k^2})/[(\tfrac{1}{4}v^2 - N\Pi(k^2)] - \frac{4}{v^2}\cdot\frac{k^\mu k^\nu}{k^2}\}$$
$$\times \exp(-ik(x-y))\; d^4k/(2\pi)^4 \tag{6.3}$$

represents an infinite series of bubbles which includes continuum contributions as well as a W or Z boson pole. The first term of $\Delta^{\mu\nu}$ is $g^2(-g^{\mu\nu} + k^\mu k^\nu/k^2)/(m_{W,Z}^2 - k^2)$ near the pole and is of $g^2 \times O(1)$. Therefore, it does not generate a harmful contribution in physical W and Z loops. On the other hand, the second term is of $g^2 \times O(N)$. If this term stays and makes a full contribution, nice low-energy results of the model would be meaningless once loop corrections are made. The dangerous longitudinal part of $\Delta^{\mu\nu}$ is separated for W as

$$\Delta_L^{\mu\nu}(x-y) = -(g^2/m_W^2)\, \partial_x^\mu \partial_y^\nu \int \frac{\exp(-ik(x-y))}{k^2}\, d^4k/(2\pi)^4. \qquad (6.4)$$

Integrating by parts in the matrix element of (6.2), we find the longitudinal contribution as

$$M_L = -i(g^2/m_W^2)\int \langle\beta|T^*(\partial^\mu J_\mu^i(x)\ \partial^\nu J_\nu^i(y)) - 2ieJ_\mu^3(x)A^\mu(x)\delta(x-y)|\alpha\rangle \\ \times \frac{1}{k^2}\exp(-ik(x-y))\, d^4k/(2\pi)^4 d^4x\, d^4y \qquad (6.5)$$

Global current conservation is broken only to O(e),

$$\partial^\mu J_\mu^i = -\varepsilon^{ij3}\, e J_\mu^j A^\mu. \qquad (6.6)$$

In the absence of electromagnetic interaction, the matrix element M_L is comletely gone and the correction is under control. If the W^0-γ mixing is included, we obtain for the W and Z mass shifts due to $\Delta_T^{\mu\nu}$

$$\delta m_W^2 = \delta m_Z^2 \cos^2\theta_W \sim (\alpha_g/4\pi)\, M^2. \qquad (6.7)$$

These shifts are of O(1/N) and do not affect the relation among m_W, m_Z and θ_W. The correction to $m_Z^2 \cos^2\theta_W = m_W^2$ is similar to that in the standard theory in which the Higgs boson mass is set equal to M.

In the presence of electromagnetic interaction, the matrix element (6.6) acquires a residual contribution of $O(e^2)$ from the longitudinal polarization. Since e^2 is to be regarded as O(1/N), e^2/m_W^2 is of O(1). In the case of W and Z mass shifts, this correction can be evaluated by the infinite series of bubble diagrams shown in Fig.3, where the broken line propagator stands for $1/k^2$ of $\Delta_L^{\mu\nu}(k)$. The result is

$$\delta m_W^2 \simeq (\alpha/2\pi)\, m_W^2 \ln M^2, \\ \delta m_Z^2 = 0. \qquad (6.8)$$

This is a small correction.

Fig.3. The contribution to $\delta m^2_{W,Z}$ in the presence of electromagnetic interaction. See Eq.(6.8).

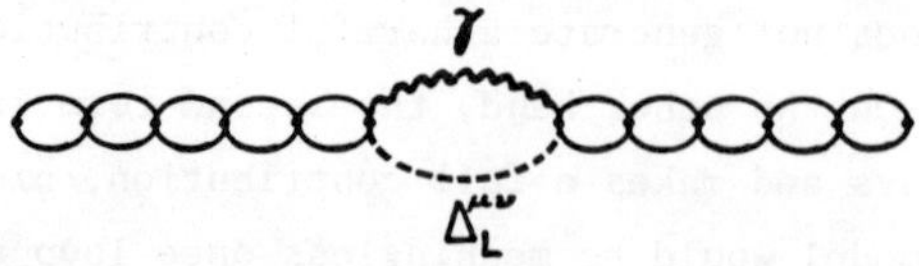

The argument for suppression of the longitudinal part of $\Delta^{\mu\nu}$ can be extended to general diagrams where more than one $\Delta^{\mu\nu}$ are involved. Take diagrams having two of them:

$$M_L = i^2 \int d^4x\, d^4y\, d^4z\, d^4w \;\; \langle\beta|T^*(J^i_\mu(x)\, J^i_\nu(y)\, J^i_\kappa(z)\, J^j_\lambda(w))|\alpha\rangle \times \Delta^{\mu\nu}(x-y)\Delta^{\kappa\lambda}(z-w). \tag{6.9}$$

Separating the longitudinal parts from $\Delta^{\mu\nu}$ and $\Delta^{\kappa\lambda}$, we can show by current algebras that the most singular contribution from $\Delta^{\mu\nu}_L$ and $\Delta^{\kappa\lambda}_L$ vanishes when electromagnetic interaction is absent. When we combine this result with the preceding argument in which only one $\Delta^{\mu\nu}$ was singled out, we can easily show that neither $\Delta^{\mu\nu}_L$ nor $\Delta^{\kappa\lambda}_L$ can survive in (6.9) when e = 0. This argument goes through in diagrams which contain an arbitrary number of $\Delta^{\mu\nu}$. Therefore, the 1/N expansion is perfectly consistent for e = 0. When electromagnetic interaction is present as an explicit breaking of symmetry, the singular terms from $\Delta^{\mu\nu}_L$ do not go away completely. A factor of

$$(\alpha_g/4\pi)(M/m_W)^2 = (M/2\pi v)^2 \tag{6.10}$$

keeps accumulating for each $\Delta^{\mu\nu}_L$ in general for $e \neq 0$. Therefore, it is necessary to require that

$$M < \sqrt{4\pi/\alpha_g}\; m_W \tag{6.11}$$

in order for perturbative calculation to be meaningful. This constraint (6.11) suggests that the preon mass M should not be much larger than 1 TeV as long as SU(2) symmetry is broken explicitly by electromagnetic interaction.

Let us turn to another interesting electroweak correction, the anomalous magnetic moment of charged lepton. Since we have not built a model for light fermions, we simply assume that quarks and leptons interact through V - A and satisfy the weak universality. If they are composites of the same preons as the constituents of W and Z, the weak universality is guaranteed by the gauge coupling of preons. For g - 2, the relevant matrix element is that of the electromagnetic current which is the very origin of symmetry breaking. The longitudinal part of $\Delta^{\mu\nu}$ still vanishes by the Ward identity or by use of an identity

$$\int d^4k/k^2 = 0, \tag{6.12}$$

if we take the lowest order in electromagnetic coupling in the matrix element

$$i\int \langle\ell|T^*(J^i_\mu(x)\ J^i_\nu(y)\ J^{em}_\lambda(0))|\ell\rangle\ \Delta^{\mu\nu}(x-y)\ d^4x d^4y. \tag{6.13}$$

Therefore, the singular part of the $\Delta^{\mu\nu}_L$ contribution in the sense of 1/N expansion disappears from loop diagrams. Once higher order electromagnetic effects are included, the situation is the same as in the process $\alpha \rightarrow \beta$ discussed before. It should be noted that in all cases disappearance of dangerous terms in 1/N occurs for e = 0 by an intricate cancellation between W/Z poles and the continuum part of the infinite series of bubble diagrams.

7. WEAK INTERACTION OF QUARKS AND LEPTONS

We have a greater motivation for building composite models of quarks and leptons than W and Z because in the standard theory there are too many parameters associated with Higgs particles whose values can not be determined from a fundamental principle. It is tempting to postulate structure in the light fermions and explain their masses and the quark mixing angles in terms of their underlying structure. Though several attempts have been made in the past, we are still in search of a right direction at present. The issue is how to make composite fermions by orders of magnitude lighter than their constituent particles. Lightness must be based on some dynamical mechanism, not by fine tuning of strength of binding force. The three basic ideas have been so far put forth as promising lines and have been explored.

The first attempt is to introduce a large chiral symmetry at the preon level and to let part of the symmetry survive spontaneous breakdown in order to realize massless composite fermions. Models along this line presume a renormalizable, confining field theory at the preon level. Then the anomaly matching condition of t'Hooft[9] eliminates a large number of attractive and realistic group theoretical frameworks.[10,11] It may be a good idea to forget about renormalizability of the preon theory and therefore discard the anomaly matching for a moment. The other line of attempts utilizes supersymmetry. It is possible to generate massless fermions as the Nambu-Goldstone fermions of supersymmetry breaking. Decoupling of the Nambu-Goldstone fermions at zero momenta can be evaded if electroweak and strong gauge interactions break supersymmetry.[12] To my knowledge, no realistic model has been built successfully along this line. There is another possibility with supersymmetry.[13] When global symmetry is broken spontaneously in a supersymmetric theory, massless fermions can appear as supersymmetric partners of the Nambu-Goldstone bosons. In some cases where global symmetry and supersymmetry are broken by the same origin, the Nambu-Goldstone fermion mechanism and the super-partner mechanism are intervined. Explicit computation was done to demonstrate the fermion partner mechanism in the supersymmetric Nambu-Jona-Lasinio model.[14] However, we have not been able to find a model applicable to the real world. One of the problems in the supersymmetric scenarios is how to keep the light fermion masses as light as several MeV in spite of a large supersymmetry breaking scale indicated by the lower limit on the scalar quark mass $\gtrsim$ 100 MeV. It appears possible to suppress some of the supersymmetric partner masses in the presence of a large supersymmetry breaking.[15]

When we attempt to introduce composite light fermions in our composite model of W and Z, chirality of the light fermions is a problem. Experiment requires that the weak isospin symmetry is left-handed for quarks and leptons. But it is difficult to build light chiral fermions from the heavy preons. Even if preons interact chirally, there is no guarantee that weak interaction of light composite fermions is also chiral. Only when the preon mass is different from the compositeness scale Λ_c and much smaller than it, can we realize a chiral weak interaction for quarks and leptons. As we can demonstrate easily in a concrete supersymmetric model of Nambu-Jona-Lasinio type, the V - A interaction of preons generally leads us to only an approximate V - A interaction for composite quarks and leptons with a deviation given by

$$(V/A) - 1 \sim M/\Lambda_c, \tag{7.1}$$

even when all interactions are chiral. It implies that approximate chiral symmetry must be embedded in advance in preon dynamics at the beginning. This observation, among others, makes introduction of our composite W and Z into composite quarks and leptons rather difficult.

Meanwhile, an interesting possibility was recently pointed out by Cohen, Georgi and Simmons.[7] Using the fact that the conjugate of a doublet is an equivalent doublet in SU(2), they give up compositeness of quarks and leptons for a moment and place the light fermions at the same footing as the preons of W and Z.

$$f_L = \begin{pmatrix} \psi \\ i\tau_2\psi^c \\ q \\ \ell \end{pmatrix}_L \tag{7.2}$$

where ψ stands for the families of preon doublets and ψ^c is the charge conjugate. W and Z are made of f_L. The quarks and leptons inside W and Z do contribute to fast damping of form factors, but such damping can be well hidden when N is large enough. Cohen et al. estimated the magnitude of N from the radiatively generated kinetic energy of W and Z in the effective Lagrangian. They found $N \approx 60$. Furthermore, they estimated the dominant contribution to the W mass shift which comes from the wave-function renormalization of W according to their 1/N expansion reasoning. They found that the W mass shift is only 2 % even for M as small as m_W itself. Although their model does not provide a new insight in the long standing problem of light fermion masses and mixing, it circumvents the chirality difficulty of heavy preons and therefore makes the composite W and Z models viable by themselves, leaving the light fermion problem as a separate issue.

REFERENCES

1. Hung, P.Q. and J.J. Sakurai, Nucl. Phys. 143B, 81, 539(E)(1978).
2. Bjorken, J.D., Phys. Rev. D19, 4257 (1974).
3. Eguchi, T. and H. Sugawara, Phys. Rev. D10, 4257 (1974).
4. Kikkawa, K., Progr. Theor. Phys. 56, 947 (1976).
   Chakrabarti, A. and B. Hu, Phys. Rev. D13, 2247 (1976).
   Eguchi, T. Phys. Rev. D14, 2755 (1976).
5. Bjorken, J.D., Ann. Phys. 24, 174 (1963).
6. Suzuki, M., Phys. Rev. D37, 210 (1988).
7. Cohen, A., H. Georgi and E.H. Simmons, Harvard University Preprint HUTP-88/A004, January 1988 (to be published in Phys. Rev.)
8. Kroll, N.M., T.D. Lee and B. Zumino, Phys. Rev. 157, 1376 (1967).
9. t'Hooft, G., "Recent Developments in Gauge Theories", edited by G. t'Hooft et al. (Plenum Press, N.Y., 1980), p.135.
   Dimopoulos, S., S. Raby and L. Susskind, Nucl. Phys. B173, 208 (1980).
10. Albright, C., Phys. Rev. D24, 1969 (1981).
    Bars, I., Nucl. Phys. B208, 77 (1982).
    Buras, A.J., S. Dawson and A.N. Schellekens, Phys. Rev. D26, 3225 (1982).
11. Peskin, M.E., Proceedings of the 1985 International Symposium on Lepton and Photon Interactions at High Energies, edited by M. Konuma and K. Takahashi (Kyoto Univ., Kyoto, Japan, 1985), p.714.
12. Bardeen, W.A. and V. Visnjic, Nucl. Phys. B194, 422 (1982).
13. Buchmüller, W., S. Love, R.D. Peccei and T. Yanagida, Phys. Lett. 115B, 233 (1982).
    Buchmüller, W., R.D. Peccei and T. Yanagida, Phys. Lett. 124B, 67 (1983).
14. Buchmüller, W. and U. Ellwanger, Nucl. Phys. B245, 237 (1984).
15. Takeshita, S. and M. Yasue, Phys. Rev. D34, 1847 (1986).

SUPERSYMMETRY IN COMPOSITE MODELS AND THE MASS SPECTRUM OF QUARKS AND LEPTONS

Hidezumi Terazawa

Institute for Nuclear Study, University of Tokyo
Midori-cho, Tanashi, Tokyo 188
JAPAN

ABSTRACT

It is shown that supersymmetry turns out to be very useful and powerful enough to determine the mass spectrum of quarks and leptons in composite models. For quarks and leptons as either almost or quasi Nambu-Goldstone fermions, many mass formulas and sum rules are derived. Among others, the mass sum rules of $m_e^{1/2}-m_{\nu_e}^{1/2} = m_d^{1/2}-m_u^{1/2}$, $m_\mu^{1/2}-m_e^{1/2} = m_s^{1/2}-m_d^{1/2}$, $m_\mu^{1/2}-m_{\nu_e}^{1/2} = m_s^{1/2}-m_u^{1/2}$ and $m_\tau^{1/2}-m_{\nu_\mu}^{1/2} = m_b^{1/2}-m_c^{1/2}$ are satisfied remarkably well by the experimental values and estimates.

1. INTRODUCTION

Supersymmetry (SUSY) was first proposed by Professor Miyazawa[1] in 1966 as a possible symmetry between mesons and baryons. The algebra of baryon-number changing currents, the hypothesis of partially conserved spinor currents and, even more ingeniously, the possible existence of scalar quarks were already suggested then. It was, however, not until the middle of seventies that SUSY was revived[2] as an additional symmetry in unified, grand unified and supergrand unified theories. Entering into eighties, SUSY has also been assumed as a central symmetry in composite models of quarks and leptons.[3] In addition, "hypersymmetry (HYSY)" was proposed by Professor Miyazawa[4] in 1981 as a possible symmetry between states with different particle numbers. In this festschrift, I shall present some latest progress in composite models, especially to show that SUSY turns out to be very useful and powerful enough to determine the mass spectrum of quarks and leptons.

One of the most important but difficult problems not only in composite models of quarks and leptons but in particle physics as a whole is how to explain the existing peculiar mass spectrum of quarks and leptons. Before investigating the detailed structure in the mass spectrum, one must first explain its gross property that quark and lepton masses are much smaller than their inverse sizes ($\gtrsim$1TeV). For such explanation, so far proposed are the following three possibilities: quarks and leptons are 1) almost chiral fermions,[5)] 2) almost Nambu-Goldstone (N-G) fermions[4),6)] or 3) quasi N-G fermions.[7)] The latter two possibilities may not be independent since quarks and leptons are taken as almost massless N-G fermions due to spontaneous breakdown of (approximate) SUSY in 2) and as almost massless super-partners of N-G bosons due to spontaneous breakdown of (approximate) flavor-color global symmetry in 3).

More precisely, let H, S and Q^A be the Hamiltonian for the unknown composite dynamics, the supercharge (of N=1) and the (vector or axial-vector) flavor-color charges, respectively. Suppose the most idealistic case in which $[H,S] = [H,Q^A] = [S,Q^A] = 0$. In the case 2), $S|0\rangle \neq 0$ and $Q^A|0\rangle = 0$ while in the case 3), $S|0\rangle = 0$ and $Q^A|0\rangle \neq 0$. In the case 2), composite quarks and leptons are taken as almost N-G fermions which behave as $Q^AS|0\rangle$ while in the case 3), they are as quasi N-G fermions which behave as $SQ^A|0\rangle$. However, it is obvious that both of these states vanish since $[S,Q^A] = 0$ and either $Q^A|0\rangle = 0$ or $S|0\rangle = 0$. The wishful existence of almost N-G fermions or quasi N-G fermions, therefore, is induced either by $[H,S] \neq 0$, $[H,Q^A] \neq 0$, $[S,Q^A] \neq 0$ or $S|0\rangle \neq 0$ and $Q^A|0\rangle \neq 0$. We will clarify the relation between almost N-G fermions and quasi ones in such a more realistic case, present a general mass formula for quarks and leptons as almost or quasi N-G fermions and, furthermore, derive simple sum rules for quark and lepton masses from the mass formula by assuming an explicit form of symmetry breaking.

2. INDUCED SUPERCURRENTS

If SUSY is exact (S=const), supercurrents with the flavor-color quantum numbers can be induced as local currents by

$$S_\mu^A = -[J_\mu^A, S] ,$$

where J_μ^A are the (vector or axial-vector) flavor-color currents, while if flavor-color symmetry is exact (Q^A=const), other supercurrents can be induced by

$$S'^A_\mu = -[Q^A, s_\mu] ,$$

where s_μ is the supercurrent (of N=1). Obviously, the charges of these supercurrents are identical to each other, *i.e.*

$$S^A = S'^A \ (= -[Q^A, S]) .$$

Let us now introduce "partially conserved induced supercurrent (PCIS) hypothesis" that the divergence of this induced supercurrent be dominated by a fermion field (f_A), which represents a quark or a lepton, as

$$\partial^\mu S_\mu^A \cong ic_{fA} f_A ,$$

where c_{fA} is a constant. This constant depends on the fermion mass (m_{fA}) as

$$c_{fA} = F_{fA}^2 m_{fA} \qquad \text{or} \qquad -F_{fA} m_{fA}^2 ,$$

where F_{fA} is the "fermion decay constant", depending on whether the fermion is close to an almost N-G fermion or a quasi one. The reason for this crucial difference of mass dependence is that

$$S_\mu^A \cong -F_{fA}^2 \gamma_\mu f_A \qquad \text{or} \qquad iF_{fA}\partial_\mu f_A$$

for an almost or quasi N-G fermion, respectively.

3. MASS FORMULAS

From the PCIS relation, it is straightforward to derive a mass formula for quarks and leptons by following Dashen[8] who derived the mass formula for pions as almost N-G bosons from the PCAC relation. The simplest derivation will be presented in the Appendix. The result is

$$c_{fA}^2 m_{fA}^{-1} \delta_{\alpha\beta} = \overline{<\{[Q^A, S_\alpha], [[Q^A, S_\beta], \Delta L]\}>_0} ,$$

where ΔL is the symmetry breaking Lagrangian density.

In order to find an explicit form of the mass formula, suppose that SUSY or flavor-color symmetry is broken not only spontaneously by the appropriate condensations of scalar subquarks (ϕ_i) but also explicitly by the masses of spinor subquarks (M_i). Then, we obtain

$$m_{fij}^p = F_{fij}^{p-5}(M_i \mp M_j)^2(M_j<\phi_i^c\phi_i>_0 + M_i<\phi_j^c\phi_j>_0) \quad \text{for} \quad p = 1 \text{ or } 3,$$

depending on whether a quark or lepton is close to an almost N-G fermion or a quasi one. The - and + signs are for the "vector" and "axial-vector" supercurrent, respectively.

This explicit mass formula indicates that the quark or lepton mass (or the cubic mass) behaves essentially as $(\Delta M)^2 M\Lambda^{p-3}$ or $(\Sigma M)^2 M\Lambda^{p-3}$ where M, ΔM, ΣM and Λ are the subquark mass, the mass difference, the mass sum and the mass scale of compositeness, respectively since the "decay constant" (F) and the vacuum-expectation value of scalar condensate ($\langle\phi_i^c\phi_i\rangle_0$) must be of order Λ and Λ^2, respectively. Therefore, it can explain, as expected, that the quark and lepton masses are much smaller than their inverse sizes (which may be of order Λ) if the subquark masses are much smaller than the mass scale of compositeness, i.e. M, ΔM, $\Sigma M \ll \Lambda$. Notice also that the masses of quasi N-G fermions (for $p=3$) may not depend on Λ but do only on the subquark mass. If this is the case, the subquark masses (or, at least, the mass differences) must be as small as the masses of quasi N-G fermions.

In the particular case that the "decay constants" and the vacuum-expectation-values of scalar-subquark condensations are almost universal or independent of their flavor-color quantum numbers, the mass formula indicates that the mass of a quark or lepton is essentially determined by

$$m_{fij}^p \cong \text{const} \times (M_i \mp M_j)^2 (M_i + M_j) \ .$$

4. MASS SUM RULES

From the last mass formula follows the mass sum rule for quarks and leptons,

$$m_{f23}^r \pm m_{f13}^r = \pm(m_{f24}^r \pm m_{f14}^r) \quad r = \frac{1}{2} \text{ or } \frac{3}{2} \text{ for the "vector"}$$

$$m_{f23}^r - m_{f13}^r = m_{f24}^r - m_{f14}^r \quad r = \frac{1}{3} \text{ or } 1 \text{ for the "axial vector",}$$

depending on whether they are almost N-G fermions or quasi ones.

In order to find whether the existing mass spectrum of quarks and leptons satisfies this type of sum rules, let us take the minimal composite model where the first generation of quarks and leptons can be identified as

$$\nu_e = f_{13}, \quad e = f_{23}, \quad u_i = f_{1i+3}, \quad d_i = f_{2i+3} \quad (i=1,2,3).$$

More precisely, they are made of a "wakem" (subquark of weak isospin doublet) and a "chrom" (subquark of color quartet) as

$$\nu_e = (w_1C_0), \quad e = (w_2C_0), \quad u_i = (w_1C_i), \quad d_i = (w_2C_i) \quad (i=1,2,3).$$

Then, the mass sum rules become in the form of

$$m_e^r - m_{\nu_e}^r = \pm(m_d^r \pm m_u^r) \ (r = \tfrac{1}{2} \text{ or } \tfrac{3}{2}) \quad \text{or} \quad m_d^r - m_u^r \ (r = \tfrac{1}{3} \text{ or } 1).$$

Obviously, since $m_e < m_u < m_d$, the possible combination of ± signs must be fixed as

$$m_e^r - m_{\nu_e}^r = m_d^r - m_u^r .$$

For r = 1/2, this mass sum rule is nothing but the one which has been found almost empirically by myself[9] and which is satisfied by the experimental value of m_e (= 0.511MeV) and the estimates of m_u (= 4.5±1.4MeV) and m_d (= 7.9±2.4MeV).[10] The ℓ.h.s. is 0.71MeV$^{1/2}$ while the r.h.s. is 0.69±0.22MeV$^{1/2}$. If this agreement is taken seriously, it may indicate that quarks and leptons (at least) in the first generation are close to the almost N-G fermions.

On the other hand, the linear mass sum rule (r=1 for quasi N-G fermions) is nothing but the one which has already been derived by us[11] in a SUSY nonlinear-interaction model of the Nambu-Jona-Lasinio type although it is not consistent with the experimental values.

5. EXTENDED MASS FORMULAS AND SUM RULES

The original mass formula for quarks and leptons can be extended if the further induced supercurrents (FIS) are introduced as

$$s_\mu^{Ah} = -[Q^h,[J_\mu^A,S]] \quad \text{or} \quad s_\mu'^{Ah} = -[Q^h,[Q^A,s_\mu]] ,$$

where Q^h are the horizontal (or generation or family) charges, and if the "partially conserved further induced current (PCFIS) hypothesis" is made as

$$s_\mu^{Ah} \cong ic_{fAh}f_{Ah} \ .$$

The extended mass formula is given by

$$c_{fAh}^2 m_{fAh}^{-1}\delta_{\alpha\beta} = \langle\{\overline{[Q^h,[Q^A,S_\alpha]],[[Q^h,[Q^A,S_\beta]],\Delta L]}\}\rangle_0 \ .$$

From this follows the mass sum rule of

$$m_\mu^{1/2}-m_e^{1/2} = m_s^{1/2}-m_d^{1/2}$$

if e, μ, d and s are identified as f_{e1}, f_{e2}, f_{d1} and f_{d2}, respectively. More precisely, they are assumed to be made of a "wachrom" (di-subquark state of a wakem and a chrom) and a "hakam" (subquark of horizontal multiplet) as

$$e = ((w_2C_0)h_1),\ \mu = ((w_2C_0)h_2),\ s_i = ((w_2C_i)h_1),\ d_i = ((w_2C_i)h_2) \qquad (i=1,2,3).$$

This sum rule is also nothing but the one which has been found almost empirically by myself[9] and which is satisfied by the experimental values of m_e and m_μ (= 105.66MeV) and the estimates of m_d and m_s (= 155±50MeV).[10] The ℓ.h.s. is 9.56MeV$^{1/2}$ while the r.h.s. is 9.64± 2.44MeV$^{1/2}$. However, from the same mass formula also follows the mass sum rule,

$$m_{\nu_\mu}^{1/2}-m_{\nu_e}^{1/2} = m_c^{1/2}-m_u^{1/2} \ ,$$

which is obviously not consistent with the experimental values. How to avoid this difficulty is a subject for future investigation.

By combining the two sum rules of $m_e^{1/2}-m_{\nu_e}^{1/2} = m_d^{1/2}-m_u^{1/2}$ and $m_\mu^{1/2}-m_e^{1/2} = m_s^{1/2}-m_d^{1/2}$, another sum rule of

$$m_\mu^{1/2}-m_{\nu_e}^{1/2} = m_s^{1/2}-m_u^{1/2}$$

can be obtained. Although not independent, this sum rule is satisfied well by the experimental values as the ℓ.h.s. is $10.28\text{MeV}^{1/2}$ while the r.h.s. is $10.33\pm2.34\text{MeV}^{1/2}$. Notice that this sum rule can be derived directly if ν_e, μ, u and s are identified as composites of a "waham" (di-subquark state of a wakem and a hakam) and a chrom, i.e.

$$\nu_e = ((w_1h_1)C_0),\ \mu = ((w_2h_2)C_0),\ u_i = ((w_1h_1)C_i),\ s_i = ((w_2h_2)C_i) \quad (i=1,2,3).$$

Similarly, if ν_μ, τ, c and b are identified as

$$\nu_\mu = ((w_1h_2)C_0),\ \tau = ((w_2h_3)C_0),\ c_i = ((w_1h_2)C_i),\ b_i = ((w_2h_3)C_i) \quad (i=1,2,3),$$

another sum rule of

$$m_\tau^{1/2}-m_{\nu_\mu}^{1/2} = m_b^{1/2}-m_c^{1/2}$$

can be obtained. This sum rule is satisfied approximately by the experimental values as the ℓ.h.s. is $42.2\text{MeV}^{1/2}$ while the r.h.s. is $36.1\pm1.4\text{MeV}^{1/2}$ or $40.1\pm1.4\text{MeV}^{1/2}$.

6. MAXIMALLY EXTENDED MASS FORMULAS

The extended mass formula can be further extended if the "maximally induced supercurrents (MIS)" are introduced as

$$s_\mu^{wCh} = -[Q^h,[Q^C,[Q^w,s_\mu]]] ,$$

where Q^W and Q^C are the weak isospin and color charges, respectively, and if the "partially conserved maximally induced supercurrent (PCMIS) hypothesis" is made as

$$S_\mu^{wCh} \cong ic_{fwCh} f_{wCh} \ .$$

The final mass formula for quarks and leptons is given by

$$c_{fwCh}^2 m_{fwCh}^{-1} \delta_{\alpha\beta} = \langle\{\overline{[Q^h,[Q^C,[Q^W,S_\alpha]]]},[[Q^h,[Q^C,[Q^W,S_\beta]]],\Delta L]\}\rangle_0 \ ,$$

which may, in principle, explain the whole mass spectrum of quarks and leptons. The procedure of deriving this mass-formula strongly suggests another type of the unified spinor model in which every quark or lepton is made from the single absolutely neutral massless spinor state (Call it "superion"), $S|0\rangle$, as $[Q^h,[Q^C,[Q^W,S]]]|0\rangle$.

7. CONCLUDING REMARKS

I have shown that supersymmetry and a way of its breakdown determine the mass spectrum of quarks and leptons in composite models and that many mass formulas and sum rules can be derived for quarks and leptons as either almost or quasi N-G fermions. I have also pointed out that among others, the mass sum rules of $m_e^{1/2}-m_{\nu_e}^{1/2} = m_d^{1/2}-m_u^{1/2}$, $m_\mu^{1/2}-m_e^{1/2} = m_s^{1/2}-m_d^{1/2}$, $m_\mu^{1/2}-m_{\nu_e}^{1/2} = m_s^{1/2}-m_u^{1/2}$ and $m_\tau^{1/2}-m_{\nu_\mu}^{1/2} = m_b^{1/2}-m_c^{1/2}$ are satisfied remarkably well by the experimental values and estimates. If one takes this agreement seriously, as I do, it may already indicate not only that quarks and leptons (at least in the first generation) are close to the almost N-G fermions but also (and more importantly) that <u>they are composites</u>.

ACKNOWLEDGEMENTS

The author cordially acknowledges Professor H. Miyazawa for his constant guidance I have had since I was a graduate student. He also wishes to thank Dr. M. Yasuè for his collaboration on the first part of this work, which has been presented in Ref.12, and for useful discussions.

APPENDIX

The simplest derivation of the mass formula for quarks and leptons as either almost or quasi N-G fermions is the following:

Let us start with the canonical commutation relation of a quark or lepton field (f) at equal time,

$$\{f^{\dagger}(x)_{\alpha}, f(y)_{\beta}\}_{x_0=y_0} = \delta_{\alpha\beta}\delta(\vec{x}-\vec{y}) \ .$$

By using the assumed partially conserved supercurrent relation of

$$\partial_{\mu}S_{f}^{\mu} \cong ic_{f}f \ ,$$

it can be transformed into

$$\{(\partial_{\mu}S_{f}^{\mu}(x)_{\alpha})^{\dagger}, \partial_{\nu}S_{f}^{\nu}(y)_{\beta}\}_{x_0=y_0} = c_{f}^{2}\delta_{\alpha\beta}\delta(\vec{x}-\vec{y}) \ .$$

Integrating both hand sides of this relation over d^3x gives the relation

$$\{(\dot{S}_{f\alpha})^{\dagger}, \partial_{\nu}S_{f\beta}^{\nu}\} = c_{f}^{2}\delta_{\alpha\beta} \ .$$

By using the Heisenberg equation of motion of $\dot{S}_{f\alpha} = i[H, S_{f\alpha}]$ and its generalized one of $\partial_{\nu}S_{f\beta}^{\nu} = i[\mathcal{H}, S_{f\beta}]$, it can be further transformed into

$$c_{f}^{2}\delta_{\alpha\beta} = \{[H, S_{f\alpha}]^{\dagger}, [\mathcal{H}, S_{f\beta}]\} \ ,$$

where H and $\mathcal{H}$ are the Hamiltonian and its density, respectively.

Taking the vacuum expectation value of both hand sides of this relation and using the relations of $H|0\rangle = 0$ and $HS_f|0\rangle = m_f S_f|0\rangle$ at rest give

$$c_f^2 \delta_{\alpha\beta} = m_f \langle\{S_{f\alpha}^\dagger,[\mathcal{H},S_{f\beta}]\}\rangle_0 .$$

This can be finally transformed into the mass formula of

$$c_f^2 m_f^{-1} \delta_{\alpha\beta} = \langle\{S_{f\alpha}^\dagger,[S_{f\beta},\Delta L]\}\rangle_0 ,$$

where ΔL is the symmetry breaking Lagrangian density. From this general form of mass formula follows various mass formulas if the supercurrent $s_{f\mu}$ is identified with either one of induced supercurrents such as $s'^{A}_{\mu} = -[Q^A,s_\mu]$, $s'^{Ah}_{\mu} = -[Q^h,[Q^A,s_\mu]]$ and $s^{WCh}_{\mu} = -[Q^h,[Q^C,[Q^W,s_\mu]]]$.

REFERENCES

1. Miyazawa, H., Prog. Theor. Phys. 36, 1266 (1966).
2. Volkov, D. V. and Akulov, V. P., Phys. Lett. 46B, 109 (1973); Wess, J. and Zumino, B., Nucl. Phys. B70, 39 (1974).
3. For a review, see for example Terazawa, H., in Proc. XXII Int. Conf. on High Energy Physics, Leipzig, July 19-25, 1984, edited by A. Meyer and E. Wieczorek (Akademie der Wissenschaften der DDR, Zeuthen, 1984), Vol.I, p.63.
4. Miyazawa, H., in Proc. III INS Winter Seminar on the Sub-Structure of Quarks and Leptons, INS, Univ. of Tokyo, Dec. 25-26, 1981, edited by M. Yasuè, INS-J-164 (INS, Univ. of Tokyo, Tokyo, 1982), p.71.
5. 't Hooft, G., in Recent Developments in Gauge Theories, edited by G.'t Hooft et al. (Plenum, New York, 1980), p.135.
6. Terazawa, H., Prog. Theor. Phys. 64, 1763 (1980), INS-Rep.-485 (INS, Univ. of Tokyo), Dec., 1983, in Flavor Mixing in Weak Interactions, edited by L.-L. Chau (Plenum, New York, 1985), p.655; Bardeen, W. A. and Višnjić, V., Nucl. Phys. B194, 422 (1982); Bardeen, W. A., Taylor, T. R. and Zachos, C. K., Nucl. Phys. B231, 235 (1984).
7. Buchmüller, W., Peccei, R. D. and Yanagida, T., Phys. Lett. 124B, 67 (1983); Barbieri, R., Masiero, A. and Veneziano, G., Phys. Lett. 128B, 179 (1983); Greenberg, O. W., Mohapatra, R. N. and

Yasuè, M., Phys. Lett. <u>128B</u>, 65 (1983)

8. Dashen, R., Phys. Rev. <u>183</u>, 1245 (1969).

9. Terazawa, H., J. Phys. Soc. Jpn. <u>55</u>, 4249 (1986).

10. For a review, see Gasser, J. and Leutwyler, H., Phys. Rep. <u>87</u>, 77 (1982).

11. Yasuè, M. and Terazawa, H., Phys. Lett. <u>197B</u>, 161 (1987).

12. Terazawa, H. and Yasuè, M., INS-Rep.-664 (INS, Univ. of Tokyo), Dec., 1987; Terazawa, H., INS-Rep.-667 (INS, Univ. of Tokyo), Jan. 1988, to be published in <u>Proc. INS Winter Seminar on Superstring, Composite Models and Anomaly Related Problems</u>, INS, Univ. of Tokyo, Dec. 21-23, 1987, edited by M. Tomiya, INS-J (INS, Univ. of Tokyo, 1988).

Chapter 4 STRING AND CONFORMAL FIELD THEORY

Conditions To Be Satisfied By String Field Theory And Third Quantized Gravity

Hirotaka Sugawara

KEK, National Laboratory for High Energy Physics
Tsukuba, Ibaraki, Japan

and

Department of Physics, University of Tokyo
Tokyo, Japan

Abstract: Conditions to be satisfied by the vertex operator for the closed string field theory are derived. They are the consequence of the Jacobi identify for the string field vertex. BRS invariance gives other conditions but they can be satisfied by the conventional method. Modular invariance is awkward to be formulated in string field theory. Application to third quantization of gravity is suggested.

1. Introduction

The difficulty of formulating string field theory stems from the limiting procedure where the Riemannian surface becomes infinitesimally narrow (fig. 1). The continous diffeomorphism retains its trace because the BRS operator can be defined even on the infinitesimally narrow surface. But the global transformation (modular transformation) cannot be defined on it.

The product of string fields is defined to satisfy either the associativity (open string) or the derivation condition (closed string)[1]. These properties, together with BRS invariance, guarantee the gauge invariance of the string field theory. Unfortunately, however, these conditions do not seem to guarantee modular invariance.[2]

Can we, then, forget about the string field theory entirely? Unfortunately, so far we have no better way. Take, for example, the problem of space time compactification. String field theory provides us the gap equation

$$Q\Psi + g\Psi * \Psi = 0 \; , \tag{1}$$

where Ψ is the string field and Q is the BRS operator. Although equation (1) states nothing but the conformal invariance of the string theory, it requires the specific form of Q and a definition of the $*$ product. The conformally invariant solutions could be classified by studying the representation of Virasoro algebra[3] but we would certainly need the gap equation such as equation (1) to actually obtain the string ground state.

Let us now turn to the problem of quantum gravity. String theory surely provides us one way of quantizing the gravity. This may actually be the only consistent theory of <u>second</u> quantized gravity. But a physically more interesting approach to quantum gravity seems to be provided by quantizing the Wheeler-DeWitt equation[4]. This corresponds to the <u>third</u> quantized gravity[5]. The action for the third quantized gravity will be given by:

$$S = \langle\Psi|\, Q \,|\Psi\rangle + g\langle\Psi, \Psi * \Psi\rangle \tag{2}$$

where Q is the gravitational BRS operator. The problem is to define a $*$ product that satisfies the condition of making equation (2) gauge invariant. The purpose of the present work is to get some insight into the method of formulating the third quantized gravity using string theory as a model. The physics content of the third quantized gravity is characterized by the fact that it describes the states that correspond to either a single elementary particle or the entire universe on an non-discriminatory basis. All existing gauge interactions are hopefully put into the single $g\langle\Psi, \Psi * \Psi\rangle$ term in equation (2).

2. Third quantized gravity and the string field

We start with the following set of field theory axioms.

(1) Let Ψ_1 and Ψ_2 be the field variables and the $*$ product be defined such that:

$$\Psi_1 * \Psi_2 = (-1)^{N(\Psi_1)N(\Psi_2)+1}\Psi_2 * \Psi_1 \tag{3}$$

where $N(\Psi)$ = even or odd (Graßmann number).

(2) There exists a nil-potent BRS operator Q which satisfies the derivation condition :

$$Q(\Psi_1 * \Psi_2) = Q(\Psi_1) * \Psi_2 + (-1)^{N(\Psi_1)N(Q)}\Psi_1 * Q(\Psi_2) \tag{4}$$

(3) adj_Ψ is also a derivation:

$$adj_\Psi(\Psi_1 * \Psi_2) = adj_\Psi(\Psi_1) * \Psi_2 + (-1)^{N(\Psi_1)N(\Psi)}\Psi_1 * adj_\Psi(\Psi_2)\ , \tag{5}$$

where adj_Ψ is defined to be

$$adj_\Psi(\Psi_1) = \Psi * \Psi_1 \tag{6}$$

(4) The inner product is defined for the pair Ψ_1 and Ψ_2 . Q and adj_Ψ are

hermitian operators in terms of this inner product:

$$(Q\Psi_1, \Psi_2) = (\Psi_1, Q\Psi_2) \,, \tag{7}$$

and

$$(adj_\Psi(\Psi_1), \Psi_2) = (-1)^{N(\Psi_1)}(\Psi_1, adj_\Psi(\Psi_2)) \,. \tag{8}$$

The above set of axioms is enough to guarantee the invariance of the action

$$\mathcal{L} = (\Psi, Q\Psi) + g(\Psi, \Psi * \Psi) \,, \tag{9}$$

under the gauge transformation

$$\Psi \to \Psi + Q\xi + f\Psi * \xi \,. \tag{10}$$

Depending on what we take as a BRS operator Q and the appropriate $*$ product, we get the corresponding gauge theory.

One example is provided by the gravitational BRS operator:

$$\begin{aligned} Q = \int d^3x(\chi_i(x)c^i(x) + H(x)c(x)) \\ - i\bar{c}_j(x)c^i(x)\partial_i c^j(x) - \frac{i}{2}\bar{c}(x)c^i(x)\partial_i c(x) \\ - i\gamma^{ij}(x)\bar{c}_j(x)c(x)\partial_i c(x) \,, \end{aligned} \tag{11}$$

where $H(x)$ and $\chi_i(x)$ $(i = 1, 2, 3)$ are local energy and momentum operators satisfying[4]:

$$[\chi_i(x), \chi_j(x')] = i\chi_j(x)\partial_i\delta(x - x') - i\chi_i(x')\partial'_j\delta(x' - x), \tag{12a}$$

$$[\chi_i(x), H(x')] = iH(x)\partial_i\delta(x - x'), \tag{12b}$$

and

$$[H(x), H(x')] = i(\gamma^{ij}(x)\chi_j(x) + \gamma^{ij}(x')\chi_j(x'))\partial_i\delta(x - x'), \tag{12c}$$

where ∂_i' stands for the derivative with respect to x' and $\gamma^{ij}(x)$ is the space-space component of the metric tensor. $c(x)(\bar{c}(x))$ is the ghost field (anti-ghost field) corresponding to the energy and $c^i(x)(\bar{c}_i(x))$ is the ghost (anti-ghost) field corresponding to the momentum. To define an appropriate $*$ product that satisfies the set of axioms we start from the BRS invariant Lagrangian which gives an explicit form for $H(x)$ and $\chi_i(x)$ in terms of the gravitational and the matter fields. We then calculate the following object:

$$V = lim \int_M e^{i\int Ld^4x} \prod dg^{\mu\nu} d\Psi \ , \tag{13}$$

where M is the four dimensional manifold with three distinct boundaries. (see fig. 2). The limit is taken in such a way that it satisfies in particular axiom (3). The equation (13) in the operator formalism will have the form

$$V = e^D \ , \tag{14}$$

where D is the quadratic form:

$$D = \int d^3\xi f_{ij}^{ab}(\xi)\Psi_a^{(i)}(\xi)\Psi_b^{(j)}(\xi) \ , \tag{15}$$

with $\Psi_a^{(i)}(\xi)$ standing for the field defined on the i-th three dimensional boundary. The suffix 'a ' stands for the field species including gravity. The coefficient $f_{ij}^{ab}(\xi)$ depends on the way the limit is taken. We are interested in what conditions the axioms place on $f_{ij}^{ab}(\xi)$ in order for the axioms to be satisfied. The commutation (or the anti-commutation) relations among $\Psi_a^i(\xi)$ depend on how we take the limit. Working directly with the gravity requires understanding of how we take into account the matter field, what the possible limiting procedures are and, eventually, what is meant by the modular invariance[6]. In this paper I would like to give a detailed discussion of only string theory, in this paper as a preparation for the future study of the gravitational theory.

In string theory the BRS operator is given by

$$Q = \int L(\sigma)c(\sigma)d\sigma + i\int \bar{c}(\sigma)\partial c(\sigma)c(\sigma)d\sigma \ , \tag{16}$$

where $c(\sigma)$, $(\bar{c}(\sigma))$ is the ghost (anti-ghost) field and $L(\sigma)$ is the Virasoro operator with the commutation relation

$$\begin{aligned}[L(\sigma), L(\sigma')] =& 2\pi i\delta'(\sigma-\sigma')(L(\sigma)+L(\sigma')) \\ &- 2\pi i\frac{N}{12}(\delta'(\sigma-\sigma')+\delta'''(\sigma-\sigma')) \ .\end{aligned} \tag{17}$$

N is known to be fixed for $Q^2 = 0$[7]. For simplicity let us consider the case when $L(\sigma)$ is given in terms of fermion fields:

$$\mathcal{L}(\sigma) = \frac{i}{2}\Psi_\mu(\sigma)\partial\Psi_\mu(\sigma) \ , \ (\mu = 1, \cdots, 2N) \ , \tag{18}$$

with

$$\{\Psi_\mu(\sigma), \Psi_\nu(\sigma')\} = 2\pi\delta(\sigma-\sigma')\delta_{\mu\nu} \ . \tag{19}$$

For this case the string field can be defined to be an element $|\Psi\rangle$ of the Fermion Fock space. The $*$ product is defined to be

$$|\Psi_1\rangle * |\Psi_2\rangle = \langle 0_2|\, \langle 0_1|\, V(3;2,1)\, |\Psi_1\rangle\, |\Psi_2\rangle\, |0_3\rangle \ , \tag{20}$$

where $|0_i\rangle$ $(i = i, 2, 3)$ denotes the vacuum state of the i-th Fock space. $V(3;2,1)$ is considered to be defined by starting with the BRS invariant action and taking the narrow Riemannian surface limit. Generically, V can be written (except for the factor with odd Graßmann number which is irrelevant for the following discussions) as:

$$V = e^{D_\Psi + D_c} \tag{21}$$

where

$$D_\Psi = \int f_{1\underline{3}}(\sigma', \sigma'')\Psi_\mu^{(1)(-)}(\sigma')\Psi_\mu^{(3)(+)}(\sigma'')d\sigma' d\sigma''$$
$$+ \int f_{11}(\sigma', \sigma'')\Psi_\mu^{(1)(-)}(\sigma')\Psi_\mu^{(1)(-)}(\sigma'')d\sigma' d\sigma''$$
$$+ \int f_{2\underline{3}}(\sigma', \sigma'')\Psi_\mu^{(2)(-)}(\sigma')\Psi_\mu^{(3)(+)}(\sigma'')d\sigma' d\sigma''$$
$$+ \int f_{22}(\sigma', \sigma'')\Psi_\mu^{(2)(-)}(\sigma')\Psi_\mu^{(2)(+)}(\sigma'')d\sigma' d\sigma''$$
$$+ \int f_{\underline{33}}(\sigma', \sigma'')\Psi_\mu^{(3)(+)}(\sigma')\Psi_\mu^{(3)(+)}(\sigma'')d\sigma' d\sigma''$$
$$+ \int f_{12}(\sigma', \sigma'')\Psi_\mu^{(1)(-)}(\sigma')\Psi_\mu^{(2)(-)}(\sigma'')d\sigma' d\sigma'' , \qquad (22)$$

with $\Psi_\mu^{(i)(\pm)}$ representing the positive or the negative frequency part of the i-th fermion. The derivation property of the BRS operator (Axiom(2)) is assumed to be valid because $f_{ij}(\sigma', \sigma'')$ are obtained from the BRS invariant Lagrangian. Our main task is to obtain those conditions on the coefficients f_{ij} that guarantee axiom (3). Ghost part D_c has the similar form:

$$D_c = \int g_{1\underline{3}}(\sigma', \sigma'')c^{(1)(-)}(\sigma')\bar{c}^{(3)(+)}(\sigma'')d\sigma' d\sigma''$$
$$+ \int g_{2\underline{3}}(\sigma', \sigma'')c^{(2)(-)}(\sigma')\bar{c}^{(3)(+)}(\sigma'')d\sigma' d\sigma''$$
$$+ \int g_{3\underline{3}}(\sigma', \sigma'')c^{(3)(+)}(\sigma')\bar{c}^{(3)(+)}(\sigma'')d\sigma' d\sigma''$$
$$+ \int g_{1\underline{1}}(\sigma', \sigma'')c^{(1)(-)}(\sigma')\bar{c}^{(1)(-)}(\sigma'')d\sigma' d\sigma''$$
$$+ \int g_{2\underline{2}}(\sigma', \sigma'')c^{(2)(-)}(\sigma')\bar{c}^{(2)(-)}(\sigma'')d\sigma' d\sigma''$$
$$+ \int g_{1\underline{2}}(\sigma', \sigma'')c^{(1)(-)}(\sigma')\bar{c}^{(2)(-)}(\sigma'')d\sigma' d\sigma''$$
$$+ \int h_{\underline{1}3}(\sigma', \sigma'')\bar{c}^{(1)(-)}(\sigma')c^{(3)(+)}(\sigma'')d\sigma' d\sigma''$$
$$+ \int h_{\underline{2}3}(\sigma', \sigma'')\bar{c}^{(2)(-)}(\sigma')c^{(3)(+)}(\sigma'')d\sigma' d\sigma'' . \qquad (23)$$

Actually, $V(3;2,1)$ must have a factor containing the odd Grassman number but the conditions for f_{ij}, g_{ij} and h_{ij} decouple from the conditions for this factor. I discuss only the conditions for f, g and h in this paper.

3. **Conditions to be satisfied by f_{ij}, g_{ij} and h_{ij}**

Axiom (3), combined with equation (20), leads to the following identity for $V(i;j,k)$:

$$\begin{aligned}&\langle 0_4|\, V(5;3,4)V(4;2,1)\,|0_4\rangle\\&+\langle 0_4|\, V(5;1,4)V(4;3,2)\,|0_4\rangle\\&+\langle 0_4|\, V(5;2,4)V(4;1,3)\,|0_4\rangle = 0\ ,\end{aligned} \tag{24}$$

where V is given explicitly in equations (21), (22) and (23). Before discussing equation (24) let us see what is imposed on f, g and h from the reality condition stated in axiom (4) (equation (8)). For this purpose let us introduce an operator $V(2,1)$ by the following equation;

$$|\Psi_1\rangle = \langle\Psi_2|\, V(2,1)\,|0_2\rangle\,|0_1\rangle\ . \tag{25}$$

$V(2,1)$ can be written in terms of the field variables $\Psi_\mu(\sigma)$ only when $|\Psi_1\rangle$ and $|\Psi_2\rangle$ satisfy certain reality conditions. The ambiguity of $V(2,1)$ emerges because of the ambiguity of the reality condition. It turns out that the BRS invariance leads to the unique from for the $V(2,1)$:

$$V(2,1) = i\exp(\frac{1}{2\pi}\int \Psi_\mu^{(2)(+)}(-\sigma)\Psi_\mu^{(1)(+)}(\sigma)d\sigma)\ . \tag{26}$$

The phase factor i multiplying the right hand side is somewhat mysterious but without it we can not satisfy axiom(2).

With this definition of $V(2,1)$, equation (8) can be written as

$$\langle 0_1|\, V(2;3,1)V(4,1)\,|0_1\rangle = V^*(3;2,4)\ . \tag{27}$$

Substituting equations (21), (22), (23) and (26) we obtain;

$$if_{12}(-\sigma,\sigma') = f^*_{1\underline{3}}(\sigma,\sigma') \tag{28a}$$
$$f_{1\underline{3}}(\sigma',\sigma'') = f^*_{1\underline{3}}(\sigma'',\sigma') \tag{28b}$$
$$f_{11}(\sigma',\sigma'') = f^*_{\underline{33}}(\sigma'',\sigma') \tag{28c}$$
$$f^*_{11}(-\sigma',-\sigma'') = f_{11}(\sigma',\sigma'')\ , \tag{28d}$$

and similar equations for the ghost part. We also note that symmetry under the exchange of suffix 1 and suffix 2 in equation(22) leads to:

$$f_{1\underline{3}}(\sigma,\sigma') = f_{2\underline{3}}(\sigma,\sigma'), f_{11}(\sigma,\sigma') = f_{22}(\sigma,\sigma') \text{ and } f_{12}(\sigma,\sigma') = -f_{12}(\sigma',\sigma)\ . \tag{29}$$

To calculate the left hand side of equation (24) we use the coherent states as the intermediate states, *e.g.*we substitute:

$$1 = \int e^{\frac{1}{2\pi}\int Z^{(+)}_\alpha(\sigma)\Psi^{(-)}_\alpha(\sigma)d\sigma}\,|0\rangle\,\langle 0|\,e^{\frac{1}{2\pi}\int Z^{(-)}_\alpha(\sigma)\Psi^{(+)}_\alpha(\sigma)d\sigma}d\Omega \tag{30}$$

between the two V's in equation (24). From now on we discuss only the Ψ part. The ghost part can be treated in a similar method. Let us define the matrix $K_{ij}(\sigma,\sigma')(i,j=1,2)$ by the following equation:

$$\int d\sigma' \begin{pmatrix} \frac{1}{(2\pi)^2}\Delta(\sigma-\sigma'), & 2f_{11}(\sigma,\sigma') \\ -2f_{\underline{33}}(\sigma,\sigma'), & \frac{1}{(2\pi)^2}\Delta^*(\sigma-\sigma') \end{pmatrix} \begin{pmatrix} K_{11}(\sigma',\sigma'') & K_{12}(\sigma',\sigma'') \\ K_{21}(\sigma',\sigma'') & K_{22}(\sigma',\sigma'') \end{pmatrix} - \begin{pmatrix} \frac{1}{(2\pi)^2}\Delta(\sigma-\sigma'') & 0 \\ 0 & \frac{1}{(2\pi)^2}\Delta^*(\sigma-\sigma'') \end{pmatrix}, \tag{31}$$

where Δ is the propagator of the Ψ field:

$$\Delta(\sigma-\sigma')=\frac{e^{-\frac{i}{2}(\sigma-\sigma')}}{1-e^{-i(\sigma-\sigma'-i\epsilon)}}\,, \tag{32}$$

for the Neveu-Schwarz boundary condition. We must use the Ramond Boundary condition for the ghost field. From equation (31) we obtain

$$K_{11}(\sigma,\sigma'')-4(2\pi)^2\int d\sigma' d\sigma''' f_{11}(\sigma,\sigma')f^*_{11}(\sigma',\sigma''')K_{11}(\sigma''',\sigma'')=\Delta(\sigma-\sigma'') \tag{33}$$

This equation for $K_{11}(\sigma,\sigma'')$ can be at least formally solved by iteration to give:

$$\begin{aligned}K_{11}(\sigma,\sigma'')&=\Delta(\sigma-\sigma'')+(2\pi)(4\pi)^2\int f_{11}(\sigma,\sigma')f^*_{11}(\sigma',\sigma'')d\sigma'\\&+(2\pi)(4\pi)^4\int f_{11}(\sigma,\sigma')f^*_{11}(\sigma',\sigma''')f_{11}(\sigma''',\sigma^{(4)})f^*_{11}(\sigma^{(4)},\sigma'')d\sigma' d\sigma''' d\sigma^{(4)}\\&+(2\pi)(4\pi)^6\int ff^*ff^*ff^*+\cdots\end{aligned} \tag{34}$$

K_{21} is given in terms of K_{11} as follows:

$$K_{21}(\sigma,\sigma'')=-4\pi\int d\sigma' f^*_{11}(\sigma,\sigma')K_{11}(\sigma',\sigma'')\,. \tag{35}$$

We also have the following relations

$$K_{21}(\sigma'',\sigma)=-K_{21}(\sigma,\sigma'')=-K_{12}(-\sigma,-\sigma'')\,, \tag{36a}$$

$$K_{12}(\sigma',\sigma'')=K^*_{21}(\sigma',\sigma'')\,, \tag{36b}$$

and

$$K_{22}(\sigma,\sigma'')=K^*_{11}(\sigma,\sigma'')=K_{11}(\sigma'',\sigma)\,. \tag{36c}$$

With this definition of $K_{ij}(\sigma,\sigma')$ the constraints imposed on $f_{ij}(\sigma,\sigma')$ by axiom (3) can be written as:

$$f_{12}(\sigma'',\sigma''')-\frac{1}{2}\int d\sigma d\sigma'(K_{12}(\sigma,\sigma')-K_{12}(\sigma',\sigma))f_{1\underline{3}}(\sigma'',\sigma)f_{1\underline{3}}(\sigma''',\sigma')=0\,, \tag{37a}$$

and

$$\int d\sigma d\sigma'(K_{11}(\sigma',\sigma)+K_{22}(\sigma,\sigma'))f_{1\underline{3}}(\sigma'',\sigma)f_{1\underline{3}}(\sigma,\sigma''')=0\ . \tag{37b}$$

These two equations are the main results of this work. Unfortunately, they are unexpectedly complicated non-linear integral equations. Directly solving these equations to obtain f_{ij} does not seem to be practical.

4. Discussion

One starts from the BRS invariant Lagrangian and, by taking the appropriate limit, one obtains the coefficient function f_{ij}. The validity of the limiting procedure can be checked by imposing the conditions expressed by equations (37a) and (37b). As is well known, there is an alternative method to obtain the coefficient function, e.g. one solves the so called 'connection condition'. Since the Lagrangian method is basically used in order to guarantee BRS invariance, an alternative method that guarantees BRS invariance is equally acceptable. The 'connection condition' replaces the conservation of the BRS charge by the conservation of other currents in terms of which the BRS charge is written. This method, combined with equations (37a) and (37b), proves to be an adequate procedure from the practical point of view. Take, for example, the Kyoto-type[1] connection condition which in our case turns out to be:

$$-(2\pi)^2 f_{1\underline{3}}(\frac{\sigma}{\alpha_1},\sigma')\frac{1}{\sqrt{\alpha_1}}+2(2\pi)^2 f_{\underline{33}}(\sigma',\frac{\sigma}{\alpha_3})\frac{1}{\sqrt{\alpha_3}}-\frac{i}{\sqrt{\alpha_3}}\Delta^*(\sigma'-\frac{\sigma}{\alpha_3})=0\ . \tag{38}$$

One has to check the compatibility of equation (38) with equations (37a) and (37b). So far I have checked only that if (38) is compatible with (37a), then, it is also compatible with (37b). Equation (38) actually has a unique solution and the compatibility can be checked using this explicit form.

Since Q is given explicitly in terms of Ψ_μ, one may try to satisfy axiom (2) without using the Lagrangian nor the connection condition. I have done this and again obtained a set of complicated non-liniar integral equations such as

$$2(2\pi)^2 \int f_{12}(\sigma,\sigma'')\partial_\sigma f_{11}(\sigma',\sigma)d\sigma' + (2\pi)^2 \int \partial_\sigma f_{12}(\sigma,\sigma'')f_{11}(\sigma',\sigma)d\sigma'$$
$$+ \int \partial_\sigma \Delta^{(-)}(\sigma-\sigma')f_{12}(\sigma,\sigma'')d\sigma' + \int \partial_{\sigma''} f_{12}(\sigma,\sigma'')f_{12}(\sigma',\sigma'')d\sigma' - [(\sigma \leftrightarrow \sigma')] = 0 \,, \tag{39}$$

where $\Delta^{(-)}(\sigma-\sigma')$ means the negative frequency part of $\Delta(\sigma-\sigma')$. Solving these equations seem to be less practical than the Lagrangian or the connection condition method.

So much for string theory. Let us now turn to the third quantized gravity. Unlike the case of string theory, where the linearized version of the field equation $Q\Psi = 0$ is relatively easy to solve, the linearized version of the gravitational field equation cannot be solved easily. $Q\Psi = 0$ in this case is essentially the Wheeler-DeWitt equation[4] $H(x)\Psi = 0$ and $\chi_i(x)\Psi = 0$. Let us briefly discuss the case of pure gravity. In this case we can show that $H(x)$ can be written as:

$$H(x) = \frac{1}{2}\gamma^{-\frac{1}{2}}(J^0(x)J^0(x) - 2J_i^j(x)J_j^i(x)) + \gamma^{\frac{1}{2}}{}^{(3)}R(x) \,, \tag{40}$$

where ${}^{(3)}R(x)$ is the intrinsic 3 dimensional curvature and $\gamma^2 = det\gamma_{ij}$ with γ_{ij} the three dimensional metric tensor. The extrinsic part (the first term) is written in terms of the current densities with the current algebra as:

$$[J^0(x), J^0(y)] = 0, \quad [J^0(x), J_i^j(y)] = 0 \tag{41a}$$

and

$$[J_i^j(x), J_k^l(y)] = \frac{-i}{2}\delta(x-y)(\delta_k^j J_i^l(x) - \delta_i^l J_k^j(x)) \,, \tag{41b}$$

with $J_i^i(x) = 0$. We notice that there are two main differences when we compare the above equations with the case of the Virasoro algebra. First is the appearance of ${}^{(3)}R(x)$ in equation(40). This shows that the finding of a representation of current algebra does not solve equation(40). We must find the representation for $\gamma_{ij}(x)$ to obtain the expression for ${}^{(3)}R(x)$. Secondly, we notice that there appears no central charge in the current algebra, unlike the case of Kac-Moody algebra. We are currently studying the Wheeler-DeWitt equation from the current algebra point of view.

(Acknowledgment)

I would like to thank Professor Miyazawa for his guidance and encouragement for the last 25 years. I deeply miss my student days when professor Miyazawa was a young, active and extremely bright teacher. I learned every branch of physics from him except for gravity. Maybe this is why I am trying to learn gravity all by myself now. Finally I would like to thank K. Higashijima and K. Ogawa for valuable discussions.

(References)

1. E.Witten Nucl.Phys. B276(1986)291
   H.Hata, K.Itoh, T.Kugo, H.Kunitomo and K.Ogawa
   Phys.Rev. D34(1987)2360
   A.Neveu and P.C.West, Phys.Lett. 168B(1986)192
2. B.Zwiebach, MIT Preprint(1988)
3. D.Gepner, Nucl.Phys. B296(1988)757
4. B.S.DeWitt, Phys.Rev. 160(1967)1113
5. S.W.Hawking, Phys.Rev. D37(1988)904
   S.Giddings and A.Strominger, (Harvard Preprint)HUTP-87/A067
   S.Coleman, (Harvard Preprint)HUTP-88/A008
   T.Banks, (Stanford Preprint)SCIPP 88/09
6. In this connection see
   E.Witten, (Princeton Preprint)IAS-PUB-IASSNS-HEP-87/77
7. M.Kato and K. Ogawa, Nucl.Phys. B212(1983)443

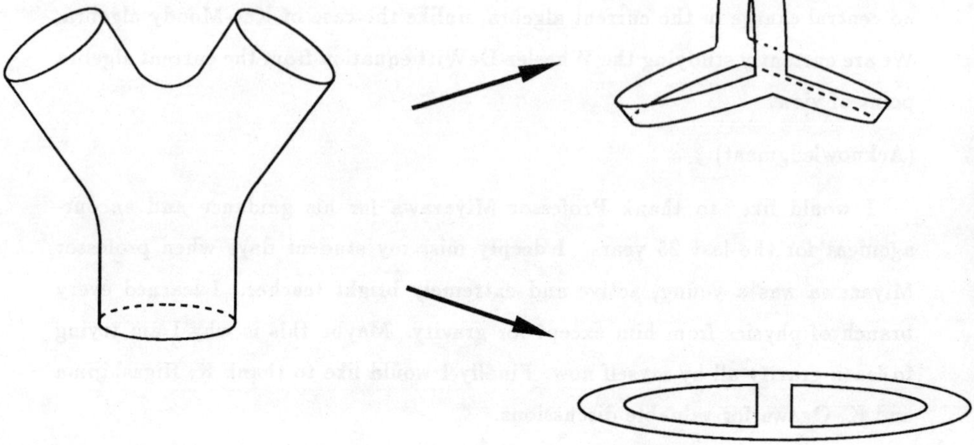

fig 1. Narrowing of Riemann Surface

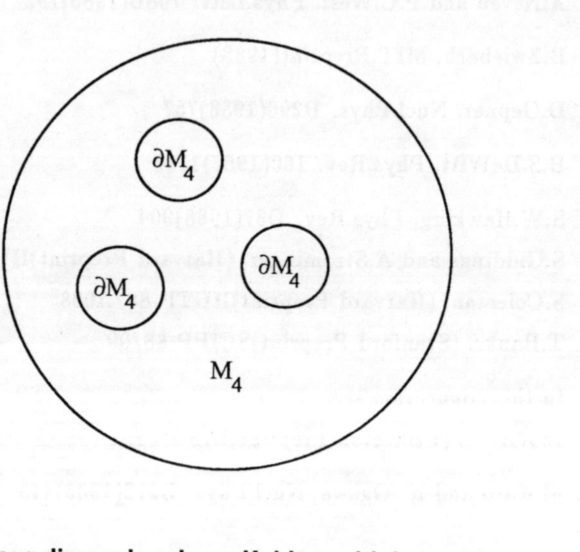

fig 2. Four dimensional manifold M_4 with boundary

Generation of OSp(9,1|2) × OSp(3,3|6) Affine Superalgebra from the Bosonic String

Z.F. Ezawa
Department of Physics, Tohoku University, Sendai 980, Japan

ABSTRACT
The compactification of the D=26 closed bosonic string into D=10 superstrings is analyzed in the covariant formalism. A mechanism is proposed for the emergence of the OSp(9,1|2) × OSp(3,3|6) affine superalgebra together with negative metric states from the "Euclidean" compactified bosonic degrees of freedom.

I. Introduction

When I was a graduate student at the University of Tokyo, I wrote several papers on the parton model under supervision of Professor Hironari Miyazawa. At that time the parton was the ultimate object in particle physics. Time has passed since then. Now, the ultimate object in particle physics is the superstring: all the particles -- quarks, gluon, leptons, gravitons, etc. -- are believed to be created from the 10-dimensional superstring. Moreover, the superstring itself may well be generated from the 26-dimensional bosonic string as a result of spontaneous compactificaton.

In this paper I present a mechanism by way of which the superstring can arise as a solution of the purely bosonic string theory. In this compactification scheme, from 16 "Euclidean" bosonic fields, 2×8 fermionic vector fields and 8 symplectic bosons are generated and they make up an $\hat{O}Sp(9,1|2) \times \hat{O}Sp(3,3|6)$ supermultiplet. Here, the OSp(9,1|2) multiplet contains the Neveu-Schwarz-Ramond (NSR) fields and the superghosts, while the OSp(3,3|6) multiplet contains the extra fields. The contribution to the conformal anomalies is -16 from the OSp(9,1|2) multiplet, but it is zero from the OSp(3,3|6) multiplet. The extra fields do not make any contribution to the conformal anomalies as should be the case. The existence of the extra fields is potentially dangerous because they create negative-norm states and may break the unitarity. The decoupling mechanism of these extra fields is formulated in analogy with the no-ghost theorem[2] in the covariant string theory.

II. Compactification of Bosonic String

In the conformal gauge the D = 26 free closed bosonic string theory is characterized by the string coordinate $X^{\hat{\mu}}(\tau,\sigma)$,

$$X^{\hat{\mu}}(\tau,\sigma) = Q^{\hat{\mu}} + P^{\hat{\mu}}\tau + 2w^{\hat{\mu}}\sigma$$

$$+ \frac{i}{2}\sum_{n\neq 0}\frac{1}{n}[\alpha^{\hat{\mu}}_{Ln}e^{-2in(\tau+\sigma)} + \alpha^{\hat{\mu}}_{Rn}e^{-2in(\tau-\sigma)}], \qquad (2.1)$$

obeying the free field equation. Here, the indices $\hat{\mu}$ are separated into the space-time indices $\hat{\mu} = \mu$ and the internal indices $\hat{\mu} = I$ to be compactified.

After compactification, two independent 16-component Fubini-Veneziano (FV) fields appear as quantum operators[3]

$$\Phi^I(z) = q^I - ip^I \log z + i\sum_{n\neq 0}\frac{1}{n}\alpha^I_n z^{-n}, \qquad (2.2)$$

where z is a complex number, I = 1,2,...,16, with

$$[q^I, p^J] = i\delta^{IJ}, \qquad [\alpha^I_n, \alpha^J_m] = n\delta(n+m)\delta^{IJ}. \qquad (2.3)$$

Here, we have suppressed the index L or R. Compactification should occur onto an even self-dual lattice for the modular invariance of the one-loop amplitudes[4]. The eigenvalue of momentum p^I may take values on the root lattice of group G, where G is either $E_8 \times E_8$ or

$Spin(32)/Z_2$. Then, from one set of the FV fields, the gauge group $E_8 \times E_8$ may be generated.[5]

We wish to use the other set of 16-component FV fields to create the space-time fermions and the superghosts. In so doing, it is crucial to generate negative-norm states from the compactified bosonic degrees of freedom. For this purpose we observe the facts that (i) the hermiticity condition on internal coordinate q^I may be modified as in (3.6a) below, and that (ii) the internal momentum p^I may be allowed to take a pure imaginary eigenvalue as in (2.4) below. Both of these lead to the introduction of negative metric states without spoiling the self-consistency of the formalism.

We propose that the internal 16 coordinates of the bosonic string are compactified onto the lattice $\Lambda = \Lambda_1 \cup \Lambda_2$ which is even and self-dual, where

$$\Lambda_1 = (\Lambda_S^8, \Lambda_S^4, \tilde{\Lambda}_C^4) \cup (\Lambda_C^8, \Lambda_S^4, \tilde{\Lambda}_S^4) \cup (\Lambda_V^8, \Lambda_C^4, \tilde{\Lambda}_R^4) \cup (\Lambda_R^8, \Lambda_C^4, \tilde{\Lambda}_V^4), \quad (2.4a)$$

$$\Lambda_2 = (\Lambda_R^8, \Lambda_R^4, \tilde{\Lambda}_R^4) \cup (\Lambda_V^8, \Lambda_R^4, \tilde{\Lambda}_V^4) \cup (\Lambda_S^8, \Lambda_V^4, \tilde{\Lambda}_S^4) \cup (\Lambda_C^8, \Lambda_V^4, \tilde{\Lambda}_C^4). \quad (2.4b)$$

Thus, 4 internal momenta are assumed to take pure imaginary eigenvalues; these are associated with the superghosts and the extra fields, as we shall see. Here, Λ_j^p stands for the sets of weight vectors of SO(2p) corresponding to the four congruence classes; Λ_R^p for the root, Λ_V^p for the vector-weight, Λ_S^p and Λ_C^p for two kinds of the spinor-weight: on the other hand, $\tilde{\Lambda}_j^4$ denotes that the lattice components are pure imaginary.

III. Neveu-Schwarz-Ramond Fields and Superghosts

[3.1] SO(2a,2b) Affine Algebra and Fermionic Vector Fields

Let us construct the SO(2a,2b) affine algebra and its vector representation based on the (a+b)-component FV fields (2.2). The index I of the FV field runs from I = 1 to I = a+b.

The SO(2a,2b) affine algebra is defined by

$$[M_m^{\mu\nu}, M_n^{\rho\sigma}] = i(g^{\mu\rho}M_{m+n}^{\nu\sigma} + g^{\nu\sigma}M_{m+n}^{\mu\rho} - g^{\mu\sigma}M_{m+n}^{\nu\rho} - g^{\nu\rho}M_{m+n}^{\mu\sigma})$$

$$+ m(g^{\mu\rho}g^{\nu\sigma} - g^{\nu\rho}g^{\mu\sigma})\delta(m+n), \qquad (3.1)$$

where the index m is an integer and $M_n^{\mu\nu} = -M_n^{\nu\mu}$. The metric $g_{\mu\nu}$ involves 2a of 1's and 2b of -1's. We also consider fermionic vector fields ψ_m^μ transforming as

$$[M_m^{\mu\nu}, \psi_n^\rho] = i(g^{\mu\rho}\psi_{m+n}^\nu - g^{\nu\rho}\psi_{m+n}^\mu), \qquad (3.2)$$

with

$$\{\psi_m^\mu, \psi_n^\nu\} = g^{\mu\nu}\delta(m+n), \qquad (3.3)$$

where the index m of ψ_m^μ is an integer (half-integer) for the R (NS) sector. These operators should obey the hermiticity: $(M_m^{\mu\nu})^\dagger = M_{-m}^{\mu\nu}$, $(\psi_m^\mu)^\dagger = \psi_{-m}^\mu$. Because generators $M_n^{\mu\nu}$ can be expressed as

$$M_n^{\mu\nu} = -i \sum {:}\psi_{n+m}^\mu \psi_{-m}^\nu{:} \qquad (3.4)$$

it is enough to construct fermionic field ψ_n^μ obeying (3.3) as vertex operators from the FV field (2.2).

The basic vertex operator is

$$A_m^{\boldsymbol{\lambda}} = \frac{1}{2\pi i}\int \frac{dz}{z}\, :\exp\{i\boldsymbol{\lambda}\cdot\boldsymbol{\Phi}(z)\}:z^m \qquad (3.5)$$

where $\boldsymbol{\lambda}$ is a (a+b)-component vector taking values on the weight lattices of SO(2a+2b). The integration contour encircles the origin once; the index m is an integer or a half-integer. We now make the following hermiticity assignment:

$$(q^I)^\dagger = q^I + \pi \qquad \text{for } I = 1,2,\ldots,b \qquad (3.6a)$$

$$(q^I)^\dagger = q^I \qquad \text{for } I = b+1,b+2,\ldots,a+b. \qquad (3.6b)$$

This is consistent with (2.3) and $(p^I)^\dagger = p^I$ since q^I is defined only modulo 2π. Then, it follows that

$$(A_m^{\boldsymbol{\lambda}})^\dagger = \prod_{I=1}^{b} (-1)^{|\lambda^I|} A_{-m}^{-\boldsymbol{\lambda}}, \qquad \boldsymbol{\lambda} = \sum_{I=1}^{a+b} \lambda^I \mathbf{e}^I \quad (\lambda^I = \pm 1), \tag{3.7}$$

where $\mathbf{e}^I$ is the basis vector of Z^{a+b}; $\mathbf{e}^I \cdot \mathbf{e}^J = \delta^{IJ}$.

Choosing a (a+b)-dimensional unit vector $\mathbf{v}$, $\mathbf{v} = (\pm 1, 0, \ldots, 0)$, etc., we set

$$B_n^{\mathbf{v}} = C(\mathbf{v}) \prod_{I=1}^{b} i^{|v^I|} A_n^{\mathbf{v}} \tag{3.8}$$

with $\mathbf{v} = \sum_{I=1}^{a+b} v^I \mathbf{e}^I$, $v^I = \pm 1$. Here, $C(\mathbf{v})$ is precisely the cocycle used to construct the SO(2a+2b) affine algebra, and its existence is well established[6,7]. The hermiticity condition is

$$C(\mathbf{v})^\dagger = C(-\mathbf{v}), \tag{3.9}$$

which yields $(B_n^{\mathbf{v}})^\dagger = B_{-n}^{-\mathbf{v}}$.

It is straightforward to prove that[8]

$$\{B_m^{\mathbf{v}}, B_n^{\mathbf{u}}\} = \theta(\mathbf{v},\mathbf{u})\delta(\mathbf{v}+\mathbf{u})\delta(m+n) \tag{3.10}$$

with

$$\theta(\mathbf{v},\mathbf{u}) = (-1)^N, \quad N = \frac{1}{2} \sum_{I=1}^{b} (|u^I + v^I| - |u^I| - |v^I|). \tag{3.11}$$

Then, by defining

$$\psi_m^{2I-1} = (B_n^{\mathbf{e}^I} + B_n^{-\mathbf{e}^I})/\sqrt{2}, \qquad \psi_m^{2I} = i(B_n^{\mathbf{e}^I} - B_n^{-\mathbf{e}^I})/\sqrt{2}, \tag{3.12}$$

the commutation relation (3.3) is derived from (3.10).

Consequently, starting with the FV field (2.4), we have constructed the SO(2a,2b) affine algebra together with the SO(2a,2b) fermionic vector fields.

[3.2] Sp(2n) Affine Algebra and Superghost Fields

Let us next construct the Sp(2n) affine algebra based on the FV fields (2.2) in association with the superghosts[9,10].

The Sp(2n) affine algebra is defined by

$$[N_m^{\alpha\beta}, N_n^{\gamma\varepsilon}] = i(J^{\alpha\gamma}N_{m+n}^{\beta\varepsilon} + J^{\beta\varepsilon}N_{m+n}^{\alpha\gamma} + J^{\alpha\varepsilon}N_{m+n}^{\beta\gamma} + J^{\beta\gamma}N_{m+n}^{\alpha\varepsilon})$$

$$- m(J^{\alpha\gamma}J^{\beta\varepsilon} + J^{\beta\gamma}J^{\alpha\varepsilon})\delta(m+n), \tag{3.13}$$

where the index m is an integer, and $N_m^{\alpha\beta} = N_m^{\beta\alpha}$; $J^{\alpha\beta}$ is the block diagonal matrix with n copies of

$$\begin{vmatrix} 0 & 1 \\ -1 & 0 \end{vmatrix}.$$

We also consider symplectic boson fields ζ_m^μ transforming as

$$[N_m^{\alpha\beta}, \zeta_n^\gamma] = i(J^{\alpha\gamma}\zeta_{m+n}^\beta + J^{\beta\gamma}\zeta_{m+n}^\alpha), \tag{3.14}$$

with

$$[\zeta_m^\alpha, \zeta_n^\beta] = iJ^{\alpha\beta}\delta(m+n), \tag{3.15}$$

where the index m of ζ_m^α is an integer (half-integer) for the R (NS) sector. These operators should obey the hermiticity $(N_m^{\alpha\beta})^\dagger = N_{-m}^{\alpha\beta}$, $(\zeta_m^\alpha)^\dagger = \zeta_{-m}^\alpha$. It is easy to see that generators $N_m^{\alpha\beta}$ are expressed as

$$N_m^{\alpha\beta} = \sum \mathop{\circ}\limits^{\circ}\zeta_{m+n}^\alpha\zeta_{-n}^\beta\mathop{\circ}\limits^{\circ} \tag{3.16}$$

in terms of the symplectic bosons.

The system of the superghosts is defined by a set of bosonic oscillators satisfying

$$[e_m,e_n] = [\bar{e}_m,\bar{e}_n] = 0, \qquad [e_m,\bar{e}_n] = \delta_{m+n,0} \tag{3.17}$$

with $e_m^\dagger = e_{-m}$, $\bar{e}_m^\dagger = -\bar{e}_{-m}$. Here the index m of oscillators e_m and $\bar{e}_m$ are integers (half-integers) for the R (NS) sector. Therefore, by setting $(\zeta_n^1,\zeta_m^2) = (e_n,i\bar{e}_m)$, the superghosts are found to be symplectic bosons and generate the Sp(2) affine algebra. Similarly, p sets of pseudo-superghosts $(e_n^k,i\bar{e}_m^k)$, $k = 1,\dots,p$, obeying the same commutation relation as (3.17) generate the Sp(2p) affine algebra.

Now, the superghost fields are constructed from the FV fields as follows. We start with the vertex operators:

$$E(z) = h(z)\eta(z), \qquad \bar{E}(z) = \bar{h}(z)z\partial_z\bar{\eta}(z). \tag{3.18}$$

where

$$h(z) = :\exp\{\Phi^I(z)\}:, \qquad \bar{h}(z) = :\exp\{-\Phi^I(z)\}:, \tag{3.19}$$

with $\Phi^I(z)$ being one component of the FV fields (2.2). Similarly, three sets of operators $(h^k, \bar{h}^k)$ are defined. On the other hand, to construct fields $\eta(z)$ and $\bar{\eta}(z)$, we create the SO(4,4) fermionic vector fields $\hat{\psi}_m^\mu$ based on 4 components of the FV fields (2.2); the index m is chosen to be integer and $\mu = 1 \sim 6$. They obey

$$\{\hat{\psi}_m^\mu, \hat{\psi}_n^\nu\} = \hat{g}^{\mu\nu}\delta(m+n), \qquad \hat{g}^{\mu\nu} = \mathrm{diag.}(-1,-1,-1,-1,1,1,1,1), \tag{3.20}$$

When we make combinations

$$\eta_m - (\hat{\psi}_m^1 + \hat{\psi}_m^6)/\sqrt{2}, \qquad \bar{\eta}_m = (-\hat{\psi}_m^1 + \hat{\psi}_m^6)/\sqrt{2}, \tag{3.21a}$$

$$\eta_m^1 = (\hat{\psi}_m^2 + \hat{\psi}_m^5)/\sqrt{2}, \qquad \bar{\eta}_m^1 = (-\hat{\psi}_m^2 + \hat{\psi}_m^5)/\sqrt{2}, \tag{3.21b}$$

and similar ones for $(\eta^2_m, \bar{\eta}^2_m)$ and $(\eta^3_m, \bar{\eta}^3_m)$ from $\hat{\psi}^3_m, \hat{\psi}^4_m, \hat{\psi}^7_m, \hat{\psi}^8_m$, they satisfy

$$\{\eta_m, \eta_n\} = \{\bar{\eta}_m, \bar{\eta}_n\} = 0, \qquad \{\eta_m, \bar{\eta}_n\} = \delta(m+n), \tag{3.22}$$

with $(\eta_m)^\dagger = \eta_{-m}$, $(\bar{\eta}_m)^\dagger = \bar{\eta}_{-m}$; similar relations hold also for η^k_m and $\bar{\eta}^k_n$. The fields $\eta(z) = \sum \eta_m z^{-m}$, $\bar{\eta}(z) = \sum \bar{\eta}_m z^{-m}$, etc., are the desired operators.

It can be proved that the oscillators

$$e_m = i(\eta_0+\bar{\eta}_0)E_m - i\bar{\eta}_0\eta_0 h_m, \qquad \bar{e}_m = i(\eta_0+\bar{\eta}_0)\bar{E}_m, \tag{3.23}$$

with

$$E_m = \int \frac{dz}{2\pi i z} E(z)z^m, \qquad \bar{E}_m = \int \frac{dz}{2\pi i z} \bar{E}(z)z^m, \tag{3.24}$$

satisfy the commutation relations (3.17) and the hermiticity of the superghosts. These are precisely the superghosts which we want. Similar construction is made for the pseudo-superghosts e^k and $\bar{e}^{-k}$ from h^k and η^k.

It is important to note that the ghost number charge is given by $j^3_0 = ip^I$ with (3.19). Hence, the eigenvalue of the zero mode p^I takes a pure imaginary value though it is a hermitian operator. This implies a compactification of the internal coordinates onto a lattice such as (2.4) that contains pure imaginary components.

[3.3] Identification of Fields and OSp(9,1|2) × OSp(3,3|6) Affine Superalgebra

After compactification, we obtain the 16-component FV fields (2.2) and hence 32 vertex operators (3.5) with 32 unit vectors $\boldsymbol{\lambda}$ such that $\boldsymbol{\lambda}\cdot\boldsymbol{\lambda}^* = 1$. The vector $\boldsymbol{\lambda}$ are assumed to contain four imaginary components. According to the lattice structure (2.4), we group 32 unit vectors $\boldsymbol{\lambda}$ as follows:

$$\boldsymbol{\lambda}(\psi) \ : \ (\mathbf{v}^8;\mathbf{0}^4;\mathbf{0}^4), \tag{3.25a}$$

$$\boldsymbol{\lambda}(\eta) \ : \ (\mathbf{0}^8;\mathbf{v}^4;\mathbf{0}^4), \tag{3.25b}$$

$$\boldsymbol{\lambda}(h) \ : \ (\mathbf{0}^8;\mathbf{0}^4;i\mathbf{v}^4). \tag{3.25c}$$

Here, $\mathbf{v}^n$ is a n-dimensional unit vector, $\mathbf{v}^n = (\pm 1,0,\ldots,0,0)$, etc., and $\mathbf{0}^n$ is the n-dimensional zero vector, $\mathbf{0}^n = (0,\ldots,0)$.

We use the vectors $\boldsymbol{\lambda}(\psi)$ to create the SO(12,4) fermionic vector fields; from them we extract the SO(9,1) vector fields $\psi^\mu(z)$ to be identified with the NSR fields ($\mu = 0 \sim 9$), and the SO(3,3) vector fields $\psi^{(k)\mu}(z)$ which are to be decoupled ($k = 1,2,3$; $\mu = 0,1$).

The vectors $\boldsymbol{\lambda}(\eta)$ are used to create the SO(4,4) fermionic vector fields, which are combined into $\eta(z)$, $\bar{\eta}(z)$, $\eta^k(z)$, $\bar{\eta}^k(z)$ ($k = 1,2,3$) as described in subsection 3.2.

The vectors $\boldsymbol{\lambda}(h)$ are used to create the field $h(z)$ and $h^k(z)$ as described also in subsection 3.2. We then combine $h(z)$ and $\eta(z)$ to define the superghost $e(z)$ and $\bar{e}(z)$; similarly, we define pseudo-superghost $e^k(z)$ and $\bar{e}^k(z)$, k=1,2,3.

It is notable that the combined system $(\psi^\mu;e,\bar{e})$ of the NSR fields and the superghosts generates the OSp(9,1|2) affine superalgebra and that the the combined system $(\psi^{(k)\mu};e^k,\bar{e}^k)$ of the extra fields generates the OSp(3,3|6) affine superalgebra. The generators are given by (3.4), (3.16) and

$$S_n^{\mu\alpha} = \sum \psi^\mu_{n+m}\zeta^\alpha_{-m}. \tag{3.26}$$

Note that

$$[S_m^{\mu\alpha},\ \psi_n^\nu] = g^{\mu\nu}\zeta^\alpha_{m+n}, \qquad [S_m^{\mu\alpha},\ \zeta_n^\beta] = J^{\alpha\beta}\psi^\mu_{m+n}. \tag{3.27}$$

The affine superalgebra is described by (3.1) and (3.13) together with

$$[M_m^{\mu\nu},\ N_n^{\alpha\beta}] = 0, \tag{3.28a}$$

$$[M_m^{\mu\nu}, S_n^{\rho\alpha}] = i(g^{\mu\rho}S_{m+n}^{\nu\alpha} - g^{\nu\rho}S_{m+n}^{\mu\alpha}), \tag{3.28b}$$

$$[N_m^{\alpha\beta}, S_n^{\mu\gamma}] = i(J^{\alpha\gamma}S_{m+n}^{\mu\beta} + J^{\beta\gamma}S_{m+n}^{\mu\alpha}), \tag{3.28c}$$

$$\{S_m^{\mu\alpha}, S_n^{\nu\beta}\} = (g^{\mu\nu}N_{m+n}^{\alpha\beta} - J^{\alpha\beta}M_{m+n}^{\mu\nu}) + img^{\mu\nu}J^{\alpha\beta}\delta(m+n). \tag{3.28d}$$

Consequently, we have presented a new compactification scheme of the internal bosonic coordinates that yields an OSp(9,1|2) × OSp(3,3|6) multiplet fields together with the associated affine superalgebra.

IV. Virasoro Generators

Compactification would occur due to a condensation of string fields, which would make a contribution ΔL_m to the Virasoro generators L_m^{BOSON} of the D=26 bosonic string, and we would have

$$L_m^{BOSON} + \Delta L_m = L_m^{NSR} + L_m^{EXTRA}. \tag{4.1}$$

Here, L_m^{BOSON} contain the terms $L(c)_m$ for the conformal ghosts; L_m^{NSR} are composed of the Virasoro generators $L(\psi)_m$ of the SO(9,1) NSR field ψ, $L(e)_m$ of the superghost e and $\bar{e}$, $L(X)_m$ of the SO(9,1) bosonic matter fields and the $L(c)_m$ of the conformal ghost fields; L_m^{EXTRA} are Virasoro generators of the extra fields which are to be decoupled. Both L_m^{NSR} and L_m^{BOSON} are free from anomaly terms, and it is reasonable to assume that L_m^{EXTRA} should also be free from anomaly. In this paper we postulate that the extra fields mimic the NSR fields and the superghost fields. Then, L_m^{EXTRA} are composed of three sets of the Virasoro generators $L(\psi^{(k)})_m$ of the SO(1,1) NSR field $\psi^{(k)\mu}$ and $L(e^k)_m$ of the superghost e^k and $\bar{e}^k$.

The Virasoro generators of the SO(p,q) fermionic vector fields are given by

$$L(\psi)_m = \frac{1}{2}\sum(\frac{1}{2}m+k):\psi_{-k}^{\mu}\psi_{m+k}^{\nu}:\eta_{\mu\nu} + A^{\psi}\delta_{m,0}, \tag{4.2}$$

with the c-number A^{ψ} being

$A^{\psi} = 0$ (NS sector), $A^{\psi} = (p+q)/16$ (R sector).

The Virasoro generators of the superghosts are given by

$$L(e)_m = \sum \{m(J^e-1)-k\}:\bar{e}_{m+k}e_{-k}: + B^e\delta_{m,0}, \tag{4.3}$$

with the c-number B^e being

$$B^e = \frac{1}{2}J^e(J^e-1) + \frac{1}{8} \quad \text{(NS sector)}, \qquad B^e = \frac{1}{2}J^e(J^e-1) \quad \text{(R sector)},$$

where J^e is the conformal weight of the anti-superghost field $\bar{e}(z)$. These c-numbers have been chosen so that the Virasoro algebra contain SL(2,**R**) as a subalgebra without anomaly terms for each $L_m(\psi)$ and $L_m(e)$. The Virasoro algebra for L_m^{NSR} is free from anomaly when J^e = 3/2 with p = 9 and q = 1. Similarly, the combined algebra for $L_m^k = L_m(\psi^{(k)}) + L_m(e^k)$ is free from anomaly if J^e =1/2 with p = q = 1 for each k.

It is straigtforward to evaluate the Virasoro generators explicitly in terms of the FV fields by substituting the Vertex operators. It is notable that we do find that

$$L_0 = L_0^{BOSON} = L_0^{NSR} + L_0^{EXTRA} = \frac{1}{2}\sum_{\mu=0}^{25} :\alpha_k^{\mu}\alpha_{-k\mu}:, \tag{4.4}$$

up to the conformal ghost terms which is common both in L_0^{BOSON} and L_0^{NSR}. There is no contribution from the condensation of the string fields to the zero mode of the Virasoro generators in our scenario. This enable us to discuss the one-loop amplitudes without knowledge of precise dynamics of compactification.

V. Decoupling Mechanism of Extra Fields

Finally, we analyze the decoupling mechanism of the extra fields $\psi^{(k)}$ and e^k, $\bar{e}^k$ (k= 1,2,3). The problem is quite similar to the no-ghost theorem in the covariant string theory[2].

We first define the physical states and show that the norm of such states are positive semidefinite. It is natural to impose that the physical states |phys> do not contain any excitations of the extra fields. This turns out to require that

$$L_0^{EXTRA}|phys\rangle = 0. \tag{5.1}$$

As we see later, (5.1) is equivalent to say that the physical states are invariant under a certain "supercharge"; see (5.4) and (5.5). When we decompose $|phys\rangle = \sum |NSR\rangle|EXTRA\rangle$, (5.1) amounts to

$$L_0^{EXTRA}|EXTRA\rangle = 0. \tag{5.2}$$

Actually, (5.2) is a sufficient condition for the unitarity as far as the extra fields concern. Indeed, it singles out the unique vacuum state |vac> of the non-zero modes of the extra fields in the Fock representation, for which <vac|vac> = 1. With respect to the zero modes, there is a subtlety how to define the norm of the state because all matrix elements become zero due to a peculiar structure[9] of the space of the pseudo-superghosts. However, with a suitable choice of the metric, it is possible to make all the states to have zero-norm except for the Fock vacuum for which the norm is one. Thus, the norms of the state satisfying (5.2) is positive semidefinite, which implies that the unitarity is not violated in the physical subspace defined by (5.1).

We next show that the extra fields do not contribute to the scattering amplitudes of the physical particles. For this purpose, we observe that there is a "supersymmetry" in this system such that one state and its partner have the opposite norms. Let us define

$$Q^k = \sum \varepsilon_{\mu\nu}\psi_{-m}^{(k)\mu}\ \beta_m^{(k)\nu}, \qquad \varepsilon_{01} = -\varepsilon_{10} = 1. \tag{5.3}$$

The following "supersymmetry" algebra is easily checked:

$$\{Q^k, Q^k\} = -2L_0^k, \qquad [L_0^k, Q^k] = 0, \tag{5.4}$$

with L_0^k being the zero mode of the Virasoro generator; $L_0^k = L(\psi^{(k)})_0 + L(e^k)_0$ of the extra fields. Note the minus sign in front of L_0^k in this algebra.

When we define $Q = \sum Q^k$, the physicality condition (5.1) may also be written as

$$Q|\text{phys}\rangle = 0. \tag{5.5}$$

Namely, the physical states are the Q-invariant states.

Unless it acts onto the Q-invariant states, the operator

$$\hat{Q} = \sum Q/(L_0^{\text{EXTRA}})^{1/2} \tag{5.6}$$

is well defined, because the eigenvalues of L_0^{EXTRA} is positive definite for such states. The remarkable properties are

$$\hat{Q}^2 = -1, \quad [L_0^{\text{EXTRA}}, \hat{Q}] = 0, \qquad \hat{Q}^\dagger = \hat{Q}. \tag{5.7}$$

Therefore, except for the Q-invariant states (5.5), all states are paired into a state $|*\rangle$ and its partner $\hat{Q}|*\rangle$, and that they have opposite norms since $\langle *|\hat{Q}^\dagger\hat{Q}|*\rangle = \langle *|\hat{Q}^2|*\rangle = -\langle *|*\rangle$. It is easy to see that the Q-invariant states $|\text{phys}\rangle$ are the vacuum states of the OSp(3,3|6) affine superalgebra : $M_n^{\mu\nu}|\text{phys}\rangle = N_n^{\alpha\beta}|\text{phys}\rangle = S_n^{\mu\beta}|\text{phys}\rangle = 0$ for $n > 0$.

Let us analyze the one-loop correlation function of the vertex operators of the NSR model, which is given by

$$A = \int \prod_j d\tau_j \; \text{TR}\{\exp(i\tau_1 L_0^{\text{BOSON}})V_1 \exp(i\tau_2 L_0^{\text{BOSON}})V_2 \cdots\cdots\cdots V_n\}, \tag{5.8}$$

where we have only displayed explicitly the left-hand side variables which are assumed to compactify. Since L_0^{EXTRA} commutes with the vertex operators V_i and L_0^{NSR}, this amplitude is rewritten as

$$A = \int \prod_j d\tau_j \; TR\{\exp(i\tau_1 L_0^{NSR})V_1 \exp(i\tau_2 L_0^{NSR})V_2 \ldots V_n\} Z^E(q), \qquad (5.9)$$

where

$$Z^{EXTRA}(q) = TR\{\exp(i\textstyle\sum \tau_j L_0^{EXTRA})\}, \qquad q = e^{i\sum\tau_j}, \qquad (5.10)$$

is the partition function for the extra degrees of freedom. Now, we use (5.7) to obtain

$$Z^{EXTRA}(q) = \sum \langle n|S(\tau_j)|n\rangle = -\sum \langle n|\hat{Q}^2 S(\tau_j)|n\rangle$$
$$= -\sum \langle\tilde{n}|S(\tau_j)|\tilde{n}\rangle, \qquad (5.11)$$

up to the Q-invariant states onto which $\hat{Q}$ cannot act, where $S(\tau_j) = \exp(i\sum\tau_j L_0^{EXTRA})$ and $|\tilde{n}\rangle = Q|n\rangle$. Because the summations over the nonzero-mode states $|n\rangle$ and over the nonzero-mode states $|\tilde{n}\rangle$ are identical, it follow that $Z^{EXTRA}(q) = 1$ when a suitable treatment of the zero-mode states is made.

We now consider heuristically multi-loop correlation functions of the vertex operators V_j of the NSR model. To write down such amplitudes we need the multi-string (multi-Reggeon) vertex. Since the multi-string vertex should not mix the different field degrees of freedom, the factorization of the type (5.9) would hold;

$$A = \int \prod d\tau_j \sum \langle m|T(\tau_j)|m\rangle \; Z^{EXTRA}(\tau_j), \qquad (5.12)$$

$$Z^{EXTRA}(\tau_j) = \sum \langle n|S(\tau_j)|n\rangle, \qquad (5.13)$$

where τ_j are the moduli parameters; $T(\tau_j)$ involves the multi-string vertices and the vertex operators V_i of the NSR model; $S(\tau_j)$ involves only the corresponding multi-string vertices of the "extra" strings. If the multi-string vertex of the "extra" strings is invariant under the Q-supercharge as defined above, it will follow that $Z(q) = 1$ precisely as in the one-loop amplitude.

V. Concluding Remarks and Discussions

In the covariant formalism we have presented a new compactification scheme of the 16 components of the D=26 bosonic string. After compactification, we obtain 2×8 fermionic vector fields and 8 symplectic boson fields, which form an OSp(9,1|2) × OSp(3,3|6) supermultiplet. The members of the OSp(9,1|2) supermultiplet are the NSR fields and the superghosts, while those of the OSp(3,3|6) supermultiplet are the extra fields. The decoupling mechanism of the extra fields was studied in analogy with the no-ghost theorem in the covariant string theory.

I would like to dedicate this paper to Professor Hironari Miyazawa on the occasion of his sixtieth birthday. I am thankful to him for his constant guidance and encouragement ever since I was a graduate student. I sincerely hope he will continue to be active and happy in many more years to come.

REFERENCES

1. P.G.O. Freund, Phys.Lett. 151B (1985) 387.
   A. Casher, F. Englert, H. Nicolai and A. Taormina, Phys.Lett. 162B (1985) 121.
   F. Englert, H. Nicolai and A. Schellekens, Nucl.Phys. B274 (1986) 315.
2. P. Goddard and C.B. Thorn, Phys. Lett. 40B (1972) 235;
   R.C. Brower, Phys. Rev. D6 (1972) 1655.
   L. Brink and D. Olive, Nucl. Phys. B58 (1973) 237.
   M. Kato and K. Ogawa, Nucl. Phys. B212 (1983) 443.
   M.D. Freeman and D. Olive, Phys. Lett. 175B (1986) 155.
3. Z.F. Ezawa, S. Nakamura and A. Tezuka, Phys.Rev. D34 (1986) 3805

4. Gross, H. Harvey, E. Martinec and R. Rohm, Phys.Rev.Lett. 54 (1985) 502; Nucl.Phys. B256 (1985) 253; B267 (1986) 75.
F. Englert and A. Neveu, Phys.Lett. 163B (1985) 349.

5. S. Raby and R. Slansky, Phys.Rev.Lett. 56 (1986) 693.

6. J.B. Frenkel and V.G. Kac, Inv.Math. 62 (1980) 23.
G. Segal, Commun.Math.Phys. 80 (1981) 301.
P. Goddard and D. Olive, in Vertex Operators in Mathematics and Physics, MSRI Publ. No 3, ed. J. Lepowsky (Springer Verlag, 1984).
P. Goddard, in Superstrings, Supergravity and Unified Theories, ICTP Ser. Vol 2, ed G. Furlan et al. (World Scientific, 1986).

7. Z.F. Ezawa, S. Nakamura and A. Tezuka, Nucl. Phys. B298 (1988) 241.

8. Z.F. Ezawa, S. Nakamura and A. Tezuka, Phys.Lett. B181 (1986) 76.

9. Z.F. Ezawa, S. Nakamura and A. Tezuka, Nucl. Phys. B291 (1987) 334; Nucl. Phys. B294 (1987) 342.

10. P. Goddard, D. Olive and G. Waterson, preprint Imperial/TP/86-87/15.

11. D. Friedan, S. Shenker and E. Martinec, Phys. Lett. 160B (1985) 55; Nucl. Phys. B271 (1986) 93.
J. Cohn, D. Friedan, Z. Qui and S. Shenker, Nucl. Phys. B278 (1986) 577.

WARD IDENTITIES IN STRING THEORY

Jananadeva MAHARANA+

Laboratoire de Physique Théorique et Hautes Energies
Bât. 211, Université Paris XI, Centre d'Orsay, 91405 Orsay
FRANCE

It is a pleasure and honour for me to participate in the celebration of the sixtieth birth day of Professor Miyazawa. I shall present our investigations of string theory in the BRS Hamiltonian formalism. The superstring theory is considered to be an attractive candidate for the unification of all fundamental forces(1). The requirement of gravitational, Yang-Mills and mixed anomalies cancellations imposes strong constraints on the choice of gauge groups(2). Furthermore, compactification of strings on specific torii has provided a mechanism to implement nonabelian gauge symmetries in the string theory (3) which is an important ingredient in the construction of the heterotic string theory (4). It has been argued that various compactification schemes for the heterotic string expose different vacua of a single underlying theory (5).

It is recognised that the first quantized approach to study the evolution of a string in a generic background field is expected to provide information about several interesting attributes of the theory (6). It might be possible, within this frame work, to understand the mechanism of compactification. In other words, the choice of the backgrounds corresponding to the ground state of the theory can be determined in a self-consistent manner. It provided a simple method to describe scattering processes of string excitations and the coupling constants associated with the string loops. If we include surfaces of non-trivial topology, it leads to topological expansion which is dual and unitary at each order of perturbation theory. The conformal invariance plays a crucial role in determining the configurations of background fields ; since the β−functions associated with the coupling constant (functions) must vanish in order to maintain tracelessness of the two dimensional energy momentum tensor.

We have envisaged the evolution of a closed bosonic string in the presence of its massless excitations [7]. The BRS Hamiltonian formalism is employed in order to construct the S-matrix generating functional using the phase space path integral approach. Subsequently, suitable canonical transformations are introduced, which reveal the local symmetry properties of the generating functional. A set of Ward identities are deduced for the S-matrix elements of the scattering of massless excitations of the closed string.

It is well known that the Lagrangian describing the dynamics of the string exhibits invariance under local symmetry transformations such as reparametrization of the world sheet coordinates and the Weyl rescaling. Consequently, the Lagrangian is a singular one and the theory requires gauge fixing. The constrained Hamiltonian formalism due to Dirac [8] together with the prescriptions of Balatin, Fradkin and Vilkovisky [9] provide a very powerful and elegant tool to quantize the theories with local symmetries. I shall briefly recapitulate some of the essential ingredients of this formalism [10] which are necessary for the Hamiltonian quantization of strings in presence of background fields.

Let us consider a mechanical system described by a singular Lagrangian. Therefore, the Legendre transformation between the generalised velocities and the conjugate momenta is singular one. Constraints manifest themselves in the canonical formulations. The method due to Dirac can be employed in order to obtain all the constraints in a consistent manner. Furthermore, the set of second class constraints can be eliminated through the introduction of Dirac brackets. Let H_0 denote the canonical Hamiltonian and are q^i and p_i, $i = 1\ldots N$, represent the generalised coordinates and the canonical momenta respectively. The set of first class constraints $\{F_\alpha(p_i, q^i)\}$, $\alpha = 1\ldots M < N$ satisfy the following Poisson bracket relations

$$\{H_0, F_\alpha\} = V^\beta{}_\alpha F_\beta \tag{1}$$

$$\{F_\alpha, F_\beta\} = C^\gamma{}_{\alpha\beta}\, \varphi_\gamma \tag{2}$$

The first class constraints $\{F_\alpha\}$ are the generators of the underlying local symmetries of the theory, whereas $C^\gamma{}_{\alpha\beta}$ are the structure constants of the Lie algebra (with Poisson bracket relations (2)) satisfied by the generators (we are considering the case where $V^\beta{}_\alpha$ and $C^\gamma{}_{\alpha\beta}$ do not depend on the phase space variables). The Hamiltonian of the system is not unique and is defined up to linear combination of the first class constraints. Thus the total Hamiltonian is

$$H_T = H_o + \lambda^\alpha F_\alpha \tag{3}$$

where $\{\lambda^\alpha\}$, $\alpha = 1 \ldots M$ are the Lagrangian multipliers. The Hamiltonian action

$$S_H = \int dt\,(p_i \dot{q}^i - H_T) \tag{4}$$

is invariant under infinitesimal transformations generated by the constraints as follows

$$\delta_\varepsilon H_o = \{ H_o, \varepsilon^\alpha(t) F_\alpha \} \tag{5}$$

and

$$\delta_\varepsilon \lambda^\alpha = \dot{\varepsilon}^\alpha(t) + \lambda^\gamma \varepsilon^\beta(t) C^\alpha{}_{\beta\gamma} - \varepsilon^\beta(t) V^\alpha{}_\beta \tag{6}$$

where $\varepsilon^\alpha(t)$ is an infinitesimal time dependent parameter. It is easy to conclude from the Hamilton's equations of motion for p_i and q^i that λ^α appears as arbitrary parameters in these equations. They cannot be determined from the initial boundary conditions on the phase space variables. It is necessary to provide extra conditions (M of them) in order to eliminate the parameters λ^α, and then determine evolutions of p_i and q^i completely. These extra conditions (which are provided from outside) are the gauge fixing conditions. The gauge fixed action, in general, requires additional Faddeev-Popov ghost terms and the full action is no longer invariant under local gauge transformations. However, it is invariant under global BRS transformations. The method of Balatin, Fradkin and Vilkovisky provides an elegant prescription to obtain the BRS charge and the gauge fixed action together with the Faddeev-Popov ghost terms. We introduce a pair of canonically conjugate anticommuting variables, $(\eta^\alpha, \mathcal{P}_\alpha)$, corresponding to each first class constraint F_α. The BRS charge is

$$Q = F_\alpha \eta^\alpha + \frac{1}{2} \mathcal{P}_\alpha C^\alpha{}_{\beta\gamma} \eta^\beta \eta^\gamma \tag{7}$$

$\{\eta^\alpha, \mathcal{P}_\beta\} = \delta^\alpha{}_\beta$, and $C^\gamma{}_{\alpha\beta}$ are the structure constants defined in (2). The classical BRS charge thus constructed is nilpotent i.e. $\{Q,Q\} = 0$ which can be proved by using the algebra of constraints. The effective Hamiltonian (canonical Hamiltonian together with Faddeev-Popov ghost terms) is obtained by introducing a gauge fixing function ψ, an odd Grassmann variable. The effective Hamiltonian is

$$H_\psi = H_o + \mathcal{P}_\alpha V^\alpha{}_\beta \eta^\beta + \{Q,\psi\} \tag{8}$$

and the corresponding effective action in the extended phase space p_i, q^i, η^α and $\mathcal{P}_\alpha$ is given by

$$S = \int (p_i \dot{q}^i + P_\alpha - H_\psi)\, dt \tag{9}$$

The BRS transformation of any phase space variable or any dynamical variable, G, is

$$\delta_\epsilon G = \{ G, \epsilon Q \} \tag{10}$$

ϵ being a constant odd Grassmann parameter. The S-matrix generating functional is

$$\Sigma_\psi = \int \mathcal{D}_\mu \, e^{iS[p, q, \eta, \mathcal{P}]} \tag{11}$$

where $\mathcal{D}_\mu$ is the measure in the extended phase space and S is the effective action [9]. The gauge invariance i.e. ψ–independence, of the generating function is proved by using the BRS transformation property [10], Jacobi identity and the nilpotency of the BRS charge. We shall see later that a similar technique will be employed to derive the Ward identity.

Now, we proceed to present the Hamiltonian approach to study the evolution of the closed string in the presence of its massless background excitations such as graviton, $G_{\mu\nu}$, antisymmetric tensor field, $B_{\mu\nu}$. The actions is

$$I = -\frac{1}{2}\int d^2\sigma \, [\, \sqrt{-g}\, g^{ab}\, \partial_a X^\mu \partial_b X^\nu G_{\mu\nu}(X)$$

$$+ \epsilon^{ab}\, \partial_a X^\mu \partial_b X^\nu B_{\mu\nu}(X)\,] \tag{12}$$

where ϵ^{ab} is antisymmetric in its indices. The action is invariant under reparametrization of the world sheet and the Weyl rescaling. The other massless excitation, dilaton, couples to the curvature scalar of the world sheet as $\int d^2\sigma \sqrt{g}\, R^{(2)} D(X)$ and this term breaks conformal invariance classically. The dilaton background couples to the ghost fields in our Hamiltonian approach [7].

The canonical momenta is

$$P_\mu \equiv \frac{\partial L}{\partial \dot{X}^\mu} = \sqrt{g}\, g^{ab}\, \partial_a X^\nu G_{\mu\nu}(X) - X'^\nu B_{\mu\nu}(X) \tag{13}$$

The two dimensional energy momentum tensor vanishes identically as a constraint when we adopt Dirac's procedure. The two constraints

$$0 = L_\pm \equiv \frac{1}{4}(P_\mu \pm X'^\rho G_{\mu\rho} + X'^\rho B_{\mu\rho})\, G^{\mu\nu}(P_\nu \pm X'^\lambda G_{\nu\lambda} + X'^\lambda B_{\nu\lambda}) \tag{14}$$

are obtained independent of any gauge choice, while $H_{can} = 0$ identically. The algebra of constraints are

$$\{L_\pm(\sigma,\tau), L_\pm(\sigma',\tau)\} = \pm\,(L_\pm(\sigma,\tau) + L_\pm(\sigma',\tau))\,\frac{d}{d\sigma}\,\delta(\sigma-\sigma') \tag{15}$$

$$\{L_\pm(\sigma,\tau), L_\pm(\sigma',\tau)\} = 0 \tag{16}$$

and they can be obtained by straight forward computation of the Poisson brackets. The BRS charge is

$$Q = \int d\sigma\,(L_+\,\eta_+ + L_-\,\eta_- + \mathcal{P}_+\,\eta_+\,\eta'_+ - \mathcal{P}_-\,\eta_-\,\eta'_-)\,; \tag{17}$$

the ghosts satisfy the canonical Poisson bracket relations

$$\{\,\eta_\pm(\sigma,\tau), \mathcal{P}_\pm(\sigma',\tau)\,\} = \delta\,(\sigma-\sigma') \tag{18}$$

and other Poisson brackets between $\eta_\pm$ and $\mathcal{P}_\pm$ vanish. The gauge fixed Hamiltonian is

$$H_\psi = \{\,\psi, Q) \tag{19}$$

in the gauge ψ. Notice that $H_{can} = 0$ and the structure constants $V^\beta{}_\alpha$ defined in eq. (1) vanish in the case the string Lagrangian. The case $\psi = \mathcal{P}_+ + \mathcal{P}_-$ corresponds to the orthonormal gauge choice and

$$H_{ON+}\,L_+ + L_- + 2\,\mathcal{P}_+\,\eta'_+ + \mathcal{P}'_+\,\eta_+ - 2\,\mathcal{P}_-\,\eta'_- - \mathcal{P}'_-\,\eta_- \tag{20}$$

The S-matrix generating functional has the following path integral representation

$$\Sigma\,(G,B) = \int \mathcal{D}\,X^\mu\,\mathcal{D}\,P_\mu\,\mathcal{D}\,\eta_\pm\,\mathcal{D}\,\mathcal{P}_\pm\,\exp\,(i\int\delta\,\sigma\,(\dot{X}\bullet P + \mathcal{P}_\pm\,\dot{\eta}_\pm - H_{ON} \tag{21}$$

The S-matrix elements for the scattering of the massless excitations of the string such as $G_{\mu\nu}$ and $B_{\mu\nu}$ can be obtained by taking functional derivatives of Σ with respect to the corresponding background fields and then setting them to their ground state values. The generating functional is BRS invariant at classical (tree) level. The BRS invariance of Σ in the quantum theory is intimately connected with the nilpotency of the quantum BRS charge [11]. The nilpotency of the quantum BRS charge imposes strong constraints [12] on the configurations of the background field and these constraints are similar to those obtained by imposing conformal invariance on the theory (equivalent to vanishing of ß-function condition).

Let us introduce canonical transformations in the phase space in order to investigate the gauge invariance properties of Σ. It is argued that suitable choice of canonical transformations on the phase space path integral are closely connected with the local symmetries of Σ under transformations of the background field $G_{\mu\nu}$ and $B_{\mu\nu}$. However, we are aware of the fact that these transformations might be plagued with anomalies which might destroy the invariance properties of Σ we are looking for.

Consider two kinds of infinitesimal canonical transformations induced by the generators

$$\text{I.} \qquad \Phi_G \equiv \int d\sigma\, P_\mu\, \xi^\mu (X) \tag{22}$$

$$\text{II.} \qquad \Phi_B \equiv \int d\sigma\, X'^\mu\, \Lambda_\mu (X) \tag{23}$$

where $\xi^\mu (X)$ and $\Lambda_\mu (X)$ are infinitesimal arbitrary functions of X. Φ_G induces following transformations on the phase space variables

$$\delta_G X^\mu = \xi^\mu (X), \quad \delta_G P_\mu = -P_\nu\, \xi^\nu{}_{,\mu} (X) \tag{24}$$

$$\delta_G \eta_\pm = 0, \quad \delta_G P_\pm = 0 \tag{25}$$

where $\xi^\nu{}_{,\mu} \equiv \dfrac{\partial}{\partial X^\mu} \xi^\nu (X)$.

The backgrounds transform as

$$\delta_G G_{\mu\nu} = G_{\mu\nu,\lambda}\, \xi^\lambda, \qquad \delta_G B_{\mu\nu} = B_{\mu\nu,\lambda}\, \xi^\lambda \tag{26}$$

since they are given functions of X. It is easy to check by explicit calculations that the action

$$S_H = \int d^2 \sigma (P.\dot{X} + \mathcal{P}_\pm \dot{\eta}_\pm - H_{ON}) \tag{27}$$

has the following property

$$\delta_G S_H = -\delta^{GCT} S_H \tag{28}$$

where $\delta_G S_H$ is the variation of the action under general coordinate transformations of $G_{\mu\nu}$ and $B_{\mu\nu}$

$$\delta^{GCT} G_{\mu\nu} = -G_{\mu\lambda}\, \xi^\lambda{}_{,\nu} - G_{\nu\lambda}\, \xi^\lambda{}_{,\mu} - G_{\nu\lambda}\, \xi^\lambda \tag{29}$$

$$\delta^{GCT}\,\Phi_B = -B_{\mu\lambda}\,\xi^{\lambda}{}_{,\nu} - B_{\nu\lambda}\,\xi^{\lambda}{}_{,\mu} - B_{\nu\lambda}\,\xi^{\lambda} \tag{30}$$

Similarly, the transformations induced on phase space variables by Φ_B gives rise to the following relations

$$\delta^{B\text{-gauge}}\,\delta_H = -\,\delta_B\,S_H \tag{31}$$

where the B-gauge transformations of the background fields are

$$\delta^{B-\text{gauge}}\,B_{\mu\nu} = \partial_\mu\,\Lambda_\nu - \partial_\mu,\ \delta^{B-\text{gauge}}\,G_{\mu\nu} = 0 \tag{32}$$

Thus the generating functional Σ, satisfies the following identity

$$\Sigma'\,(G,B) = \Sigma\,(G + \delta^{GCT}\,G, B + \delta^{GCT}\,B + \delta^{B.g}\,B) \tag{33}$$

if we assume that the phase space measure remains invariant under the canonical transformations induced by Φ_G and Φ_B .

Now we are in a position to derive the Ward identities for the S-matrix elements generated by Σ using the invariance properties (33), as follows

$$\delta^{GCT}\,\Sigma = <\int d^D X\,[\,\frac{\delta\,S_H}{\delta\,G_{\mu\nu}\,(X)}\,\delta^{GCT}\,G_{\mu\nu}\,(X) + \frac{\delta\,\Sigma_H}{\delta\,G_{\mu\nu}\,(X)}\,\delta^{GCT}\,B_{\mu\nu}\,(X)] >_{G,B} \tag{34}$$

where $< \ >_{G,B}$ is to be interpreted as the functional integration with measure $\exp(i\,S_H)$. The action, S_H, is a functional of the background fields as well as the string coordinates, X^μ , the canonical momenta, P_μ, and the ghosts, $\eta_\pm$ and $\mathcal{P}_\pm$. Thus

$$S_H = \int d^2\,\sigma\,\mathcal{L}_H\,(X, P, \eta_\pm, \mathcal{P}_\pm, G,B) \tag{35}$$

and

$$\frac{\delta\,S_H}{\delta\,G_{\mu\nu}\,(X)} = \int d^2\,\sigma\,\,\delta(X-X(\sigma))\,\,\frac{\delta\,\mathcal{L}_H}{\delta\,G_{\mu\nu}\,(X)} = \int d^2\,\sigma\,\,\delta(X-X(\sigma))\,\,V^{\mu\nu}{}_G\,(X) \tag{36}$$

$$\frac{\delta\,S_H}{\delta\,B_{\mu\nu}\,(X)} = \int d^2\,\sigma\,\,\delta(X-X(\sigma))\,\,V^{\mu\nu}{}_B\,(X) \tag{37}$$

$V^{\mu\nu}{}_G(X)$ and $V^{\mu\nu}{}_B(X)$ can be identified as the graviton and antisymmetric tensor field vertices in a background, their explicit forms are

$$V^{\mu\nu}{}_G = \frac{1}{2} X'^{\mu} X'^{\nu} - \frac{1}{2} G^{\mu\rho} G^{\nu\lambda} P_\rho P_\lambda - P_\rho G^{\mu\rho} G^{\nu\lambda} B_{\lambda\sigma} X'^{\sigma}$$

$$- \frac{1}{2} X'^{\rho} B_{\alpha\rho} B_{\lambda\sigma} X'^{\sigma} G^{\mu\alpha} G^{\nu\lambda} \tag{38}$$

and

$$V^{\mu\nu}{}_B = \frac{1}{2} (P_\rho G^{\mu\rho} X'^{\rho} + X'^{\nu} B_{\rho\lambda} X'^{\lambda} G^{\mu\rho} - \mu \leftrightarrow \nu) \tag{39}$$

using the transformation properties of $G_{\mu\nu}$ and $B_{\mu\nu}$ under GCT (29) and (30) in (34) and the definitions of the vertices $V^{\mu\nu}{}_G$ and $V^{\mu\nu}{}_B$ eq. (36), (37) in (34) we arrive at

$$< \int d^2\sigma \, [V^{\mu\nu}{}_G(X) \{ G_{\mu\lambda}(X) \xi^{\lambda}{}_{,\nu}(X) + G_{\nu\lambda}(X) \xi^{\lambda}{}_{,\mu}(X) + G_{\mu\nu,\lambda}(X) \xi^{\lambda}(X)$$

$$+ V^{\mu\nu}{}_B \{ B_{\mu\lambda}(X) \xi^{\lambda}{}_{,\nu} - B_{\nu\lambda}(X) \xi^{\lambda}{}_{,\mu}(X) + B_{\mu\nu,\lambda} \xi^{\lambda}(X) \}]_{G,B} > = 0 \tag{40}$$

Since $\xi^{\lambda}(X)$ is an arbitrary infinitesimal function, we can take functional derivatives of (40) with respect to $\xi^{\mu}(Z)$ and then set $\xi^{\mu}(Z) = 0$. Consequently,

$$< \int d^2\sigma \, [V^{\mu\nu}{}_G(X) \{ G_{\mu\lambda}(X) \partial_\nu \delta(X\text{-}Z) + \mu \leftrightarrow \nu + G_{\mu\nu,\lambda} \, \delta(X\text{-}Z) \}$$

$$+ V^{\mu\nu}{}_B(X) \{ B_{\mu\lambda}(X) \partial_\nu \delta(X\text{-}Z) - B_{\nu\lambda}(X) \partial_\mu \delta(X\text{-}Z)) + B_{\mu\nu,\lambda} \, \delta(X\text{-}Z) \}]_{G,B} > = 0 \tag{41}$$

The Ward identities for multigraviton and antisymmetric field processes can be derived by taking arbitrary number of functional derivatives of (41) with respect to $G_{\mu\nu}$ and $B_{\mu\nu}$ and then setting the background fields to their ground state values. Let us take the vacuum background to be flat, $\eta_{\mu\nu}$, $D = 26$ and set $B_{\mu\nu} = 0$ for sake of simplicity. The n-graviton amplitude, obtained by taking n-derivatives of (41), satisfies

$$\frac{\delta^n}{\delta G_{\mu_1\nu_1}(X_1) .. \delta G_{\mu_n\nu_n}(X_n)} < \int d^2\sigma \, V^{\mu\nu}{}_G \{ G_{\mu\lambda}(X) \partial_\nu \delta(X\text{-}Z) + \mu \leftrightarrow \nu$$

$$+ G_{\mu\nu,\lambda} \, \delta(X\text{-}Z) \} >_{G=\eta} = 0 \tag{42}$$

The operator $\dfrac{\delta}{\delta G_{\mu_i\nu_i}(X_i)}$ acts three ways. It acts on V_G or on the metric $G_{\mu\lambda}$, $G_{\mu\nu,\lambda}$ or on the action S_H which is implicitly present in the definition of the path integral. We get contact terms in the first two cases and we get extra vertices in the last case. Therefore, we find the following relations hold (after taking Fourier transform)

$$< \prod_{i=1}^{n} \int d^2\sigma_i\, V_G^{\mu_i\nu_i}(X(\sigma_i))\, e^{iK_i X(\sigma_i)} \int d^2\sigma\, 2K_\nu\, V_G^{\lambda\nu}(X(\sigma))\, e^{iK.X(\sigma)} >_{G=\eta}$$

$$= \sum_{i=1}^{n} K^\lambda{}_i < \int d^2\sigma_i\, V_G^{\mu_i\nu_i}(X(\sigma_i))\, e^{i(K+K_i)X(\sigma_i)} \prod_{j\neq i} \int d^2\sigma_j\, V_G^{\mu_j\nu_j}\, e^{iK_j X(\sigma_j)} >_{G=\eta}$$
$$+ \text{ other contact terms} \qquad (43)$$

This is the desired WI. Noted that the divergence of the (n+1) graviton amplitude on LHS of eq. (43) is related to the lower point functions [13-16]. We have deduced (43) through formal manipulations and therefore, anomalies might creep in while the canonical transformation is being performed. Furthermore, we do not expect (43) to hold for arbitrary backgrounds field configurations. Presumably WI are valid only if the backgrounds preserve DRS invariance and the fluctuations considered correspond to on-shell external particles.

We would like to mention that similar WI can be obtained for the compactified closed bosonic string [17,18]. We would like to present another interesting result in this context. Consider the action

$$I = -\frac{1}{2}\int d^2\sigma\, [\sqrt{-g}\, g^{ab}\, \partial_a X^\mu\, \partial_b X^\nu\, G_{\mu\nu} + \epsilon^{ab}\, \partial_a X^\mu\, \partial_b X^\nu\, B_{\mu\nu}(X)$$
$$+ i e\, \psi^i{}_L\, e^a{}_+\, \partial_a \psi^i{}_L + i e\, \psi^i{}_R\, e^a{}_-\, (\partial_a \delta_{ij} + T^M{}_{ij}\, A^\mu{}_\mu\, \partial_a X^\mu)\, \psi^j{}_R\,] \qquad (44)$$

Here we have ferminised the compacified bosonic coordinates denoted by $\psi^i{}_{L,R}$, i = 1...2d, d being the number of compactified dimensions.

The theory has massless gauge bosons denoted by $A^M{}_\mu$; $T^M{}_{ij}$ are the generators of the gauge group in the fundamental representation to which $\psi^{'s}{}_{R,L}$ belong. The action is invariant under nonabelian gauge transformations $\psi^i \rightarrow \exp(i\,\theta_M\, A^M)$ and $A^M{}_\mu \rightarrow \partial_\mu \theta^M + f^{abc} A^b{}_\mu\, \theta^c$ where f^{abc} are the structure constants of the gauge group. Note that we have coupled the gauge background to the right handed spinors only exposing therefore half of the actual gauge symmetry.

We find that the fermionic measure picks up a nontrivial Jacobian factor under the above gauge transformations [18]. This extra factor can be eliminated if the antisymmetric tensor field transforms as

$$\delta B_{\mu\nu} = \partial_{[\mu} \quad \theta^M(X)\, A^M{}_{\nu]} \tag{45}$$

under the nonabelian gauge transformation mentioned above. However, it is interesting to note that the transformation (45) is necessary to make the modified field strenght, $\widetilde{H}_{\mu\nu\lambda}$, of associated with $B_{\mu\nu}$ to be gauge invariant ; where

$$\widetilde{H}_{\mu\nu\lambda} = \partial_{[\mu}\, B_{\mu\nu\,]} - \omega_{\mu\nu\lambda} \tag{46}$$

where $\omega_{\mu\nu\lambda} = T_r (A_{[\mu} F_{\nu\lambda]} - \frac{1}{3} A_{[\mu} A_{\mu} A_{\nu} A]$ (47)

is the Yang-Mills Chern-Simons term. In super string theory cancellation of Yang-Mills anomalies [18] requires that the field strenght $H_{\mu\nu\lambda}$ should be modified to $\widetilde{H}_{\mu\nu\lambda}$ given by (46) and $B_{\mu\nu}$ must transform according to (45) to make (46) gauge invariant. It is heartening to note that although the gauge transformation produced an anomaly in path integral measure this anomaly can be eliminated by requiring a transformation property of $B_{\mu\nu}$ which is required for a very fundamental reason in string theory.

Therefore, in view of our results on compactified closed string, it will be interesting to investigate whether the canonical transformations generated by Φ_G and Φ_B induce any anomaly in derivation of WI. Another important issue is to explore the consequences of the problem of operator ordering. Notice, that, in H_{ON}, the quadratic term in P_μ is multiplied by $G^{\mu\nu}(X)$ and in general gives extra terms (to order h^2) when we define the path integral properly [20]. This problem is currently under investigation.

To summarize, we have presented a BRS path integral representation of the S-matrix generating functional for closed bosonic string in the background of its massless excitations. The symmetry property of Σ is exhibited under suitable canonical transformations and WI are derived for scattering amplitudes.

Most of the work presented in this article was done in collaboration with G. Veneziano. I express my gratitude to him for sharing his deep insight in string theory with me. I would like to thank Professor K. Chadan and the members of the Laboratoire de Physique Théorique et Hautes

Energies, Orsay, for their gracious hospitality. A research associateship from CNRS is gratefully acknowledged.

I had the privilege of being associated with Professor Miyazawa and I had the pleasant experience of collaborating with him while I was a Post-doctoral fellow at Tokyo. He has made a profound impact on my academic carreer. I would like to express my indebtedness to Professor Miyazawa and I wish him many happy returns of the day.

REFERENCES

+ Permanent address : Institute of PHysics, Bhubaneswar 751005,, India

(1) Green, M.B., Schwarz, J.H. and Witten, E., Superstring Theory, Cambridge University Press (1987).

(2) Green, M.B. and Schwarz, J.H., Phys. Lett. 136B, 367 (1984), Phys. Lett. 149B (1984) Schwarz, J.H. in "Geometry, Anomaly and Topology", Editors W.A. Bardeen and A. White, World Scientific, Singapore.

(3) Frenkel, I.B. and Kac, V.G., Inv. Math 62, 23 (1980),
Segal, G., Commun. Math. Phys. 80, 301 (1982),
Goddard, P. and Olive, D., Vertex operators in Mathematics and Physics,
Eds. J. Lepowsky et al. (Springer, Berlin, 1984).

(4) Gross, D.J., Harvey, J.A., Martinec, E. and Rohm, R., Nucl. Phys. B256, 253 (1985).

(5) Schwarz, J.H.., Proc. of International Workshop on Superstrings, Composite Structures and Cosmology, Maryland, (1987) (to be published) and references therein.

(6) Fradkin, E.S. and Tseytlin, A.A., Phys. Lett. 158B, 316 (1985), Nucl.Phys. B261, 1 (1986).
Callan, C.G., Lectures at Workshop on High Energy Physics and Cosmology ICTP, Trieste (1986).
Sen, A., Lectures at Workshop on High Energy Physics and Cosmology, ICTP, Trieste (1986) and references therein.

(7) Maharana, J. and Veneziano, G., Nucl. Phys. B283, 126 (1987).

(8) Dirac, P.A.M., Can. J. Math. 2, 124d (1950).

(9) Fradkin, E.S. and Vilkovisky, G.A., Phys. Lett. 55B, 224 (1975),
Balatin, I.A. and Vilkovisky, G.A. Phys. Lett. 69B, 409 (1977).

(10) For a review see Henneaux, M., Phys. Rep. 126C, 1 (1986) and references therein.

(11) The nilpotency of BRS charge for a string in flat background gives critical dimension, D=26 and unit intercept. See Kato, M. and Ogawa, K., Nucl. Phys. B212, 443 (1983), Huang, S. Phys. Rev. D28, 2614 (1983).

(12) Maharana, J. and Veneziano, G., unpublished.

(13) Maharana, J. and Veneziano, G., réf. 7.

(14) Veneziano, G., Phys. Lett. 167B, 338 (1986).

(15) Yoneya, T., Phys. Rev. Lett. 55, 1822 (1985).

(16) Capper, D.M. and Roman Madrano, M., Phys. Rev. D9 1641 (1974).

(17) Maharana, J. and Veneziano, G., Phys. Lett. 169B, 177 (1986).

(18) Das, A., Maharana, J. and Panigraki, P.K., Rochester Preprint, UR-1002/ER 13065-489. Maharana, J., Proceedings of 1st Asia Pacific Conference on High Energy Physics, Singapore (1987) (World Scientific Pub.).

(19) Green, M.B. and Schwarz, J.H., ref. 2. There is a Lorentz C-S term which can also be obtained using our technique.

(20) Mizrahi, M., J. Math. Phys. 16, 2201 (1975),
Gervais, J.L. and Jevicki, A., Nucl. Phys. B110, 93 (1976).
Schulman, L.S., "Techniques and applications of path integrals", J. Wiley (1981).
Lee, T.D., "Particle Physics and isntroduciton to field Theory", Harward (1981).

REDUCING THE RANK OF GAUGE GROUPS IN ORBIFOLD COMPACTIFICATION

Hikaru SATO

Department of Physics, Hyogo University of Education
Yashiro-cho, Hyogo 673-14, Japan

ABSTRACT

The Wilson-line mechanism in orbifold compactification is investigated for both Abelian and non-Abelian embedding of the Z_3 group in the $E_8 \times E_8$. We give general argument in the fermionic formulation for the gauge degrees of freedom and show that the rank of the gauge group is reduced by introducing nondiagonal Wilson-line matrix in the fermionic boundary conditions.

1. INTRODUCTION

Heterotic string theories are promising candidates for unified theories of all known interactions.[1] However if they are to make contact with the real world it is necessary to understand how the enormous symmetry of these theories is broken down to the symmetry which we observe at low energies. A powerful method to implement symmetry breaking in string theories is to consider string propagation on orbifold.[2] In the standard Z orbifold model the gauge group in the $E_8 \times E_8$ heterotic string theory breaks down to $E_6 \times SU(3) \times E_8$ and there are 36 chiral generations.[2] An effective method to break further the gauge symmetry and to reduce the number of generations is an application of the Wilson-line mechanism in the framework of orbifold compactification.[3,4]

In ref.(4) we have made a systematic approach to this problem and given a complete classification of the possible gauge symmetry breaking by Wilson lines on the standard Z orbifold. There we have embedded the action of the Z_3 discrete group in the internal degrees of freedom in an

Abelian way so that the rank of the gauge group could not be reduced. String propagation on orbifold is expressed by imposing twisted boundary conditions on string variables. In this paper we introduce most general twisted boundary conditions in the fermionic formulation and show that a non-Abelian embedding is possible when background gauge field is present on orbifold. This may lead to the reduction of the rank of the gauge group.

2. BOUNDARY CONDITIONS ON ORBIFOLD

In what follows we focus on the heterotic string in the fermionic formulation. We will work in the light-cone gauge and use the NSR formulation to describe the superstring part of the heterotic string. We describe the bosonic variables in the six compactified dimensions in the complex notation as $z^\alpha (\alpha=1,2,3)$ and their complex conjugate $\bar{z}^\alpha$. Then the string propagation on the Z orbifold is described by imposing the following boundary conditions on the bosonic variables:

$$z^\alpha(\sigma_1+\pi,\sigma_2) = \exp(2\pi ik/3)\, z^\alpha(\sigma_1,\sigma_2) \ , \tag{1}$$

$$z^\alpha(\sigma_1,\sigma_2+\pi) = \exp(2\pi ih/3)\, z^\alpha(\sigma_1,\sigma_2) \ , \tag{2}$$

$$h,k=1,2,3$$

where σ_1 and σ_2 parametrize a torus of a closed string theory at one-loop level and they run between 0 and π. The center of mass coordinate of z^α must be at one of the fixed points for the Z_3 transformation $z^\alpha \to \exp(2\pi i/3)\, z^\alpha$.

The group action on the right-moving NSR fermions $\lambda^a (a=1,\cdots,4)$ is determined by the requirement of preserving world sheet supersymmetry and the boundary conditions amount to

$$\lambda^a(\sigma_1+\pi,\sigma_2) = -(-1)^n \exp(2\pi ik\xi^a)\, \lambda^a(\sigma_1,\sigma_2) \ , \tag{3}$$

$$\lambda^a(\sigma_1,\sigma_2+\pi) = -(-1)^m \exp(2\pi ih\xi^a)\, \lambda^a(\sigma_1,\sigma_2) \ , \tag{4}$$

$$3\,\xi^a = \text{integer}$$

where n,m=0,1 specifies the spin structure.

Now let us consider the gauge degrees of freedom which are represented by two sets of 16 fermionic coordinates ψ^I and $\tilde{\psi}^I$ (I=1,···,16) transforming as thr vector of O(16). We embed the Z_3 group action on $E_8 \times E_8$ in an $O(16) \times O(16)$ subgroup. If we introduce background gauge fields A_μ^{IJ} which transform as the adjoint representation of O(16), the left-moving gauge fermion $\psi^I(\tilde{\psi}^I)$ have a coupling to A_μ^{IJ} as $i\bar{\psi}^I A_\mu^{IJ} \bar{\partial} x^\mu \psi^J$ with $\partial = \partial_z$, $z=(\sigma_1+\tau\sigma_2)/2\pi$ and τ is the complex parameter defining the torus. Under the group action a fixed point of Z_3 must be accompanied by a shift on the Γ_6 lattice defining the six torus T^6, since different fixed points correspond to different elements of the space group although they correspond to the same element of the point group. This shift picks up a following phase factor to $\psi^I(\tilde{\psi}^I)$ in order to preserve gauge invariance:

$$\Theta_{f,k} = \exp[2\pi i\, \theta_{f,k}^{IJ}\, T^{IJ}] \ , \tag{5}$$

where T^{IJ} are generators of O(16) and

$$\theta_{f,k}^{IJ} = n_{f,k}^i (e_i)^\mu A_\mu^{IJ} \equiv n_{f,k}^i a_i^{IJ} \quad (n_{f,k}^i = \text{integer}) \ . \tag{6}$$

Here e_i(i=1,···,6) is a set of basis vectors for the Γ_6 lattice and $n_{f,k}^i(e_i)^\mu$ represents a shift necessary for the fixed point f to return to the starting point under the rotation $z^\alpha \to \exp(2\pi ik/3)\, z^\alpha$. The factor (5) is nothing but a Wilson line.

The Z_3 group action on $\psi^I(\tilde{\psi}^I)$ is represented by the rotation matrix of O(16). By choosing appropriate basis we may diagonalize the rotation matrix as follows:

$$\Omega = \exp[2\pi i\, \omega^\ell\, H_\ell] \ , \tag{7}$$

where $\Omega^3 = 1$ and H_ℓ (ℓ=1,···,8) are Cartan subalgebra of O(16). The boundary conditions for the gauge fermions are given by

$$\psi(\sigma_1+\pi,\sigma_2) = -(-1)^n \Omega^k\, \Theta_{f,k}\, \psi(\sigma_1,\sigma_2) \quad , \tag{8}$$

$$\psi(\sigma_1,\sigma_2+\pi) = -(-1)^m \Omega^h\, \Theta_{f,h}\, \psi(\sigma_1,\sigma_2) \quad , \tag{9}$$

and matrix notation is understood here. These boundary conditions are imposed at each fixed point f of the Z orbifold.

3. SYMMETRY BREAKING BY WILSON LINES

The rotation matrix Ω and the Wilson-line matrix $\Theta_{f,k}$ are incommutable in general. Commutability of these matrices depends on the choice of the background gauge fields A_μ^{IJ} introduced on the Z orbifold. First we consider the case where these two matrices are commutable.

(i) $[\Omega, \Theta_{f,k}] = 0$: This case corresponds to the Abelian embedding. The two matrices Ω and $\Theta_{f,k}$ are diagonalized simultaneously. By choosing an appropriate complex basis we can express $(\psi^I, \tilde{\psi}^I)$ by Ψ^I (I=1,$\cdots$, 16) which is $(\underline{1},\underline{8}) + (\underline{8},\underline{1})$ of the two U(8) subgroup of O(16) × O(16). Then the boundary conditions (8),(9) are rewritten as

$$\Psi^I(\sigma_1+\pi,\sigma_2) = -(-1)^n \exp[2\pi i(k\zeta^I + \theta^I_{f,k})]\, \Psi^I(\sigma_1,\sigma_2) \ , \tag{10}$$

$$\Psi^I(\sigma_1,\sigma_2+\pi) = -(-1)^m \exp[2\pi i(h\zeta^I + \theta^I_{f,h})]\, \Psi^I(\sigma_1,\sigma_2) \ , \tag{11}$$

$$3\,\zeta^I = \text{integer}$$

where the rotation matrix is described by a diagonal matrix $\{\exp(2\pi i k\zeta^I)\}$ and the Wilson-line matrix is written by a diagonal matrix $\{\exp(2\pi i\theta^I_{f,k})\}$. Since we are considering Z_3 symmetry, $3\,\theta^I_{f,k} = \theta^I_{f,3k}$ = integer. This implies

$$\theta^I_{f,k} = k\,\theta^I_f = k\, n^i_f\, a^I_i \quad (n^i_f = \text{integer mod } 3) \ , \tag{12}$$

and

$$3\, a^I_i = \text{integer mod } 3 \ . \tag{13}$$

Thus the background gauge field A^I_μ which lies in the Cartan subalgebra in this case must be constrained so as to obey the Z_3 symmetry and $3(\zeta^I + \theta^I_f)$ must be a vector in the $E_8 \times E_8$ lattice.

We should notice that this embedding of Z_3 in the fermionic formulation is equivarent to the embedding of the shift in the $E_8 \times E_8$ lattice in the bosonic formulation. Under the boundary conditions (10),(11) we have calculated the one-loop partition function in ref.(5) and derived the condition of modular invariance in the presence of the background gauge field as follows:

$$3 \sum_a \xi^a = 0 \mod 2, \tag{14}$$

$$3 \sum_I (\zeta^I + \theta_f^I) = 0 \mod 2, \tag{15}$$

$$3 \{ \sum_I (\zeta^I + \theta_f^I)^2 - \sum_a (\xi^a)^2 \} = 0 \mod 2. \tag{16}$$

The string states on the orbifold must be invariant under Z_3. The group invariant condition has been obtained in ref.(5) by constructing the projection operator onto the invariant subspace of the string Hilbert space. The condition in the (k,h) sector is given by

$$(V^I + k v_f^I/2) v_f^I + (K^a - k\xi^a/2) \xi^a + \frac{1}{3} m_k = \text{integer} , \tag{17}$$

where V^I is the vector in the $E_8 \times E_8$ lattice and K^a is the vector in the SO(8) vector or spinor lattice. The shift vector v_f^I is defined by $v_f^I = \zeta^I + \theta_f^I$ and m_k is the eigenvalue of the operator

$$\hat{m}_k = \sum_{n>0} \{ \frac{1}{n+1-k/3} \beta^\alpha_{-n-1+k/3} \bar{\beta}^\alpha_{n+1-k/3} - \frac{1}{n+k/3} \bar{\beta}^\alpha_{-n-k/3} \beta^\alpha_{n+k/3} \} \tag{18}$$

where β^α is the oscillator of z^α. The twist operator $\hat{g}_k$ for z^α is given by $\hat{g}_k = \exp[(2\pi i/3)\hat{m}_k]$, which acts on the string variable in the k-twisted sector as

$$\hat{g}_k z^\alpha \hat{g}_k^{-1} = e^{2\pi i/3} z^\alpha . \tag{19}$$

The gauge symmetry and the generation structure of the massless fermions in the case of Abelian embedding of Z_3 are obtained by taking into account the condition of modular invariance (14)-(16) and the group invariant condition (17). More detailed discussions and explicit examples are found in refs.(4,5), where the rank 22 gauge group obtained by the toroidal compactification breaks down to smaller groups but the total rank of the gauge group is not reduced. We give here brief description why the rank is not reduced in the Abelian embedding. The gauge bosons are contained in the untwisted sector and the gauge symmetry is determined by the group invariant condition (17), which turns out $V^I v_f^I =$

integer for the gauge sector. This condition restricts only nonzero roots of the $E_8 \times E_8$. In the bosonic formulation the gauge degrees of freedom corresponding to the zero roots or Cartan subalgebra come from the oscillator part in the partition function, $a^I_{-1}|0\rangle$ which are not touched by the group action so that the rank of the gauge group is not altered.

(ii) $[\Omega, \Theta_{f,k}] \neq 0$: In this case Ω and $\Theta_{f,k}$ are not diagonalized simultaneously. We have chosen the basis such that Ω is diagonal. Then the Wilson-line matrix $\Theta_{f,k}$ is nondiagonal in this basis and can be expressed as

$$\Theta_{f,k} = \exp[2\pi i\, \theta^{\alpha}_{f,k}\, E_{\alpha}] \,, \tag{20}$$

where E_α stands for the generator corresponding to the root α. Now let us examine the condition of group transformation $z^\alpha \to \exp(2\pi i/3)\, z^\alpha$. Under this transformation the basis vector in the complex z plane (fig.1) transforms as $e_1 \to e_2 - e_1$, $e_2 \to -e_1$ so that the gauge field $a_i \equiv (e_i)^\mu A_\mu$ transform as $a_1 \to a_2 - a_1$, $a_2 \to -a_1$. For other complex planes we will obtain the same relation for the corresponding gauge components.

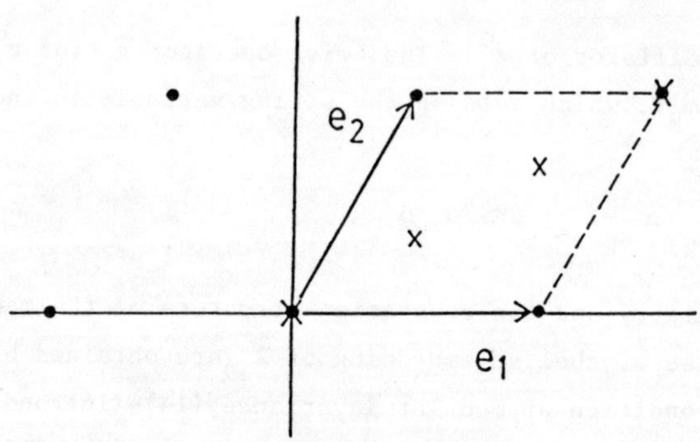

Fig.1 The lattice to define the torus, a direct product of which gives a six torus T^6. The Z orbifold is obtained by T^6/Z_3. The fixed points are marked by ×. The basis vectors are denoted by e_1 and e_2.

The Wilson-line matrix associated with the fixed point $f=(1,0,0)d$ with $d = \sqrt{1/3}\exp(i\pi/6)$ in the $k=1$ sector is given by

$$\Theta_{(100),1} = \exp[2\pi i\, a_1^{\alpha} E_{\alpha}] \ , \tag{21}$$

where under the $2\pi/3$ rotation, the shift e_1 is necessary for the fixed point (1,0,0)d so that the contribution of the gauge field a_1^{α} is picked up in (20). The group transformation for the Wilson-line matrix then reads

$$\Omega^{-1}\Theta_{(100),1}\Omega = \exp[2\pi i(a_2 - a_1)^{\alpha} E_{\alpha}] = \Theta_{(100),2} \ , \tag{22}$$

which is the Wilson-line matrix cprresponding to the $4\pi/3$ rotation. Finally we should note that $\Omega^k\Theta_{f,k}$ and $\Omega^h\Theta_{f,h}$ is commutable. This condition is necessary in order for the two boundary conditions with respect to σ_1 and σ_2 are compatible.

4. REDUCTION OF THE RANK OF THE GAUGE GROUP

Now let us examine the concequence of the boundary conditions (8) and (9) specified by the nondiagonal Wilson-line matrix (20). First we consider the untwisted sector k=0, h=0,1,2, since the gauge symmetry is determined by this sector. The one-loop partition function for the gauge fermion is given by

$$\frac{1}{3}\sum_f \sum_{h=0}^{2} \mathrm{Tr}[\,\Theta_{f,h}\,\Omega^h q^{\hat{H}_0}\,]$$

$$= \frac{1}{3}\sum_f \sum_{h=0}^{2} \mathrm{Tr}[\exp(2\pi i\,\theta_{f,h}^{\alpha}\, E_{\alpha})\exp(2\pi i h\,\omega^{\ell} H_{\ell})\, q^{\hat{H}_0}] \ , \tag{23}$$

for each spin structure where $q = \exp(2\pi i\tau)$ and $\hat{H}_0$ is the Hamiltonian of the fermionic string in the untwisted sector.

The projection operator onto the invariant subspace of the string Hilbert space is given by

$$P_f = \frac{1}{3}\sum_{h=0}^{2} \exp[2\pi i\,\theta_{f,h}^{\alpha}\, E_{\alpha}]\exp[2\pi i h\,\omega^{\ell} H_{\ell}] \ , \tag{24}$$

and the physical states should obey $|\mathrm{phys}\rangle = P_f|\mathrm{phys}\rangle$ for all f. The adjoint representation <u>248</u> of E_8 are represented by the generators E_β and H_ℓ (ℓ=0,•••,8). The state corresponding to the root β obeys

$$P_f|\beta\rangle = \exp(2\pi i h\,\omega\cdot\beta)\,|\beta\rangle \ , \tag{25}$$

if $[\theta^{\alpha}_{f,h} E_\alpha, E_\beta] = 0$. For the states corresponding to the Cartan subalgebra H_ℓ, we obtain

$$P_f |\ell> = |\ell> , \qquad (26)$$

if $[\theta^{\alpha}_{f,h} E_\alpha, H_\ell] = 0$. Thus the conditions which determine the gauge group with the roots β and the Cartan subalgebra H_ℓ are

$$\omega^\ell \beta_\ell = 0 \mod 1 , \qquad (27)$$

$$[\theta^{\alpha}_{f,h} E_\alpha, E_\beta] = 0 , \qquad (28)$$

$$[\theta^{\alpha}_{f,h} E_\alpha, H_\ell] = 0 , \qquad (29)$$

for all the fixed point f and h=0,1,2.

Here we shall give a simple example. We embed the Z_3 in the first E_8 of $E_8 \times E_8$ by choosing $\omega = (1/3,1/3,-2/3,0^5)$ such that

$$\omega^\ell H_\ell = \frac{1}{3}T^{12} + \frac{1}{3}T^{34} - \frac{2}{3}T^{56} . \qquad (30)$$

This corresponds to the standard embedding and the symmetry which obeys (27) is $E_6 \times SU(3)$. We introduce the background gauge field with only one nonzero component $A^{23}_\mu (e_i)^\mu$ for i=1,2. Then the contribution to the Wilson line turns out

$$\theta^{\alpha}_{f,h} E_\alpha = a^{23}_1 T^{23} \quad \text{for } h=1 , \qquad (31)$$

$$= a^{23}_2 T^{23} \quad \text{for } h=2 , \qquad (32)$$

at the fixed point f = (1,0,0)d. For f = (2,0,0)d we have just two times (31). The SU(3) generators $\{H_\ell; E_\beta\} = \{T^{12},T^{34};T^{13},T^{24},T^{25},T^{26},T^{35},T^{36}\}$ do not obey (28) and (29) for the Wilson lines (31) and (32) so that the remaining symmetry turns out E_6. In this example the rank of the original E_8 group is reduced by two units and the symmetry has been broken down to $E_6 \times E_8$.

The possible Z_3 embedding in $E_8 \times E_8$ is constrained by the condition of modular invariance.[2] The level-matching condition[6] restricts the $E_8 \times E_8'$ breaking to (i) $E_6 \times SU(3) \times E_8'$, (ii) $E_7 \times U(1) \times SO(14)' \times U(1)'$,

(iii) $E_6 \times SU(3) \times E_6' \times SU(3)'$ and (iv) $SU(9) \times SO(14)' \times U(1)'$, when the Z_3 is embedded in the Cartan subalgebra of $E_8 \times E_8'$. Further symmetry breaking is achieved by introducing the background gauge fields. When the Wilson-line matrix has no component in the Cartan subalgebra as in (20), the background gauge fields have no influence on the fractionally moded quantum numbers so that the level-matching conditions are not changed by the introduction of Wilson lines. In this case the Wilson-line matrix (20) works as rotation in $O(16) \times O(16)$. The boundary condition (8) in the twisted sector reduces the number of the independent oscillators of the string variables. Under the modular transformation $\tau \to -1/\tau$, this condition turns into the projection onto the invariant subspace of the string Hilbert space realized by the boundary condition (9). More detailed discussion on modular invariance and explicit calculation of the one-loop partition functions will be given elsewhere.[7]

Massless fermions are obtained from the untwisted as well as the twisted sector. In the untwisted sector they are combined with the right-moving ground states with helicity 1/2 in 8_s of SO(8), for which $K^a\xi^a = 2/3 \bmod 1$. Then the left-moving ground states associated with the gauge fermions must have the eigen value of P_f such that $\langle P_f \rangle \exp(i4\pi/3) = 1$, in order to obey the group invariant condition. In the k-twisted sector the projection must be done according to

$$\langle P_f^{(k)} \rangle \exp[2\pi i(K^a - k\xi^a/2)\xi^a] = 1 \ , \tag{33}$$

where $P_f^{(k)}$ is the projection operator onto the k-twisted subspace of the string Hilbert space spanned by the left-moving gauge fermions.

5. CONCLUSIONS AND DISCUSSIONS

Orbifold compactification with Wilson lines is a powerful method to break the gauge symmetry associated with the string theory. We have shown in this paper that when the Wilson lines lie in the Cartan subalgebra, the gauge group is broken down in such a way that the rank of the gauge group is not changed. On the other hand when the Wilson lines do not lie in the Cartan subalgebra and the Wilson-line matrix is not commutable with the Z_3 rotation, the rank of the gauge group is reduced. We have derived the conditions (27)-(29) which determine the unbroken symmetry on the Z orbifold with Wilson lines.

Our formulation is based on the fermionic representation of the gauge degrees of freedom. Another approach to the symmetry breaking by Wilson lines is to consider in the bosonic formulation.[8] In this formalism the Z_3 rotation is embedded in the $E_8 \times E_8$ lattice by the Weyl rotation and the Wilson lines are represented by shifts in the $E_8 \times E_8$ lattice. These two formalisms may be equivalent and the difference is in the choice of the representation of the gauge degrees of freedom. In the fermionic formulation, however, we can treat both the Abelian and non-Abelian embedding in a unified way by adjusting the nonzero components of the background gauge fields introduced on the orbifold.

ACKNOWLEDGMENT

It is my great pleasure to dedicate this paper to Professor Hironari Miyazawa on the occasion of his sixtieth anniversary. I have learned physics from him since my undergraduate studies. I wish him many happy returns and many more fruitful years to come.

REFERENCES

1. Gross,D.J.,Harvey,J.A.,Martinec,E.and Rohm,R., Phys.Rev.Lett.54,502 (1985); Nucl.Phys.B256,253(1985), B267,75(1986)
2. Dixon,L.,Harvey,J.A.,Vafa,C.and Witten,E., Nucl.Phys.B261,651(1985), B274,285(1986)
3. Ibàñez,L.E.,Nilles,H.P.and Quevedo,F., Phys.Lett.187B,25(1987); Ibàñez,L.E.,Kim,J.E.,Nilles,H.P.and Quevedo,F., Phys.Lett.191B,282 (1987)
4. Sato,H., Phys.Lett.201B,78(1988)
5. Sato,H., Phys.Rev.D37,368(1988)
6. Vafa,C., Nucl.Phys.,B273,592(1986)
7. Sato,H., Hyogo Univ. of Education Preprint HUE015 (1988)
8. Ibàñez,L.E.,Nilles,H.P.and Quevedo,F., CERN Preprint TH.4673 (1987); Ibàñez,L.E.,Mas,J.,Nilles,H.P.and Quevedo,F., CERN Preprint TH.4859 (1987)

Superconformal Field Theory with Extended Supersymmetry[1]

Satoshi Matsuda

Department of Physics, College of Liberal Arts and Sciences
Kyoto University, Kyoto 606

We present a brief review on superconformal invariance and its realization in the form of superconformal algebras for models of two dimensions in physics. It seems to me that it is quite appropriate to discuss the subject at this ceremonial occasion for Professor Hironari Miyazawa, who first gave a concrete introduction of Lie superalgebra in a physical model of hadrons, attempting an approximate unified model of mesons and baryons[2].

1. Introduction

Conformal invariance has been recognized as a significant symmetry when one deals with models of two dimensions in physics. Its supersymmetric extension with supercharge N is realized in the form of superconformal algebras with extended supersymmetry of N = 1, 2, and 4. The representation of N = 1 and 2 superconformal algebras was studied extensively[3] and their application to the study of superstrings has been under current intensive study. Recently, the detailed study of the representation of N = 4 superconformal algebras has just begun to be seriously attempted[4,5].

In view of these developments we shall discuss some aspects of superconformal invariance with a brief introduction to the subject. We shall also present some historical overview on the intimate interplay between mathematics and physics on the present subject.

2. Conformal Algebra

Conformal symmetry is associated with invariance under general coordinate transformations $x^\mu \to y^\mu(x)$. The conformal transformation is specified by the following transformation of the metric tensor $g_{\mu\nu}(x)$:

$$g_{\mu\nu}(x) \to g'_{\mu\nu}(x) = \frac{\partial y^\tau}{\partial x^\mu} \frac{\partial y^\rho}{\partial x^\nu} g_{\tau\rho}(y) = \lambda(x) g_{\mu\nu}(x) \tag{2.1}$$

Then the distance $ds^2 = g_{\mu\nu}(x)dx^\mu dx^\nu$ undergoes the local scale transformation

$$ds^2 \to ds'^2 = g'_{\mu\nu}(x)dx^\mu dx^\nu = \lambda(x)\, ds^2 \tag{2.2}$$

For the sake of simplicity we restrict ourselves to the Euclidean flat metric

$$g_{\mu\nu}(x) = \delta_{\mu\nu} \tag{2.3}$$

Then, considering an infinitesimal coordinate transformation with $|\varepsilon| \ll 1$

$$y^\mu = x^\mu + \varepsilon\, \xi^\mu(x), \tag{2.4}$$

the conformal Killing vector ξ^μ now satisfies the Killing condition

$$\partial_\mu \xi_\nu + \partial_\nu \xi_\mu = \frac{2}{d} \delta_{\mu\nu}\, \partial_\lambda \xi_\lambda \tag{2.5}$$

It is a well-known fact that for $d > 2$ there are only a finite number of solutions to (2.5) corresponding to translation, rotation, dilatation, and special conformal transformation. The case of $d = 2$ is special in that there are an infinite number of solutions with arbitrary power of x. This can be seen by use of analytic arguments. For the

Euclidean two dimensions ($\mu, \nu = 1, 2$) the relation (2.6) reads $\partial_1 \xi_1 = \partial_2 \xi_2$, $\partial_1 \xi_2 = -\partial_2 \xi_1$, which is nothing but the Cauchy-Riemann relation if one defines a complex variable z and a complex function f(z):

$$z = x_1 + ix_2,\ f(z) = \xi_1 + i\xi_2. \tag{2.6}$$

The Cauchy-Riemann relation guarantees that any holomorphic function f(z) can be a solution to (2.5) in the regular domain. Similarly one can consider a complex conjugate variable z and anti-analytic function $\bar{f}(\bar{z})$:

$$\bar{z} = x_1 - ix_2,\ \bar{f}(\bar{z}) = \xi_1 - i\xi_2 \tag{2.7}$$

If one regards $\bar{z}$ as an independent complex variable and $\bar{f}(\bar{z})$ a complex analytic function of $\bar{z}$, the whole argument we will make for the z space is also equally valid for the $\bar{z}$ space. One can recover the real space of (x_1, x_2) and the corresponding relations (2.6) and (2.7), restricting ourselves to the appropriate real sections of the (z, $\bar{z}$) space by putting $\bar{z} = z^*$. In the following we present our arguments by confining ourselves to the z space only, unless it is necessary to consider the $\bar{z}$ space explicitly.

Now one easily finds that under the conformal transformation $z \to w = f(z)$ and $\bar{z} \to \bar{w} = \bar{f}(\bar{z})$, dz and $d\bar{z}$ transform as $dz \to dw = (df/dz)dz$ and $d\bar{z} \to d\bar{w} = (d\bar{f}/d\bar{z})d\bar{z}$. Consequently, the distance $ds^2 = dzd\bar{z}$ undergoes the change

$$ds^2 \to ds'^2 = dwd\bar{w} = \lambda(z, \bar{z})ds^2 \tag{2.8}$$

with the local scaling parameter $\lambda(z,\bar{z}) = (df/dz)(d\bar{f}/d\bar{z})$.

Choosing $f_n(z) = \varepsilon_n z^{n+1}$, one obtains an infinite set of generators

$$l_n = z^{n+1}\frac{d}{dz},\quad n = 0, \pm 1, \pm 2, \cdots \tag{2.9}$$

which generates the transformation $z \to w = z + \varepsilon_n z^{n+1}$. l_n's form an infinite dimensional Lie algebra

$$[l_m, l_n] = (n - m)\, l_{m+n} \tag{2.10}$$

Similarly one has the generators $\bar{l}_n = \bar{z}^{n+1} \frac{d}{d\bar{z}}$ which satisfy the same form of the algebra as (2.10). The quantized form of the "classical" algebra (2.10) will be discussed at the end of the next section.

3. Supersymmetric Extension of Conformal Algebra

We consider a superspace $Z = (z, \theta_i)$ with N Grassmann variables θ_i $(i = 1, 2, \cdots, N)$ which satisfy $\theta_i \theta_j = -\theta_j \theta_i$, in particular $\theta_i^2 = 0$. We also have the conjugate superspace $\bar{Z} = (\bar{z}, \bar{\theta}_i)$ $(i = 1, 2, \cdots, N)$. Our arguments to be presented are quite general to accommodate higher supersymmetry of $N > 1$, but, to save space, we only consider here the case of $N = 1$.

The difference between the two points $Z_1 = (z_1, \theta_1)$ and $Z_2 = (z_2, \theta_2)$ is defined by $Z_1 - Z_2 = z_1 - z_2 - \theta_1 \theta_2$. Thus the infinitesimal difference dZ is given by

$$dZ = dz + \theta\, d\theta \tag{3.1}$$

As in the case of $N = 0$, the superconformal transformation is defined such that one has

$$dZ \to dZ' = (\text{a local scale factor}) \times dZ \tag{3.2}$$

The resulting superconformal transformation is explicitly given by

$$z' = z + \varepsilon(z) + \theta \beta(z) = z + \delta z \tag{3.3a}$$

$$\theta' = \theta + \beta(z) + \frac{1}{2} \partial_z \varepsilon(z) \theta = \theta + \delta \theta \tag{3.3b}$$

which gives $dZ \to dZ' = (1 + \partial_z \varepsilon + 2\theta \partial_z \beta)\, dZ$ where $\beta(z)$ is a Grassmannian function of z. The superspace distance $dS^2 = dZd\bar{Z}$ is simply rescaled as

$$dS^2 \rightarrow dS'^2 = dZ' d\bar{Z}' = \Lambda(Z,\bar{Z})\, dS^2 \tag{3.4}$$

where $\Lambda(Z,\bar{Z}) = (1 + \partial_z \varepsilon + 2\theta\partial_z\beta)(1 + \partial_{\bar{z}}\bar{\varepsilon} + 2\bar{\theta}\partial_{\bar{z}}\bar{\beta})$. The generators $\delta \equiv l(\varepsilon) + g(\beta)$ corresponding to the transformations with $\varepsilon(z) = \varepsilon_n z^{n+1}$ and $\beta(z) = \beta_m z^m$ are given by

$$l(\varepsilon) = \varepsilon_n l_n = \varepsilon_n z^n \{ z\partial_z + \frac{1}{2}(n+1)\theta\partial_\theta \} \tag{3.5a}$$

$$g(\beta) = \beta_m g_m = \beta_m z^m (\partial_\theta - \theta\partial_z) \tag{3.5b}$$

which form a "classical" superconformal algebra of N = 1:

$$[l_m, l_n] = (n-m)\, l_{m+n} \tag{3.6a}$$

$$\{g_m, g_n\} = -2 l_{m+n} \tag{3.6b}$$

$$[l_m, g_n] = (n - \frac{m}{2})\, g_{n+m} \tag{3.6c}$$

The quantized form of the algebra emerges when one considers the transformation of fields under the conformal transformation. The primary conformal field $\phi_{h,\bar{h}}(z,\bar{z})$ with conformal weights h and $\bar{h}$ is defined by the transformation[6]

$$\phi_{h,\bar{h}}(z,\bar{z}) dz^h d\bar{z}^{\bar{h}} = \tilde{\phi}_{h,\bar{h}}(w,\bar{w}) dw^h d\bar{w}^{\bar{h}} \tag{3.7}$$

under $z \rightarrow w = z - \varepsilon f(z)$ and $\bar{z} \rightarrow \bar{w} = \bar{z} - \bar{\varepsilon}\bar{f}(\bar{z})$. for the infinitesimal change $|\varepsilon|, |\bar{\varepsilon}| \ll 1$ one finds

$$\begin{aligned} \delta\phi_{h,\bar{h}}(z,\bar{z}) &= \tilde{\phi}_{h,\bar{h}}(z,\bar{z}) - \phi_{h,\bar{h}}(z,\bar{z}) \\ &= \varepsilon\{ f\partial_z\phi_{h,\bar{h}} + h(\partial_z f)\phi_{h,\bar{h}} \} \\ &+ \bar{\varepsilon}\{ \bar{f}\partial_{\bar{z}}\phi_{h,\bar{h}} + \bar{h}(\partial_{\bar{z}}\bar{f})\phi_{h,\bar{h}} \} \end{aligned} \tag{3.8}$$

Restricting ourselves to the z space and making the choice of $f(z) = f_n(z) - z^{n+1}$, one obtains

$$\delta_{f_n}\phi_h(z) = \varepsilon [l_n + h(n+1)z^n]\, \phi_h(z) \tag{3.9}$$

The quantized form of l_n, i.e. the Virasoro operators L_n is defined in the Hilbert space which is spanned by the conformal fields and is given such that

$$[L_n, \phi_h(z)] = \{L_n + h(n+1)z^n\} \phi_h(z) \tag{3.10}$$

It turns out that L_n should form an algebra similar to what l_n satisfies, i.e. (2.10), but with the anomalous constant term:

$$[L_m, L_n] = (m-n)L_{m+n} + \frac{c}{12}(m^3 - m)\,\delta_{m+n,0} \tag{3.11}$$

This is the Virasoro algebra with the central extention. c is called the central charge.

Similarly one has the supersymmetric extention of the Virasoro algebra corresponding to the classical expression (3.6):

$$[L_m, L_n] = (m-n)L_{m+n} + \frac{\hat{c}}{8}(m^3 - m)\,\delta_{m+n,0} \tag{3.12a}$$

$$\{G_m, G_n\} = 2L_{m+n} + \frac{\hat{c}}{2}(m^2 - 1)\,\delta_{m+n,0} \tag{3.12b}$$

$$[L_m, G_n] = (\frac{m}{2} - n)\,G_{m+n} \tag{3.12c}$$

The N = 1 supersymmetry incorporates boson-fermion correspondence, and the coefficient of the central extention in (3.12a) carries that information such that $c/12 = (1 + 1/2)\,\hat{c}/12 = \hat{c}/8$. The fermion contributes half as much as the boson to the term of central extention.

4. Representation of Conformal Algebra

The representation space of the Virassro algebra is spanned on the vacuum $|0\rangle$ which is specified by the condition:

$$L_n |0\rangle = 0, \quad n = -1, 0, 1, 2, \cdots \tag{4.1}$$

The first three relations for n = -1, 0, 1 are representing the invariance of the vacuum under the rotations, dilatations, and special conformal transformations. What is special here is the additional condition with n = 2. All the remaining relations for $n \geq 3$ follow as a consequence of the Virasoro algebra.

The ground state $| h \rangle$ with conformal dimension h is constructed by operating the conformal field $\phi_h(z)$ on the vacuum $| 0 \rangle$:

$$| h \rangle = \lim_{z\to 0} \phi_h(z) | 0 \rangle \qquad (4.2)$$

As a consequence of the relation (3.10) the state $| h \rangle$ satisfies

$$L_0 | h \rangle = h | h \rangle, \quad L_n | h \rangle = 0, \quad n \geq 1, \qquad (4.3)$$

which are refereed to as the highest weight conditions of $| h \rangle$.

The representation of the Virasoro algebra corresponding to the highest weight state $| h \rangle$ is spanned by those states which are obtained by operating the raising Virasoro operators on $| h \rangle$:

$$| h; \{n_l\} \rangle = (L_{-k})^{n_k} (L_{-(k-1)})^{n_{k-1}} \cdots (L_{-1})^{n_1} | h \rangle \qquad (4.4)$$

The level N of a given excited state is defined by $N = \sum_{l=1}^{N} l n_l$. Obviously

$$L_0 | h ; \{n_l\} \rangle = (h + N) | h; \{n_l\} \rangle \qquad (4.5)$$

One can introduce secondary fields

$$\phi_h^{\{n\}}(z) = \phi_h^{\{\overbrace{-k,\cdots,-k}^{n_k};\cdots;\overbrace{-1,-1,\cdots,-1}^{n_1}\}}(z) \qquad (4.6)$$

in one to one correspondence to each excited state such that

$$| h; \{n_l\} \rangle = \lim_{z\to 0} \phi_h^{\{n\}}(z) | 0 \rangle \qquad (4.7)$$

For example, $\phi_h^{(-1)}(z) = \partial_z \phi_h(z)$, etc. The primary and secondary conformal fields saturate the representation with highest weight h. Their transformation property is known by construction. In other words, the central charge c and the conformal dimensions h and $\bar{h}$ completely characterize the theory of fields involving $\phi_{h,\bar{h}}(z,\bar{z})$.

As a simple example, let us consider a two-point function

$$G_{12}(z_1,\bar{z}_1,;z_2,\bar{z}_2) = \langle \phi_1(z_1,\bar{z}_1)\phi_2(z_2,\bar{z}_2) \rangle \qquad (4.8)$$

Naturally, the correlator must be a function of $z_1 - z_2$ and $\bar{z}_1 - \bar{z}_2$.

Its expression is furthermore restricted from the invariance of the theory to the form G_{12} = constant x $(1/(z_1 - z_2)^{\alpha}(\bar{z}_1 - \bar{z}_2)^{\beta})$. Now we operate the conformal transformation $z \to w$, $\bar{z} \to \bar{w}$ on G_{12}. Then the relation

$$\left(\frac{dw_1}{dz_1}\right)^{-h_1}\left(\frac{d\bar{w}_1}{d\bar{z}_1}\right)^{-\bar{h}_1}\left(\frac{dw_2}{dz_2}\right)^{-h_2}\left(\frac{d\bar{w}_2}{d\bar{z}_2}\right)^{-\bar{h}_2} G_{12}(z_1,\bar{z}_1;z_2,\bar{z}_2)$$

$$= \frac{\text{constant}}{(w_1 - w_2)^{\alpha}(\bar{w}_1 - \bar{w}_2)^{\beta}} \tag{4.9}$$

holds. Suppose we consider the transformation $w = z + \varepsilon z^2$, $\bar{w} = \bar{z} + \bar{\varepsilon}\bar{z}^2$, then we get

$$-2(\varepsilon h_1 z_1 + \bar{\varepsilon}\bar{h}_1\bar{z}_1 + \varepsilon h_2 z_2 + \bar{\varepsilon}\bar{h}_2\bar{z}_2)$$

$$= -\alpha\,\varepsilon(z_1 + z_2) - \beta\,\bar{\varepsilon}(\bar{z}_1 + \bar{z}_2) \tag{4.10}$$

From this we conclude that $G_{12} = 0$ unless $h_1 = h_2 = h$, $\bar{h}_1 = \bar{h}_2 = \bar{h}$, and that $\alpha = 2h$, $\beta = 2\bar{h}$. Introducing the radial coordinates

$z_1 - z_2 = r_{12}e^{i\theta_{12}}$, $\bar{z}_1 - \bar{z}_2 = r_{12}e^{-i\theta_{12}}$, we obtain

$$G_{12}(z_1,\bar{z}_1;z_2,\bar{z}_2) = \frac{\text{constant}}{(z_1 - z_2)^{2h}(\bar{z}_1 - \bar{z}_2)^{2\bar{h}}}$$

$$= \frac{\text{constant}}{r_{12}^{2\Delta}\, e^{2i\Sigma\theta_{12}}} \tag{4.11}$$

where the scaling dimension Δ and conformal spin Σ is generally defined by

$$\Delta = h + \bar{h}, \qquad \Sigma = h - \bar{h} \tag{4.12}$$

Σ takes only half-integers. This example typically exemplifies how the theory is determined in the conformal framework.

When one applies the "minimal"[6] representations of the conformal theory to statistical models of critical phenomena, their critical exponents such as Δ can be determined to fractional values which are unambignously given as the mathematical numbers of the representation theory of the conformal algebra.

5. Vertex Operators

Vertex operators were first invented in the operator formalism of dual resonance models (DRM)[7]. It is a non-polynomial operator of a scalar field:

$$V(t,z) = :e^{it\phi(z)}: \tag{5.1}$$

where

$$\phi(z) = q - i(p-\lambda)\,ln\,z + i\sum_{n\neq 0}\frac{a_n}{n}z^{-n} \tag{5.2}$$

with $p \equiv a_0$,

$$[q,p] = i, \quad [a_m, a_n] = m\delta_{m+n,0} \tag{5.3}$$

and λ a c-number. Let us define $A(z) = i\partial_z\phi(z)$. Then

$$L(z) = :(\frac{1}{2}A(z)^2 - \lambda\partial_z A(z)): = \sum_{n=-\infty}^{\infty} L_n z^{-n-2} \tag{5.4}$$

and

$$L_n = \frac{1}{2}\sum_{k=-\infty}^{\infty} :a_k a_{n-k}: + \lambda n a_n - \frac{\lambda^2}{2}\delta_{n,0} \tag{5.5}$$

which is the oscillator representations of the Virasoro algebra (3.11) with

$$c = 1 - 12\lambda^2 \tag{5.6}$$

In this representation the vertex operator V(t,z) plays the role of primary conformal field with conformal dimension

$$h = h(t) = \frac{t^2}{2} + \lambda t \tag{5.7}$$

It turns out that this representation is quite powerful for the generic construction of null-states as well as the general proof of Kac-formulas in the N = 0 conformal and the N = 1 and 2 superconformal algebras, which have been rigorously accomplished by Kato and Matsuda [3].

The vertex operators could also play an important role in constructing the basic representation of a Kac-Moody Lie algebra. The development was sparked by the paper of Frenkel and Kac[8]. In the first page of their paper we can find an impressive description about the remarkable interplays between mathematics and physics which we have come across in the past history of science. In particular, we note here as the most recent one[8] the so-called Frenkel-Kac construction of the affine Lie algebra based on the vertex operators of DRM.

6. Comments on N = 1, 2, and 4 Superconformal Algebras

The Virasoro algebra of N = 0 is supersymmetrically extended to the N = 1, 2, and 4 superconformal algebras. The N = 1 algebras corresponding to the Neveu-Schwarz and the Ramond boundary conditions were introduced as the realization of the world-sheet supersymmetry of the string. The superstring variable [9]

$$X^{\mu}(z) = x^{\mu}(z) + \theta\, \psi^{\mu}(z), \qquad \mu = 1, 2, \cdots, d \tag{6.1}$$

which carries boson ($x^{\mu}(z)$) and fermion ($\psi^{\mu}(z)$) fields maps the two dimensional string manifold to the target space of d dimensions:

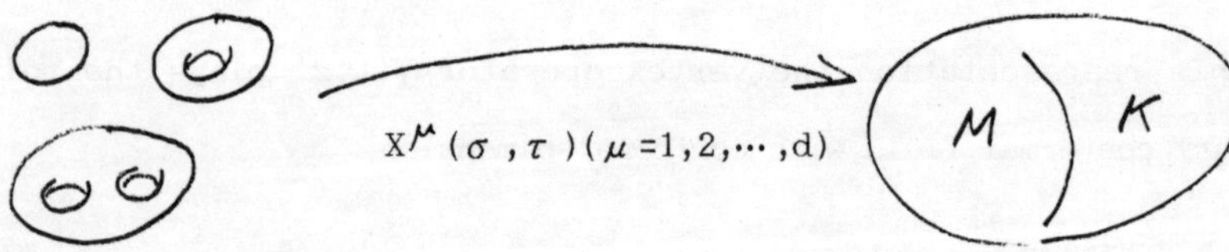

string manifold target space

The target space is compactified to $M_{even} \times K_{even}$ space, where M_{even} stands for the non-compactified open space of even space-time dimensions while K_{even} the compactified closed space representing internal symmetries. The critical dimension of the superstring is d = 10. Thus the target space is decomposed into $M_4 \times K_6$, $M_6 \times K_4$, and $M_2 \times K_8$. the manifold structures of each compactified space are Calabi-Yau (K_6), K3(K_4), and "Calabi-Yau like" (K_8) with extended supersymmetry of N = 2, N = 4, and N = 4, respectively[10]. Thus the higher symmetry of N = 2 and N = 4 are realized as the space-time structure of the superstring. In terms of n complex dimensions K_{2n} lives in odd (N = 2) and even (N = 4) dimensions. The geometrical correlation between the complex manifold and the extended supersymmetry is shown in the Table.

	Manifold (M)	Grassmann (θ_i)	Kac-Moody	Ult.viol.div. behavior
N = 1	Riemann	real		4 loop div.
N = 2	Kahler	complex	U(1)	4 loop div.
N = 4	hyper-Kahler	quaternion	SU(2)	on-shell ult. viol.finite to all orders

Table

The Grassmann variables are real, complex, and quaternionic for N = 1, 2, and 4, respectively. Furthermore, the embedded Kac-Moody algebraic structure for N = 2 and 4 is ableian U(1) and non-abelian SU(2). We also have added some comments on the ultraviolet behavior of the non-linear σ model with the corresponding extended supersymmetry.

I am currently working with T. Uematsu on the superfield representation of the N = 4 superconformal algebra[5]. The N = 4 structure obeys the symmetry reduction of SO(4) ~ $SU(2)_{local}$ x $SU(2)_{global}$ where $SU(2)_{local}$ is identified to be the SU(2) Kac-Moody algebra. In our formulation $SU(2)_{global}$ plays a crucial role in constructing a consistent superfield formalism for the N = 4 algebra.

7. Historical Overview

Finally I shall close my talk by presenting a diagrammatic historical overview of the interplay between physics and mathematics concerning the development of conformal approach. A concrete example of infinite dimensional conformal Lie algebras was first given by the Virasoro algebra which was actually invented by the collaboration of many people who were involved in the study of DRM in late 60's. The Virasoro operators were first recognized as the generalized gauge operators corresponding to SL(2,C) invariance and its extended symmetry of the n-point Veneziano amplitude. In the string picture of DRM, i.e. dual string model (DSM), the symmetry is just the invariance under the general coordinate transformation of the string coordinates. Furthermore, DRM triggered the invention of supersymmetry in the modern form, of superstrings as TOE (theory of everything), of vertex operators for the Frenkel-Kac construction, and thus has opened up a generic approach based on conformal invariance to the problems of physics and mathematics, including particle physics, statistical physics, and representation theory of infinite dimensional Lie algebras.

It is quite interesting and suggestive to note that when Polyakov[11] proposed the idea of local scale transformation for studying critical phenomena in 1970, the existence of infinite dimensional conformal Lie algebras in the field of DRM seemed not to have been recognized by him.

Historical Overview

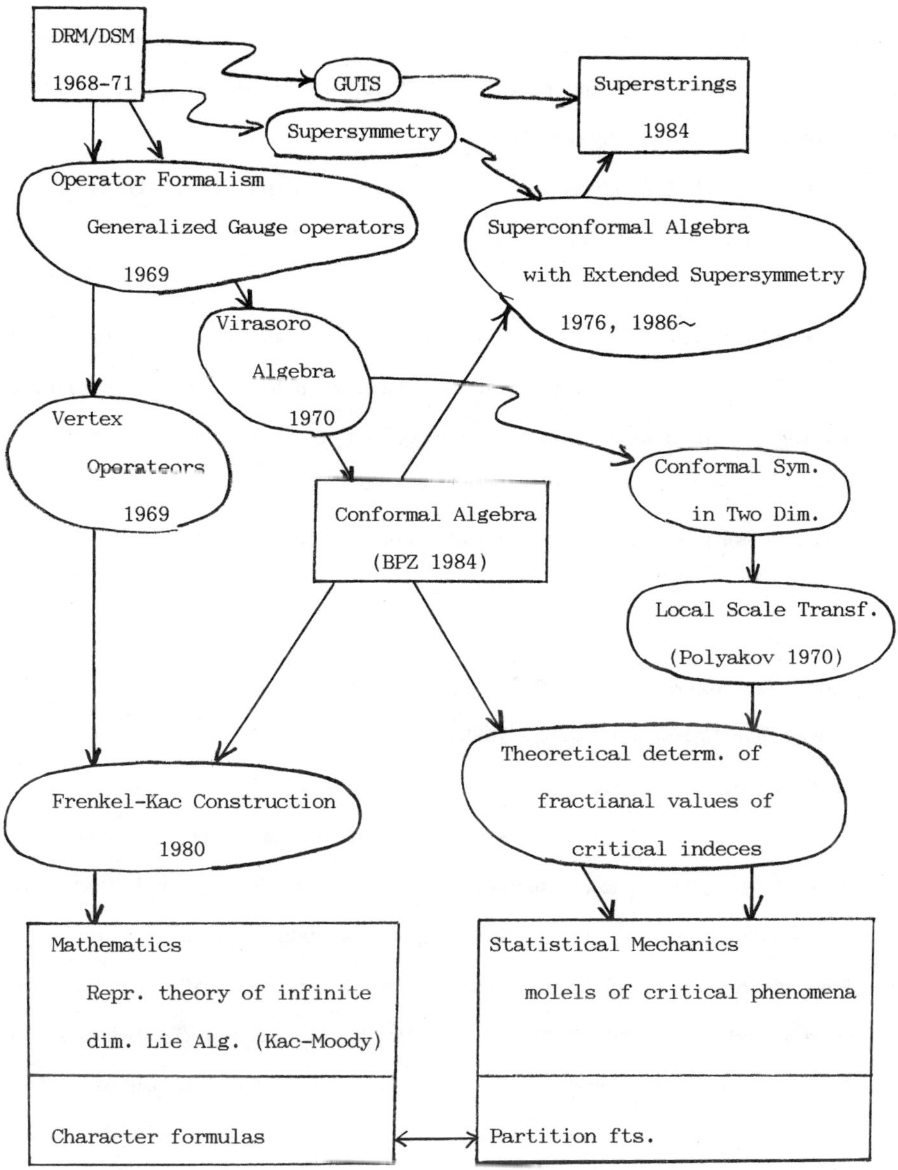

It was only after the mathematicians picked up the machinery of vertex operators from DRM for the study of the basic representation of the affine Kac-Moody algebras that the remarkable development of conformal approach was initiated by the paper of BPZ[6] and the fruitful interplay between mathematics and physics had began.

I presume that conformal approach in physics will be as fruitful as ever in the future.

Finally I would like to express my gratitude and respect to my life-long teacher, Professor Hironari Miyazawa on his sixtieth birthday, for having provided enthusiasm, encouragement and constant support to my carrier as a physicist throughout hard times as well as good times. I do wish him many happy returns and as lively life as ever.

References

1. See, for example, P.G.O. Freund, Introduction to Supersymmetry (Cambridge University Press 1986).
2. H.Miyazawa, Phys.Rev.170, 1586(1968).
3. S.Matsuda and M.Kato, Proceedings of the Workshop on Superstrings, June 9-11, 1987, KEK, Tsukuba, Japan, ed. by M.Kato and Y.Kazama, KEK Report 87-23, p.112; M.Kato and S.Matsuda, Advanced Studies in Pure Mathematics 16: Conformal Field theory and Solvable Lattice Models, eds. M.Jimbo, T.Miwa and A.Tsuchiya, p.205-p.254 (Kinokuniya Co.Ltd., Tokyo , Japan, April 1988). See also references cited therein.
4. T.Eguchi and A.Taorumina, Phys.Lett.B196, 75(1987); A.Kent and H.Riggs, Phys.Lett.B198, 491(1987).

5. S.Matsuda and T.Uematsu, in preparation.

6. A.A.Belavin, A.M.Polyakov and A.B.Zamolodchikov, Nucl.Phys.B241, 333(1984).

7. M.Jacob(ed.): Dual Theory (North-Holland, Amsterdam 1975).

8. I.B.Frenkel and V.G.Kac, Inv.Math.62, 23(1980).

9. M.G.Green, J.H.Schwarz and E.Witten, Superstring Theory Vol. 1 and 2 (Cambridge University Press 1987).

10. T.Eguchi, private communication.

11. A.M.Polyakov, JETP Lett.12, 381(1970).

MONOPOLES FROM SUPERSTRING THEORY WITH COMPACTIFIED EXTRA DIMENSIONS

Mitsuyoshi Tomiya

Institute for Nuclear Study, University of Tokyo

Tanashi, Tokyo 188

JAPAN

ABSTRACT

The non-abelian magnetic monopoles associated with symmetry breaking by the compactification of extra dimensions is considered. We find the monopole solution of the SU(2) gauge theory on $M^4 \times S^1$. It is shown that the compactified space may blow up inside the monopoles. It is also discussed the relation betweeen the Higgs mechanism and the compactification briefly.

1. INTRODUCTION

One of the most mysterious fact in physics is that we cannot find any monopole on earth until today.[1] Gauge theory succeeds to understand the strong interaction and the electro-weak interaction. Even the general relativity is also a "gauge theory" in general. In mathematical notion, both are described in the fibre bundle theory. Therefore we expect monopole to exist, because it is a natural by-product of gauge theory. But we have no experimental proof of its existence. Physicists have discussed various theories that try to explain the absence of monopoles.[2] This paper is another trial of it from the view of superstring unification. In GUTs the Higgs field with which gauge theories is associated, plays essential role to break down the symmetries from very high energy region to our world's energy scale. The idea of non-local elements (i.e. superstring, super-membrane $\cdots$)[3,4] are now seriously considered as the candidate for the grand unified theory. The heterotic string is the most favorable model which is made of a left-moving bosonic string in 26 dimensions and a right-moving superstring in 10 dimensions [5]. The gauge group is obtained by the skillful compactification of extra 16 dimensions(= rank of the gauge group) of the bosonic string. Therefore the heterotic string theory is 10-dimensional theory with $E_8 \times E_8$ (or SO(32),...). In spite of miraculous success in 10 dimension super-string theories are still lack of reliable dynamics of reduction of the dimension of space-time to 4 dimension.(There are also another type of superstring theories which are not necessary by the route of 10 dimensions[6], but it will not discussed here.) Anyhow a low-energy effective theory must be 4-dimensional. It is discussed by the compactification of extra dimensions:

$$M^{10} \to M^4 \times K^6 \quad . \tag{1}$$

without "physical" dynamics[7]. The structure of compact manifold K^6 depends on the pattern of the compactification[6,7].

Orbifold and Calabi-Yau manifold are mostly adopted as K^6 from the requirement of supersymmetry in the low-energy effective theory. In both case K^6 is constructed from a simply-connected manifold K_0^6 divided by discrete symmetries $D \subset K_0^6$: K_0^6/D. The non-triviality of the fundamental group is necessary to break the gauge symmetries($E_8 \times E_8$, SO(32) ...) down to some realistic one. The Wilson loop around the fixed point would not be trivial, because the loop cannot always shrink to a point. We adopt $M^4 \times S^1$ as a model of $M^4 \times K^6$ [8], because $M^4 \times S^1$ plays an essential role in expanding gauge configuration on M^4 to higher dimensional space [9].

2. THE HIGGS MECHANISM AND THE COMPACTIFICATION

At first let us discuss the relation between the Higgs mechanism and the compactification briefly. For example we consider SU(2) Yang-Mills theory on $M^4 \times S^1$. We introduce a metric on $M^4 \times S^1$:

$$ds^2 = (dx^0)^2 - (dx^1)^2 - (dx^2)^2 - (dx^3)^2 - R^2 d\theta^2 . \tag{2}$$

The action is defined as

$$S = \int d^5x\sqrt{g}\,\mathcal{L}(x)$$

$$= \int d^5x\sqrt{g}\left[-\frac{1}{4}\,\mathrm{Tr}(F_{mn}F^{mn})\right] \tag{3}$$

$$m,n = 0,1,2,3,5$$

R is the radius of S^1 and assumed to be constant for a while. We shall rewrite (3) by using the following new variables:

$$A_\mu = A^3_\mu, \tag{4a}$$

$$D_\mu = \partial_\mu - ieA_\mu, \tag{4b}$$

$$F_{\mu\nu} = \partial_\mu A_\nu - \partial_\nu A_\mu, \tag{4c}$$

$$W^\pm_\mu = \frac{1}{\sqrt{2}}\,(A^1_\mu \mp A^2_\mu), \tag{4d}$$

$$\phi^\pm = \frac{1}{\sqrt{2}}\,(A^1_5 \mp A^2_5), \tag{4e}$$

$$v = \frac{1}{eR}\,\left(\frac{e\Phi}{2\pi} + m\right), \tag{4f}$$

$$\xi + v = A^3_5, \tag{4g}$$

where μ,ν run from 0 to 3. Then we have

$$\mathcal{L} = -\frac{1}{4}\,F_{\mu\nu}F^{\mu\nu}$$

$$-(D_\mu W^+_\nu)(D^{\mu *}W^{-\nu}) + (D_\nu W^+_\mu)(D^{\mu *}W^{-\nu}) + e^2v^2W^+_\mu W^{-\mu}$$

$$+\frac{1}{2}\,(\partial_\mu\xi)(\partial^\mu\xi) + (D_\mu\phi^+)(D^\mu\phi^-)$$

$$+\ (\text{interaction terms}), \tag{5}$$

This shows that W has a mass of $m_W = \frac{1}{R}(\frac{e\Phi}{2\pi} + m)$, and $A_5^i (i=1,2,3)$ plays the role of a Higgs boson in the adjoint representation of the SU(2) group. Note that symmetries is broken down even if the boundary condition is periodic. Only we need is non-zero vacuum expectation value of the Wilson loop.

3. TWISTED BOUNDARY CONDITIONS IN SUPERSTRING THEORY

The obstruction theory tells us the necessary condition, when we try to extend the fibre bundle defined on n-dimensional manifold M to (n+1) dimensional manifold K: $M \subset K$[10]. Wen and Witten derived the condition that the monopole charge must be $Z = \ell ne$ (ℓ:integer), when $\pi_1(K^6) = Z_n$[9]. They keep in mind E_6 as the gauge group of low energy effective theory. Therefore the process $M^4 \to M^4 \times S^1$ is an only possible "obstruction" for the extension, because $\pi_3(E_6) \neq 0$ and $\pi_i(E_6) = 0$, $4 \leq i \leq 6$. But the necessary condition says almost nothing about the property of actual monopole solutions except the charge.

We embed S^1 in K^6 such that S^1 surrounds the fixed point of K^6 ($S^1 \subset K^6$). Hence we get the twisted boundary conditions (T.B.C.) is obtained for fermions;

$$\Psi(x,g(x^5)) = U_g \Psi(x,x^5) \tag{6}$$

where $x \; M^4$, $x^5 \; S^1$, $x^5 = R\theta$; $0 \leq \theta < 2\pi$, R is a radius of S^1 and

$$g(x^5) = x^5 + 2\pi R \; .$$

U_g is the Wilson loop and has a property;

$$U_g{}^n(x) = 1 \tag{7}$$

It is a consequence of $\pi_1(K^6) = Z_n$. U_g becomes a representation of D in a gauge group G. Smaller gauge group G' on M^4 is defined:

$$[a, U_g] = 0, \qquad \forall a \in G' \subset G \; . \tag{8}$$

This time radius of S^1 must be seriously considered to find out monopole solutions. We assume that fields have no dependence on x^5(a few more assumptions are needed to find one-monopole solutions later.);

$$S = \int d^4x \mathcal{L}^{4\text{-dim}}(x)$$

$$\mathcal{L}^{4\text{-dim}}(x) = 2\pi R\{-\frac{1}{4}\,\mathrm{Tr}(F_{\mu\nu}F^{\mu\nu}) + \frac{1}{2}R^2(D_\mu\Phi)(D^\mu\Phi)\} \tag{9}$$

$$D_\mu = \partial_\mu + eA_\mu\times, \quad \mu,\nu = 0,1,2,3$$

where a part a vector potential $A^5(x)$ turns out to be an adjoint scalar field Φ; $\Phi = A^5$. From (9) the equations of motion are derived

$$D_\mu(RF^{\mu i}) + eR^3\Phi\times(D^i\Phi) = 0 \tag{10a}$$

$$D_\mu(R^3D^\mu\Phi) = 0 \tag{10b}$$

and we obtain the constraint equations as

$$D_\mu(RF^{\mu 0}) + eR^3\Phi\times(D^0\Phi) = 0.$$

On this model the vacuum expectation value of Φ generates the non-trivial loop on S^1:

$$U_g(x) = e^{ie\Phi\tau_r}$$

$$\Phi = \int dx^5 \langle A_5(x)\rangle = \int_0^{2\pi} d\theta\; g_{55}\langle\Phi(x)\rangle \tag{11}$$

$$= 2\pi R^2\langle\Phi(x)\rangle\ .$$

Then the gauge symmetry on M^4 breaks down to U(1) from SU(2). If we adopt $A_5(x)$ as $\Phi(x)$, $\Phi = 2\pi\langle\Phi(x)\rangle$. Therefore $\langle\Phi(x)\rangle$ become constant and we cannot make monopole solutions.

Later we consider spherically symmetric monopole. Here we must check the relation between $U_g(x)$ and spherically symmetric ansatz. It derives a stringent constraint on $U_g(r)$. The T.B.C. becomes

$$\psi(\theta+2\pi) = e^{ie\Phi\tau_r}\psi(\theta)$$

where $\tau_r = \frac{\vec{\tau}\cdot\vec{X}}{r}$ and Φ is determined so that $U_g(r) = e^{ie\Phi\tau_r}$ satisfied $U_g^n(r) = 1$. The constraint that $U_g(r)$ must be regular at r=0, namely $\lim_{r\to 0}[\cos(e\Phi)+i\sin(e\Phi)] = \text{const.}$, requires

$$U_g(0) = \pm 1 \ . \tag{12}$$

To accomplish (12), we have two choices:

(i) $\lim_{r\to 0} e\Phi(r) = \pi k$ (k: integer)

(ii) $U_g(r) = \pm 1$ everywhere. Hence $e\Phi = \pi k$ (k: integer).

In the case (i) where Φ is a function of r, it is impossible to put the condition $U_g^n = 1$ for fixed integer n. Since $\pi_1(K^6) = Z_n$, varying n means the change of topology of K. We want to keep the topology of K unchanged. Thus we shall not consider this case further. On the other hand, the case (ii) is acceptable. $U_g = +1$ and $U_g = -1$ correspond to periodic and anti-periodic boundary conditions respectively.

4. MONOPOLE SOLUTIONS ON $M^4\times S^1$

Next, let us find the monopole solutions on $R^3\times S^1$. At least, in the theory of a (anti-)self-dual monopole, which is constructed from the twistor theory, proved that various solutions of one (anti-)monopole are essentially equivalent [11]. It means that all the solutions are gauge transformed each other. Therefore we afterwards omit time (x^0)-dependence of the fields and adopt the spherically symmetric ansatz [12]:

$$A_i^a(x) = \varepsilon_{abi}x^b\left(\frac{K(r)-1}{er}\right)$$

$$\phi^a(x) = H(r)\,\frac{\hat{X}^a}{e} \tag{13}$$

$$A_0^{a}(x) = 0 \qquad : \text{unitary gauge}$$

and there is no x^0-dependence. From eq.(11) we get a constraint;

$$R^2(r)H(r) = \text{const.} \tag{14}$$

Therefore the equations of motion become

$$r^2K'' = K^3 - \frac{r^2H'K'}{2H} + e(m + \frac{\ell}{2})r^2HK - K \tag{15a}$$

$$r^2H'' = \frac{3r^2H'^2}{2H} - 2rH' + 2HK^2 . \tag{15b}$$

where we use the relation from eq.(14) $\left(\frac{R'}{R} = -\frac{H'}{2H}\right)$.
If the regularity at the origin (the center of monopole) is physically expected, the Taylor expansion tells us;

$$\begin{aligned} K(r) &= 1 + \alpha r^k + \cdots \quad (k>0) \\ & \qquad\qquad\qquad\qquad\qquad (r\approx 0) \\ H(r) &= \beta r^h + \cdots \quad (h>0) . \end{aligned} \tag{16}$$

We substitute these for eq.(15b) and we get in the lowest order

$$h^2 - 2h + 4 = 0 . \tag{17}$$

Note that the cases of h=0 and h=1 must be checked independently, because eq.(15b) has a second order derivative term of H(r). Eq.(17) does not have real solutions ($h = 1\pm i\sqrt{3}$) and the cases of h=0,1 has no solutions which are consistent with (15a) and (15b). Therefore we have to abandon the regularity and admit the solution of the type;

$$K(r) = ar^1 + br^2 + \cdots . \tag{18}$$

Then we cannot avoid a singularity of the vector potential: $A_i^{a}(x) \propto \frac{1}{r}$ $(r\approx 0)$. It implies the monopole has infinite electro-magnetic energy, if it is a precise solution through the core. But, inside the core of some scale $O(1/\mu)$ such as the Planck mass, we suppose some unknown mechanism will be dominant as usual discussion of monopoles[13]:

$$E_{em} = \int d^3x\ B^aB^a \propto 4\pi\int_{1/\mu}^{\infty} dr\ r^2\left(\frac{1}{r^4}\right) \propto 4\pi\mu \ . \tag{19}$$

The electromagnetic energy E_{em} may not necessarily diverge. (But the mass of monopole is crucially dependent on μ and also v.) Hence we look for the solution of the type (18).

Let us try $K(r) = 0$. It is the simplest case which admit the singularity $A_i{}^a(x) \propto \frac{1}{r}$ ($r \approx 0$) (type (18)). This time (15b) becomes

$$H^{3/2}\ \frac{d}{dr}\left(\frac{r^2H'}{H^{3/2}}\right) = 0 \ . \tag{20}$$

At $r \to \infty$ we require the same boundary conditions as an well-known self-dual one-monopole solutions[11]:

$$(R(r)\cdot r^2H'(r))|_{r\to\infty} = 1$$
$$H(r)|_{r\to\infty} = v \quad (\neq 0) \tag{21}$$

because we disallow manifest physical difference between our solution and ordinary one. Then (20) is exactly solved:

$$H(r) = v\left(\frac{r}{\sqrt{\frac{e}{2kV}} + r}\right)^2 \tag{22}$$

and

$$A_i{}^a(x) = -\varepsilon_{abi}x^b/er$$
$$\phi^a(x) = \frac{v}{e}\left(\frac{r}{\sqrt{\frac{e}{2kV}} + r}\right)^2 x^a \tag{23}$$

$$R(x) = \frac{1}{2v}\cdot\frac{1}{r} + \sqrt{\frac{k}{2eV}}) .$$

The radius of the compactified space becomes larger when we get closer to the core. This property is almost inevitable by the inspection that all the Prasad-Sommerfield type solutions take $(\mathrm{tr}\Phi^2)^{1/2}(=H(r))\to 0$ $(r\to 0)$[11,14] and we also have the condition (14).

5. DISCUSSION

The most significant result is that the compactified space blows up inside the spherically symmetric monopoles. As a consequence, the SU(2) is restored inside the core. If there exist such monopoles, it has been shown that entropy is pumped into the effective four dimensional world by the shrinkage of extra dimensions[15]. Hence, annihilation of a monopole and an anti-monopole release more energy than that expected from the GUT's monopoles. We have also shown that spherically symmetric solutions permit only periodic and anti-periodic boundary conditions of (fermion) fields: $U_g^2 = 1$. If we understand the Witten mechanism as the framework of (5), even T.B.C. does not seem to be necessary for the symmetry break down. If we forbid the supposition of the unknown mechanism at very high energy, then superstrings theories seems to prohibit the existence of monopole. However more definite arguments rely on a dynamics of the compactification.

The author wish to express my cordial thanks to Prof. H. Miyazawa for continous encouragement and constructive suggestion since I was a graduate student of him. He would like to thank Prof. H. Terazawa and Dr. S. Midorikawa for encouragement and much useful advice which is directly related to this work,and also to all the other members of the theory division at INS for stimulating discussions.

References

1. Ebisu, T. and Watanabe, T., Phys. Rev. D36, 3359 (1987).
   Groom, D. E., Phys. Rep. 140, 323 (1986).
2. Guth, A. H., Phys. Rev. D23, 347 (1980).

Linde, A. D.,Phys. Lett. 114B, 431 (1982).
Albrecht, A. and Steinhardt, P. J., Phys. Rev. Lett. 48, 1220 (1982).

3. Green, M. B. and Schwartz, J. H., Nucl. Phys. B181, 502 (1981) ;Nucl. Phys. B198, 252, 441 (1982);Phys. Lett. 109B, 444 (1982).
   Green, M. B., Schwartz, J. H. and Brink, L., Nucl. Phys. B198, 474 (1982).
4. Dirac, P. A. M., Proc. Roc. Soc. A268, 57 (1982).
   Collins, P. A. and Tucker, P. W., Nucl. Phys. B112, 150 (1976).
   Howe, P. S. and Tucker, P. W., J. Phys. A10, L155 (1977).
   Sugamoto, A., Nucl. Phys. B215, 385 (1983).
   Kikkawa, K. and Yamasaki, M., Prog. Theor. Phys. 76, (1986) 1379.
   Bergshoeff, E., Sezgin, E. and Townsend, P. K., Phys. Lett. 189B, 75 (1987).
   Duff, M. D., Howe, P. S., Inami, T. and Stelle, K. S., Phys. Lett. 191B, 70 (1987).
5. Gross, D. J., Hurvey, J. A., Martinec, E. and Rohm, R., Phys. Rev. Lett. 54, 502 (1985);Nucl. Phys. B256, 253 (1985);Nucl. Phys. B267, 75 (1986).
6. Narain, K. S., Phys. Lett. 169B, 41 (1986).
   Narain, K. S., Sarmadi, M. H. and Witten, E., Nucl. Phys. B279, 369 (1987).
   Kawai, H., Lewellen, D. C. and Tye, S.-H. H., Phys. Rev. Lett. 57, 1832 (1986);Nucl. Phys. B288 1 (1987).
7. Candelas, P., Horowitz, G. T., Strominger, A., and Witten, E., Nucl. Phys. B258, 46 (1985).
   Witten, E., Nucl. Phys. B258, 75 (1985).
   Dixon, L., Harvey, J. A., Vafa, C., and Witten, E., Nucl. Phys. B261, 678 (1985).
8. Midorikawa, S., and Tomiya, M., INS-Rep.-597, 603 (1986); INS-Rep.-620 (1987).
   Tomiya, M., INS-Rep.-665 (1987) to be published in J. Phys. G.
9. Wenn, X-G. and Witten, E., Nucl. Phys. B261, 651 (1985).

10. Steenrod, N., "The Topology of Fibre Bundles"(Princeton Math. Series 14) (Princeton:Princeton UP)(1951).
11. Hitchin, N. J., Comm. Math. Phys. 83, 579 (1982).
12. Julia, B. and Zee, A., Phys. Rev. D11, 2227 (1975).
13. Coleman, S., in "Gauge Theories in High Energy Physics" ed. Gaillard, M. K. and Stora, R., (North-Holland:Amsterdam) (1983).
14. Prasad, M. K. and Sommerfield, C. M.,Phys. Rev. 35, 760 (1975).
15. Alvarez, E. and Gavela, M. B., Phys. Rev. Lett. 51, 931 (1983). Barr, S. M. and Brown, L. S.,Phys. Rev. D29, 2779 (1984).

Supersymmetric Generalization of Extended Conformal Algebras

Samir K. PAUL , Ryu SASAKI[⋆] [†] and Hisahiro YOSHII
Research Institute for Theoretical Physics
Hiroshima University, Takehara, Hiroshima 725, Japan

ABSTRACT

Integer and half odd integer spin super primary fields and their possible operator algebras with super Energy-Momentum tensor in two dimensional $N = 1$ superconformal field theories (Feigin-Fuchs construction) are studied. A general method for obtaining (half) integer spin super primary fields is given. In this framework it is shown that the super generalization of the spin-3 algebra of Fateev and Zamolodchikov does not exist.

1. Introduction

Two dimensional supersymmetric conformal field theories (SCFT)[1] [2] containing integer or half odd integer spin super primary fields are studied. Our main interest lies in whether some of these (half) integer spin fields, being generators of infinite dimensional symmetries,[3] form a closed operator algebra with the super Energy-Momentum tensor. Let us call such algebras supersymmetric generalization of extended conformal algebras in order to avoid confusion with the extended ($N \geq 2$) super conformal algebras. Such algebras, if found, would provide a very

---

⋆ Work supported in part by the Grant-in-Aid from the Ministry of Education, Science and Culture (#63540220).

† Address after Sep. 1988 Department of Mathematical Science, University of Durham, England.

interesting examples of infinite dimensional symmetry algebras relevant to two dimensional critical phenomena as well as to super strings and their possible vacua. In the non supersymmetric CFT several explicit examples of extended conformal algebras are known and well studied.[3][4][5] One simple and systematic way of investigating the possibility of extended conformal algebras is the Feigin-Fuchs construction,[6] which was adopted by one of the present authors in a previous paper.[7] (This paper will be referred to as I hereafter.) Namely, starting from a free bosonic CFT (with $c = n - 1$, integer) one constructs various primary fields of integer spin and try to find closed operator algebras among them including the Energy-Momentum tensor. If closed operator algebras are found then one deforms the Energy-Momentum tensor to make the CFT non-trivial ($c = n - 1 - 12\alpha_0^2\rho^2$) and check whether the operator algebras survive the deformation. One advantage of this method is that since all the operators are explicitly constructed the postulates of CFT, *i.e.*associativity etc., are automatically satisfied except for the unitarity.

In the present paper we generalize the method of I to the $N = 1$ supersymmetric CFT. The result is negative, namely no supersymmetric generalization of extended conformal algebra is found except for the known extended ($N \geq 2$) super conformal algebras. In particular, the supersymmetric version of the W_3-algebra of Fateev and Zamolodhikov,[5] which was given in terms of Feigin-Fuchs construction, is found to be non-existent. This paper is organized as follows. In Chapter 2 basic concepts and notation of SCFT are introduced in order to set the stage. In Chapter 3 we present a general method for finding (half) integer spin super primary fields, which is a simple generalization of the results of I. Explicit forms of general super primary fields for $\hat{c} = 2$ (*i.e.*two super fields U_1 and U_2) SCFT are given. Then the possibility of forming closed operator algebras for $s \leq 2$ is examined. In Chapter 4 the supersymmetric extension of the W_3-algebra ($s = 5/2$ fields) is examined in some detail. It is shown that supersymmetric extension of the W_3-algebra does not close even before deformation.

2. $N = 1$ Super Conformal Algebra

The super Energy-Momentum tensor of a two dimensional ($N = 1$) super conformal field theory is decomposed into the holomorphic $W(\hat{z})$ and the anti-holomorphic $\overline{W}(\hat{\overline{z}})$ parts, in which $\hat{z} = (z,\theta), \hat{\overline{z}} = (\overline{z},\overline{\theta})$. The operator product expansion (OPE) is given by,

$$W(\hat{z}_1)W(\hat{z}_2) = \frac{\hat{c}}{4z_{12}^3} + \frac{3\theta_{12}}{2z_{12}^2}W(\hat{z}_2) + \frac{1}{2z_{12}}DW(\hat{z}_2) + \frac{\theta_{12}}{z_{12}}\partial W(\hat{z}_2) + \ldots, \qquad (2.1)$$

in which $\hat{c}$ is the central charge ($\hat{c} = 3c/2$) and $\ldots$ denotes regular terms. The notation is standard:

$$z_{12} = z_1 - z_2 - \theta_1\theta_2, \theta_{12} = \theta_1 - \theta_2, D = \frac{\partial}{\partial\theta} + \theta\frac{\partial}{\partial z}, D^2 = \partial. \qquad (2.2)$$

The explicit form of the super Energy-Momentum tensor W we are going to discuss in the present paper is the so called Feigin-Fuchs type,

$$\begin{aligned} W &= W_0 + i\alpha_0 W_1, \\ W_0 &= -\frac{1}{2}\sum_{i=1}^{n-1} U_i DU_i, \\ W_1 &= \sum_{i}^{n-1} \rho_i \partial U_i, \\ \hat{c} &= n - 1 - 8\alpha_0^2\rho^2. \end{aligned} \qquad (2.3)$$

Here the fundamental super field U (Grassmann odd) is given by

$$U_i = DX_i = \psi_i + \theta\partial\phi_i, \quad i = 1, \ldots, n-1. \qquad (2.4)$$

The two point functions are

$$\left\langle U_i(\hat{z}_1)U_j(\hat{z}_2)\right\rangle = -\frac{\delta_{i,j}}{z_{12}}. \qquad (2.5)$$

In the above and hereafter polynomials in the fields and their derivatives are always understood as normal ordered. A fundamental building block of SCFT, a super primary field of dimension h is characterized by the following OPE with W :

$$W(\hat{z}_1)\Phi(\hat{z}_2) = h\frac{\theta_{12}}{z_{12}^2}\Phi(\hat{z}_2) + \frac{1}{2z_{12}}D\Phi(\hat{z}_2) + \frac{\theta_{12}}{z_{12}}\partial\Phi(\hat{z}_2) + \ldots, \tag{2.6}$$

As in the case of CFT the product of two super primary fields Φ_1 (dimension h_1) and Φ_2 (h_2) is a super primary field of dimension $h_1 + h_2$, *provided that the product at the same point develops no singularity* . In other words , at the classical level (*i.e.*at the Poisson bracket level) the product of two primary fields is always primary.

Suppose we have found a super primary field P_{h0} w.r.t. the free W_0, a super primary field with respect to the full super Energy Momentum tensor W can be found by deformation. We start from the following *ansatz*,

$$P_h = P_{h0} + i\alpha_0 P_{h1} + (i\alpha_0)^2 P_{h2} + \ldots + (i\alpha_0)^l P_{hl}, \tag{2.7}$$

in which $P_{h1}, \ldots, P_{hl}$ are as yet undetermined differential polynomials in U. The condition for super primarity Eq(2.6)gives the following set of equations,

$$\begin{aligned} &W_1(\hat{z}_1)P_{hj-1}(\hat{z}_2) + W_0(\hat{z}_1)P_{hj}(\hat{z}_2) \\ &\qquad = h\frac{\theta_{12}}{z_{12}^2}P_{hj}(\hat{z}_2) + \frac{1}{2z_{12}}DP_{hj}(\hat{z}_2) + \frac{\theta_{12}}{z_{12}}\partial P_{hj}(\hat{z}_2) + \ldots, \quad j = 1, \ldots, s-1, \end{aligned} \tag{2.8}$$

If the above set of equations admits a solution, we call P_h the deformation of P_{h0}.

3. Super Primary Fields with respect to W_0

In this chapter we derive general conditions characterizing (half) integer primary fields w.r.t. W_0, the free theory. Our starting point is that U_i is a super primary field of dimension 1/2 w.r.t W_0,

$$W_0(\hat{z}_1)U_i(\hat{z}_2) = \frac{\theta_{12}}{2z_{12}^2}U_i(\hat{z}_2) + \frac{1}{2z_{12}}DU_i + \frac{\theta_{12}}{z_{12}}\partial U_i + \ldots. \tag{3.1}$$

Then from the remark in the previous chapter we know that

$$U_{i_1} \ldots U_{i_s}(\hat{z}), \quad i_1 < i_2 < \ldots < i_s. \quad s < n-1, \tag{3.2}$$

is a super primary field of dimension $s/2$. In order to find more general forms of (half) integer spin super primary fields let us consider normal ordered polynomial in U and DU of homogeneous order s,

$$P_{s0} = \sum const.(U_1)^{a_1} \dots (U_{n-1})^{a_{n-1}} (DU_1)^{b_1} \dots (DU_{n-1})^{b_{n-1}}, \tag{3.3}$$

$$a_i = 0, 1.$$

$$\sum_i (\frac{1}{2} U_i \frac{\partial}{\partial U_i} + DU_i \frac{\partial}{\partial (DU_i)}) P_s = s P_s. \tag{3.4}$$

Double contractions in the OPE $W_0(\hat{z}_1) P_s(\hat{z}_2)$ are easily evaluated,

$$\begin{aligned} &W_0(\hat{z}_1) P_s(\hat{z}_2) \Big|_{double\, contraction} \\ &= \frac{\theta_{12}}{2z_{12}^4} \sum_i \frac{\partial^2}{\partial (DU_i)^2} P_s(\hat{z}_2) + \frac{1}{2z_{12}^3} \sum_i \frac{\partial^2}{\partial U_i \partial (DU_i)} P_s(\hat{z}_2) + \dots . \end{aligned} \tag{3.5}$$

Therefore the absence of these higher singularities gives the conditions for super primarity

$$\sum_i \frac{\partial^2}{\partial (DU_i)^2} P_s = 0, \tag{3.6}$$

$$\sum_i \frac{\partial^2}{\partial U_i \partial (DU_i)} P_s = 0. \tag{3.7}$$

These conditions could be interpreted as the "super harmonicity" in parallel with the "harmonicity" condition in bosonic case, see[7] . Contributions from simple contractions are, after Taylor expansion, given by

$$\begin{aligned} &W_0(\hat{z}_1) P_s(\hat{z}_2) \Big|_{simple\, contraction} \\ &= \frac{\theta_{12}}{z_{12}^2} \sum_i (\frac{1}{2} U_i \frac{\partial}{\partial U_i} + DU_i \frac{\partial}{\partial (DU_i)}) P_s(\hat{z}_2) \\ &+ \frac{1}{2z_{12}} \sum_i (DU_i \frac{\partial}{\partial U_i} + D^2 U_i \frac{\partial}{\partial (DU_i)}) P_s(\hat{z}_2) \\ &+ \frac{\theta_{12}}{z_{12}} \sum_i (\partial U_i \frac{\partial}{\partial U_i} + D^3 U_i \frac{\partial}{\partial (DU_i)}) P_s(\hat{z}_2) \\ &+ \frac{1}{2z_{12}^2} \sum_i U_i \frac{\partial}{\partial (DU_i)} P_s(\hat{z}_2) + \dots . \end{aligned} \tag{3.8}$$

By using Eq(3.4) and simple algebra, we arrive at

$$\begin{aligned}W_0(\hat{z}_1)P_s(\hat{z}_2)\Big|_{simple\ contraction} & \\ &= \frac{\theta_{12}}{z_{12}^2} s P_s(\hat{z}_2) + \frac{1}{2z_{12}} D P_s(\hat{z}_2) + \frac{\theta_{12}}{z_{12}} \partial P_s(\hat{z}_2) \\ &\quad + \frac{1}{2z_{12}^2} \sum_i U_i \frac{\partial}{\partial(DU_i)} P_s(\hat{z}_2) + \dots .\end{aligned} \tag{3.9}$$

The last term is unwanted. So we get another condition

$$\sum_i U_i \frac{\partial}{\partial(DU_i)} P_s = 0\,, \tag{3.10}$$

which reflects the fact that DU is not a super primary field. If P_s satisfy these three conditions Eqs(3.6) , (3.7) and (3.9) then it is a super primary field.

If we slightly generalize the form of P_s by including powers of ∂U we get similar conditions for super primarity. P_s should have homogeneous order s

$$\sum_i \left(\frac{1}{2} U_i \frac{\partial}{\partial U_i} + DU_i \frac{\partial}{\partial(DU_i)} + \frac{3}{2} \partial U_i \frac{\partial}{\partial(\partial U_i)}\right) P_s = s P_s\,. \tag{3.11}$$

Vanishing of double contractions give condition Eq(3.7) and

$$\sum_i \left(\frac{\partial^2}{\partial(DU_i)^2} - \frac{\partial^2}{\partial(\partial U_i)\partial U_i}\right) P_s = 0\,, \tag{3.12}$$

$$\sum_i \frac{\partial^2}{\partial(\partial U_i)\partial(DU_i)} P_s = 0\,. \tag{3.13}$$

Absence of extra terms in simple contractions give rise to the conditions

$$\sum_i \left(DU_i \frac{\partial}{\partial(\partial U_i)} + U_i \frac{\partial}{\partial(DU_i)}\right) P_s = 0\,, \tag{3.14}$$

$$\sum_i U_i \frac{\partial}{\partial(\partial U_i)} P_s = 0\,. \tag{3.15}$$

It should be remarked that not of all these conditions are independent. In fact, in the above example, they reduce to two independent conditions Eq.(3.14), and

Eq.(3.7). They correspond to the vanishing of $J_{1/2}$ and $J_{3/2}$ respectively, which characterize a highest weight representation in the Neveu-Schwarz sector. (J_r is the moment of the fermionic part of the super stress tensor W_0.) Similarly two independent conditions for super primarity can be easily obtained when higher order derivatives of the field U are involved.

Examples

As solutions to the above algebraic equations we find plenty of super primary fields of (half) integer spin. First we consider the case $\hat{c} = 2$, *i.e.*a free SCFT consisting of two superfields U_1 and U_2. In the corresponding bosonic CFT we know that[7] the real and imaginary parts of $(\partial\phi_1 + i\partial\phi_2)^s$, *s integer* , being harmonic polynomials of two variables , are spin s primary fields. Their supersymmetric counter parts are : the real and imaginary parts of

$$(U_1 + iU_2)(DU_1 + iDU_2)^s, \quad s; integer, \tag{3.16}$$

are spin $s + 1/2$ super primary fields. Similarly the real and imaginary parts of

$$(U_1 + iU_2)(\partial U_1 + i\partial U_2)(DU_1 + iDU_2)^{s-2} + \frac{1}{s}(DU_1 + iDU_2)^s, \tag{3.17}$$

are spin s super primary fields. Therefore one could call them "super harmonic polynomials of two variables". For general $\hat{c}$ we give explicit examples of lower spin super primary fields .

i) $s = 1/2$. $U_i (i = 1, \ldots, n-1)$ are the only primary. Since we do not discuss the super Kac-Moody algebra in this paper, possible closed operator algebras of U's are not considered here.

ii)$s = 1$. $Q_{10} = U_i U_j$ $(i \neq j)$. In the case of $n - 1 = 2$ (two free super fields U_i, $i = 1, 2$), this forms a closed operator algebra with W_0,

$$Q_{10}(\hat{z}_1)Q_{10}(\hat{z}_2) = -\frac{1}{z_{12}^2} - \frac{\theta_{12}}{z_{12}} 2W_0(\hat{z}_2) + \ldots,. \tag{3.18}$$

More generally for $n - 1 =$ even

$$Q_{10} = U_1 U_2 + \ldots + U_{n-2} U_{n-1}$$

satisfy the same closed operator algebra and the algebra remains the same after deformation. This is nothing but the Feigin-Fuchs construction of N=2 Superconformal algebras.

iii)$s = 3/2$. Three types of super primary fields are possible.

$$(a) \quad U_iU_jU_k, i<j<k, (b) \quad U_iDU_i - U_jDU_j, (c) \quad U_iDU_j + U_jDU_i. \tag{3.19}$$

(a) does not lead to a closed operator algebra with W_0. (b) and (c) are essentially same, since they transform to each other by a 45° rotation in the (U_i, U_j)plane

$$U_i = \frac{U_i' + U_j'}{\sqrt{2}}, \qquad U_j = \frac{-U_i' + U_j'}{\sqrt{2}}, \tag{3.20}$$

which leaves W_0 and the two point function $< U_iU_j >$ unchanged. (b) forms a closed operator algebra with W_0. However, it is reduced to a direct sum of super conformal algebras by certain recombination of (b) and W_0.

iv) $s = 2$. Some of the $s = 2$ super primary fields are derived directly form those of $s = 3/2$;

$$U_iU_jU_kU_l, \quad (U_iDU_j - U_jD_i)U_l, \quad (U_iDU_j + U_jDU_i)U_l, \tag{3.21}$$

in which we are not interested. The others are

$$(DU_i)^2 + 2U_i\partial U_i, \quad DU_iDU_j + U_i\partial U_j + U_j\partial U_i. \tag{3.22}$$

Again these two are related to each other by a 45° rotation and are essentially same. We show that the OPE of S with itself, is not closed due to the last term.

$$S = \frac{1}{\sqrt{32}}\sum_i((DU_i)^2 + 2U_i\partial U_i), \tag{3.23}$$

$$\begin{aligned} S(\hat{z}_1)S(\hat{z}_2) =& \frac{3}{8}\frac{\hat{c}}{2z_{12}^4} + \frac{3}{2}\frac{\theta_{12}}{z_{12}^3}W_0(\hat{z}_2) + \frac{1}{z_{12}^2}(\frac{1}{2}DW_0 + \frac{1}{\sqrt{2}}S) \\ &+ \frac{\theta_{12}}{z_{12}^2}(\partial W_0 + \frac{1}{2\sqrt{2}}DS) + \frac{1}{z_{12}}(\frac{1}{4}D^3W_0 + \frac{1}{2\sqrt{2}}\partial S) \\ &+ \frac{\theta_{12}}{z_{12}}\frac{3}{4}\sum_i(\partial^2U_iDU_i + U_iD^5U_i - D^3U_i\partial U_i) + \dots . \end{aligned} \tag{3.24}$$

4. Super version of the W_3-Algebra?

A possible closed operator algebra of an $s = 5/2$ field with the super Energy-Momentum tensor corresponds to a possible supersymmetric version of the W_3-algebra of Fateev-Zamolodchikov[5] . So we discuss $s = 5/2$ super primary fields in some detail. To be more specific we consider $n-1 = 2$ case, since the closed operator algebra of spin 3 field with the Energy-Momentum tensor is known only for two boson case[5][7] . Super primary fields are

$$U_1((DU_2)^2 + 2(U_2\partial U_2)), \tag{4.1}$$

$$P_{5/2} = i\sqrt{\frac{3}{80}}(U_1(DU_1)^2 - U_1(DU_2)^2 - 2U_2(DU_1)(DU_2)), \tag{4.2}$$

and those obtained by exchanging the indices 1 and 2. As is obvious the first one is derived from the $s = 2$ super primary field and the second one is new. It is interesting to note that $P_{5/2}$ can be expressed in terms of $\vec{h}_i$ vectors introduced in Chapter 3 of I, Eq(I,3.1).

$$P_{5/2} \propto (h_1U)(h_2DU)(h_3DU) + \ldots \tag{4.3}$$

in analogy with the spin 3 field P_{30} in W_3-algebra, Eq.(I.2.26) is expressed as

$$P_{30} \propto (h_1\partial\phi)(h_2\partial\phi)(h_3\partial\phi). \tag{4.4}$$

The analogy goes one step further and the θ_{12}/z_{12}^4 term of the OPE $P_{5/2}(\hat{z}_1)P_{5/2}(\hat{z}_2)$ is in fact proportional to W_0. However, there the parallelism ends and the OPE is not closed in contrast with the bosonic case

$$\begin{aligned} P_{5/2}(\hat{z}_1)P_{5/2}(\hat{z}_2) =& \frac{3}{8}\frac{\hat{c}}{s}\frac{1}{z_{12}^5} + \frac{3}{2}\frac{\theta_{12}}{z_{12}^4}W_0(\hat{Z}_2) + \frac{1}{z_{12}^3}(\frac{1}{2}DW_0 - \frac{1}{20}S) \\ &+ \frac{\theta_{12}}{z_{12}^3}(\partial W - \frac{1}{40}DS) + \frac{1}{z_{12}^2}(\frac{1}{4}D^3W_0 - \frac{1}{40}\partial S) \\ &+ lower\, order\, terms\,, \end{aligned} \tag{4.5}$$

In which S is the spin 2 super primary field discussed in the previous chapter and $\hat{c} = 2, s = 5/2$. Since we know that the OPE of S is not closed, we find that the operator algebra of $P_{5/2}$ with W_0 is not closed even when S is included. The situation does not improve when the other $s = 5/2$ super primary fields is added to $P_{5/2}$. Therefore we conclude that the supersymmetric generalization of the W_3 algebra does not exist. Increasing the number of participating superfields U_i from two will not improve the situation. For, we know that the corresponding pure boson theories do not have a closed operator algebra[7] . Similarly we would conclude that a closed operator algebra of spin s ($s \geq 3$) superprimary fields with super Energy-Momentum tensor cannot be constructed.

ACKNOWLEDGEMENTS

One of the authors (R.S) would like to thank Professor H. Miyazawa for guidance and encouragements over many years in the most exciting branches of theoretical physics. Another author (S.K.P) happens to be a former student of J. Maharana , who also worked with Professor Miyazawa some fifteen years ago.

REFERENCES

1. M.A.Bershadsky, V.G.Knizhnik and M.G.Teitelman, *Phys.Lett.* **151B** (1985) ,31.
2. D.Friedan, Z.Qiu and S.Shenker, *Phys.Lett.* **151B** (1985) ,37; D.Friedan. E.Martinec and S.Shenker, *Nucl.Phys.* **B271** (1986) 93.
3. A. B. Zamolodchikov, *Theor. Math. Phys.* **65** (1985) ,1205.
4. V. G. Knizhnik and A. B. Zamolodchikov, *Nucl. Phys.* **B247** (1984) ,83.
5. V. A. Fateev and A. B. Zamolodchikov, *Nucl. Phys.* **B280 [FS18]** (1987) , 644.
6. B. L. Feigin and D. B. Fuchs, *Funkts. Anal. Pril.* **16** (1982) , 47.
7. R. Sasaki, Hiroshima preprint RRK 88-21.

Chapter 5 FIELD THEORY

SUPERCONVERGENT PROPAGATORS

Reinhard Oehme

Department of Physics, University of Tokyo, Tokyo 113
and
Enrico Fermi Institute and Department of Physics
University of Chicago, Chicago, Illinois, 60637**

ABSTRACT

Exact properties of propagators can be obtained on the basis of the general principles of gauge field theories and the renormalization group. The results make it possible to give arguments for the confinement of transverse gauge field excitations for theories with asymptotic freedom and with a limited number of matter fields. The superconvergence relation obtained under these conditions has possible implications for quark confinement.

This article is dedicated to my friend Hironari Miyazawa on the occasion of his sixtieth birthday. Ever since our first encounter at the University of Chicago during the Fermi Era, I have admired the imagination and courage which is characteristic of his scientific work. Miyazawa has introduced and understood new concepts well before they became topical, and thereby he has made important contributions to particle physics.

** Permanent address.

1. INTRODUCTION

The short range properties of gluons and quarks are well described by the weak-coupling limit of quantum chromodynamics. This is a consequence of the asymptotic freedom of the theory. But important questions concerning the phase structure and the related problem of the confinement of gluons and quarks are non-perturbative in character. Within the framework of the continuum theory, these questions can be approached only by more indirect methods. In this connection, the study of the exact propagator functions of the non-Abelian gauge theory has been very helpful. Even though these structure functions are gauge dependent,and therefore do not directly represent physical observables, they nevertheless contain important information about the structure of theory. Results derived in a particular gauge, if used in gauge invariant setting, can have physical significance. So we will find, for example, that, under certain circumstances, transverse gluon states do not contribute to the unitarity relations of the S-matrix, and hence are not physical states in this sense; they are confined. Since the unitarity relations are invariant, we have a gauge dependent argument for a physical result.

In this talk I consider mainly the structure function of the gauge field propagator. Analytic properties, as obtained from general principles, together with the consequences of renormalization group invariance, are used in order to show that the discontinuity of the structure function satisfies a superconvergence relation, provided the number of matter fields is restricted (less than ten flavors for QCD).[1] Furthermore, under these circumstances, the sign of the discontinuity is negative for values of $k^2 > K^2(g^2)$, where $K^2(g^2)$ decreases exponentially for $g^2 \to +0$.[2] These features lend themselves to a dipole-type representation of the transverse gluon propagator. The dipole representation is of a form so that we may relate it to an approximate, linearly rising potential between static quark charges, but only up to a range $R \ll 1/\sqrt{K^2}(g^2)$.[3,4]

Of particular interest is the extension of the renormalization group results to projected propagators, where the projection does not introduce a new, dimensionful parameter.[1,2] Under these circumstances our formulae imply that, for theories with a limited number of matter fields, ($N_F \leq 9$ for QCD), the projected structure functions involving only positive norm states either vanish identically or diverge in the limit $g^2 \to +0$. For a subspace containing all positive norm states generated by the gauge field operator in application to the vacuum, only the divergent solution is possible. For a subspace containing physical states, however, we can exclude the divergent solution. Then we conclude that the norm of transverse gluon states must vanish, so that they are not contained in the physical subspace. This situation corresponds to gluon confinement.[5,6]

In this talk I consider the gauge field theory by itself and ignore the embedding into a more comprehensive scheme. We may expect that such an embedding is of importance at least near the Planck length, but we assume that it is not a dominant feature as far as confinement is concerned.

The general structure analysis of propagators and of projected propagators has been done in collaboration with Wolfhart Zimmermann.[1,2] The implications of our results for the confinement problem have been discussed extensively by Kazuhiko Nishijima,[3,5] and also in references (4) and (6).

2. STRUCTURE ANALYSIS

Let me first give a brief description of the main features of the gluon structure function[1,2] which are relevant for the later discussion of confinement. We assume the general axioms of quantum field theory as applied to non-Abelian gauge theories with a state space ϑ of indefinite matrix. Important are simple spectral conditions, like non-negative eigenvalues of $P_\mu P^\mu$ and P^0, where P^μ is the energy-momentum operator, and Lorentz covariance, although the latter way be relaxed in certain situations.[5] Finite quark masses can be accomodated as long as it is possible to use a mass-independent renormalization scheme.[7]

Exact Green's functions are assumed to be connected with the corresponding expressions of the formal perturbation expansion in the weak coupling limit $g^2 \to +0$, at least for the first few terms. This assumption is natural in view of the asymptotic freedom of the theory considered, and the physical relevance of the weak coupling, short distance limit. We note that this connection with perturbation theory is assumed only for unprojected, complete Green's functions and within the framework of the full state space with indefinite metric. Projections into subspaces may not commute with the weak coupling limit $g^2 \to +0$, in particular in the confined phase.

We consider the transverse gauge field propagator

$$\langle 0|T A_a^{\mu\nu}(x) A_b^{\rho\sigma}(x')|0\rangle, \quad A_a^{\mu\nu} = \partial^\mu A_a^\nu - \partial^\nu A_a^\mu . \tag{2.1}$$

The state with indefinite metric is assumed to be non-degenerate, and the Poincaré group is realized by isometric transformations U(a, Λ). We require $U(a, \Lambda)|0\rangle = |0\rangle$ for the vacuum state. The structure function of the Fourier transform is then defined by

$$\begin{aligned} -i D^{ab}_{\mu\nu\rho\sigma}(k) = \delta^{ab}(&k_\mu k_\rho g_{\nu\sigma} - k_\mu k_\sigma g_{\nu\rho} \\ &+ k_\nu k_\sigma g_{\mu\rho} - k_\nu k_\rho g_{\mu\sigma}) D_c(k^2) . \end{aligned} \tag{2.2}$$

Using Lorentz covariance and the spectral conditions, we find that $D_c(k^2)$ is the boundary value of an analytic function $D(k^2)$ which is regular in the complex k^2-plane, cut along the positive real axis. $D_c(k^2) = D(k^2 + i0)$ for real $k^2 \geq 0$. With the space-time expression (2.1) being a tempered distribution, the function $D(k^2)$ is bounded by a polynomial for $k^2 \to \infty$.

More detailed information about the asymptotic behavior of $D(k^2)$ can be obtained with the help of the renormalization group. We consider the fields as functions of x, g, κ^2 and α, where $\kappa^2 < 0$ is the renormalization point and α the gauge parameter for covariant gauges. We normalize the structure function at κ.

$$-k^2 D(k^2, \kappa^2, g, \alpha) = 1 , \quad k^2 = \kappa^2 < 0 , \tag{2.3}$$

and define $-k^2 D(k^2, \kappa^2, g, \alpha) \equiv R(\frac{k^2}{\kappa^2}, g, \alpha)$ with $R(1, g, \alpha) = 1$. We then have the renormalization group relation

$$R\left(\frac{k^2}{\kappa^2}, g, \alpha\right) = R\left(\frac{\kappa'^2}{\kappa^2}, g, \alpha\right) R\left(\frac{k^2}{\kappa'^2}, g', \alpha'\right), \tag{2.4}$$

where (κ^2, g, α) and (κ'^2, g', α') are equivalent parameters:

$$g' = \bar{g}(u, g), \quad \bar{g}(1, g) = g; \quad u = \exp \int_{g^2}^{\bar{g}^2(u,g)} dx\, \beta^{-1}(x); \tag{2.5}$$
$$\alpha' = \bar{\alpha}(u, g, \alpha) = \alpha R^{-1}(u, g, \alpha),$$

with $u = \kappa'^2/\kappa^2$. Writing $\kappa'^2 = -|k^2|$ and $k^2 = -|k^2|e^{i\phi}$, $|\psi| \leq \pi$, eq. (2.4) implies

$$R\left(\frac{k^2}{\kappa^2}, g, \alpha\right) = R\left(\left|\frac{k^2}{\kappa^2}\right|, g, \alpha\right) R\left(e^{i\phi}, \bar{g}, \bar{\alpha}\right) \tag{2.6}$$

with $\bar{g}$ and $\bar{\alpha}$ given by eq. (2.5) with $u = \left|\frac{k^2}{\kappa^2}\right|$. From these relations, we obtain the asymptotic behavior $k^2 \to \infty$ in any direction[8,1,2]:

$$R\left(\frac{k^2}{\kappa^2}, g, \alpha\right) \simeq \frac{\alpha}{\alpha_0} + C_V(g, \alpha)\left(\lg\left|\frac{k^2}{\kappa^2}\right|\right)^{-\gamma_{00}/\beta_0} + O\left(\left(\lg\left|\frac{k^2}{\kappa^2}\right|\right)^{-1}\right), \quad \frac{\gamma_{00}}{\beta_0} > 0; \tag{2.7}$$

$$R\left(\frac{k^2}{\kappa^2}, g, \alpha\right) \simeq C_V(g, \alpha)\left(\lg\left|\frac{k^2}{\kappa^2}\right|\right)^{-\gamma_{00}/\beta_0} + \frac{\alpha}{\alpha_0} + O\left(\left(\lg\left|\frac{k^2}{\kappa^2}\right|\right)^{-\gamma_{00}/\beta_0 - 1}\right), \quad \gamma_{00}/\beta_0 < 0. \tag{2.8}$$

Here we write $\beta(g^2) \simeq \beta_0 g^4 + \beta_1 g^6 + \cdots$ and $\gamma(g, \alpha) \simeq (\gamma_{00} + \alpha\gamma_{01})g^2 + \cdots$ in the limit $g^2 \to +0$ for the renormalization group function and the anomalous dimension of the gauge field respectively. The constant α_0 is given by $\alpha_0 = -\gamma_{00}/\gamma_{01}$. We have assumed that $\beta_0 < 0$ corresponding to asymptotic freedom and have $\bar{g}^2(u, g) \simeq 1/-\beta_0 \lg u$

for $u \to \infty$. With eqs. (2.7) and (2.8), it follows that $D(k^2)$ satisfies an unsubtracted dispersion relation

$$D(k^2, \kappa^2, g, \alpha) = \int_{-0}^{\infty} dk'^2 \, \frac{\rho(k'^2, \kappa^2, g, \alpha)}{k'^2 - k^2} \,, \tag{2.9}$$

with

$$\rho(k^2, \kappa^2, g, \alpha) \simeq -\frac{\gamma_{00}}{\beta_0} C_V(g, \alpha) k^{-2} \left(\lg \frac{k^2}{|\kappa|^2} \right)^{-\gamma_{00}/\beta_0 - 1} \tag{2.10}$$

for $k^2 \to \infty$. We see that for $\gamma_{00}/\beta_0 > 0$ the function $D(k^2, \kappa^2, \alpha) + k^{-2} \frac{\alpha}{\alpha_0}$ vanishes faster than k^{-2} for $k^2 \to \infty$, so that we have the superconvergence relations

$$-\frac{\alpha}{\alpha_0} + \int_{-0}^{\infty} dk^2 \rho(k^2, \kappa^2, g, \alpha) = 0 \,. \tag{2.11}$$

The essential k^2-dependence of the asymptotic expressions is determined by γ_{00}/β_0 and is independent of α.

We note that all discussions involving discontinuities should be interpreted in the sense of distributions.[1.2)]

3. SUPERCONVERGENCE AND DIPOLE STRUCTURE

In the following we restrict ourselves to the Landau gauge $\alpha = 0$, but the considerations can be generalized to other covariant gauges. For $\alpha = 0$, we have generally[1,2)]

$$D(k^2, \kappa^2, g) \simeq -C_V(g) k^{-2} \left(\lg \left| \frac{k^2}{\kappa^2} \right| \right)^{-\gamma_{00}/\beta_0} \tag{3.1}$$

for $k^2 \to \infty$, with the coefficient given by

$$C_V(g) = \left(g^2 |\beta_0| \right)^{-\gamma_{00}/\beta_0} \exp \int_{g^2}^{0} dx \, J(x) \,, \tag{3.2}$$

where

$$\tau(g^2) = \frac{\gamma(g^2, 0)}{\beta(g^2)} - \frac{\gamma_{00}}{\beta_0 g^2} \tag{3.3}$$

is finite for $g^2 = 0$. Now the function $D(k^2)$ itself decreases faster than k^{-2} for $k^2 \to \infty$ if $\gamma_{00}/\beta_0 > 0$, and we have the superconvergence relation

$$\int_{-0}^{\infty} dk^2 \rho(k^2, \kappa^2, g) = 0 . \tag{3.4}$$

For the SU(3) gauge theory (QCD), we have $\gamma_{00}/\beta_0 > 0$, i.e. $\beta_0 < 0$ and $\gamma_{00} < 0$, for $N_F \leq 9$, where N_F is the number of flavors. We note that, for a reasonable theory, the relation (3.4) relies on the indifinite metric of the state space. It is clearly a non-perturbative result, which is not valid order by order in perturbation theory. "Improved" perturbation theory, like summed-up one loop contributions, results in a structure function

$$D_{\text{impr. pert.}}(k^2) = -k^{-2}\left(1 - \beta_0 g^2 \lg \frac{k^2}{\kappa^2}\right)^{-\gamma_{00}/\beta_0} . \tag{3.5}$$

This expression has an asymptotic behavior in accordance with eq.(3.1), but at the cost of a singularity at $k^2 = \kappa^2 \exp(1/\beta_0 g^2) < 0$ on the negative real k^2-axis, which is not allowed in the general theory. There is a superconvergence relation, but it involves all singularities on the real k^2-axis, and it is not of the form (3.4).

In the Landau gauge, the asymptotic limit of the discontinuity is given by

$$\rho(k^2) \simeq -\frac{\gamma_{00}}{\beta_0} C_V(g) k^{-2}\left(\lg \frac{k^2}{|\kappa^2|}\right)^{-\gamma_{00}/\beta_0 - 1} . \tag{3.5}$$

Irrespective of the sign of γ_{00}/β_0, it vanishes sufficiently fast for $k^2 \to \infty$ so that we can rewrite the Lehmann representation (2.9) in the form [3,4]

$$D(k^2) = \int_{-0}^{\infty} dk'^2 \frac{\sigma(k'^2)}{(k'^2 - k^2)^2}, \tag{3.6}$$

with

$$\sigma(k^2) = \int_{-0}^{k^2} dk'^2 \rho(k'^2), \tag{3.7}$$

and

$$\sigma(k^2) \simeq C_V(g)\left(\lg \frac{k^2}{|\varkappa^2|}\right)^{-\gamma_{00}/\beta_0} \tag{3.8}$$

for $k^2 \to \infty$.

We first consider the case $\gamma_{00}/\beta_0 > 0$ (corresponding to $N_R \leq 9$ for QCD). Since $C_V(g) > 0$, the discontinuity ρ is then <u>negative</u> for large values of k^2. Since it satisfies the superconvergence relation (3.4), it should become positive for some $k^2 = K^2 \geq 0$. The position K^2 of the zero of ρ must be invariant under the renormalization group, i.e. $K^2 = K^2(\kappa^2, g) = K^2(\kappa'^2, g')$ for equivalent parameters, so that

$$K^2(\varkappa^2, g) = C_K(g_0)\varkappa^2 \exp\int_{g^2}^{g_0^2} dx\, \beta^{-1}(x), \tag{3.9}$$

and

$$K^2(\varkappa^2, g) \simeq C_K(g_0)\varkappa^2 \exp\left(1/\beta_0 g^2\right)(g^2)^{\beta_1/\beta_0^2} + \cdots \tag{3.10}$$

for $g^2 \to +0$. We see that, except for an exponentially vanishing interval, or possibly only a δ-function contribution at $k^2 = 0$, the discontinuity ρ is dominated by negative norm states. More explicitly, the relationship between ρ and the norm of the states

$$A(k)|0\rangle \equiv C^a_{\mu\nu}(k)\tilde{A}^{\mu\nu}_a(k)|0\rangle \tag{3.11}$$

is given by

$$\langle 0|\bar{A}(k')A(k)|0\rangle = 2\pi c^2 \delta(k'-k)\Theta(k^0)\rho(k^2), \tag{3.12}$$

with

$$c^2 = -4C^a_{\mu\nu}(k)k^\mu C^a_{\rho\sigma}(k)k^\rho g^{\nu\sigma} > 0.$$

Here $\tilde{A}^{\mu\nu}(k)$ is the Fourier transform of the corresponding Heisenberg operator for the gauge field.

Returning to the dipole representation (3.6), we see that for $\gamma_{00}/\beta_0 > 0$ we have $\sigma(\infty) = 0$ and $\sigma(k^2) > 0$, $\sigma'(k^2) = \rho(k^2) \leq 0$ for $k^2 \geq K^2$, while at $k^2 = K^2$ we have $\sigma'(k^2) = \rho(k^2) = 0$ and $\sigma''(K^2) = \rho'(K^2) < 0$ per definition. Hence $\sigma(k^2)$ has its first maximum from above at $k^2 = K^2 \geq 0$. While K^2 vanishes exponentially for $g^2 \to +0$, $\sigma(K^2)$ increases according to

$$\sigma(K^2) = \sigma(K_0^2)\left(\frac{g^2}{g_0^2}\right)^{-\gamma_{00}/\beta_0} \exp\int_{g^2}^{g_0^2} dx\, \tau(x), \tag{3.13}$$

where $K_0 = K(g_0, \kappa^2)$.

As a very crude estimate, let us assume that $\sigma(k^2)$ has only the one maximum at $k^2 = K^2$, and let us approximate the contribution of the peak by a δ-function. Writing then

$$D(k^2) \approx \frac{C_\sigma(g)K^2(\varkappa^2, g)}{(K^2(\varkappa^2,g)-k^2)^2} + \int_{K_1^2}^{\infty} dk'^2 \frac{\sigma(k'^2)}{(k'^2-k^2)^2}, \tag{3.14}$$

where

$$C_\sigma(g)K^2(\varkappa^2, g) \sim O\left[\exp(1/\beta_0 g^2)(g^2)^{-\gamma_{00}/\beta_0+\beta_1/\beta_0^2}\right] \tag{3.15}$$

for $g^2 \to +0$. With $K_1^2 > K^2$ chosen well above the center of the peak at K^2, the second term in eq. (3.14) contains mainly the perturbative contributions. Using eq. (3.14) in the usual approximate expression for the Wilson loop, we get an approximately linear potential for $R << 1/\sqrt{K^2}$. Of course, it is possible that $K^2 \equiv 0$ and ρ has a δ-function contribution at $k^2 = 0$ with a positive coefficient. Then the "maximum" is at $k^2 = 0$, since $\sigma(-0) = 0$.

In contrast to the situation described above for $\gamma_{00}/\beta_0 > 0$ ($N_F \leq 9$ for QCD), the shape of $\sigma(k^2)$ is very different for $\gamma_{00}/\beta_0 < 0$ ($N_F \geq 0$ for QCD). In the latter case we have no superconvergence, $\sigma(\infty) = + \infty$ and $\sigma'(k^2) = \rho(k^2) > 0$, at least for sufficiently large values of k^2. Hence $\sigma(k^2)$ is more likely to have a minimum at some point $K^2 \geq 0$ which vanishes like $\exp(1/\beta_0 g^2)$ for $g^2 \to +0$. In this case the perturbative short distance expansion is the dominant feature. However, we cannot exclude the possibility that, also for $\gamma_{00}/\beta_0 < 0$, the weight-function $\sigma(k^2)$ rises to a maximum at some $K^2 < K^2$, or has a δ-function singularity at $k^2 = 0$ corresponding to a k^{-4}-term in $D(k^2)$. Even for $\gamma_{00}/\beta_0 > 0$, there may be an additional term of the form $b(g)\delta(k^2)$ in $\sigma(k^2)$, which corresponds to a dipole contribution $b(g)\delta'(k^2)$ in $\rho(k^2)$ and which would not contribute to the superconvergence relation (3.4).

Summing up, we see that the behavior of $D(k^2)$ for $\gamma_{00}/\beta_0 > 0$ ($N_F \leq$ for QCD) suggests an approximate confinement potential as a consequence of the general structure we have obtained. On the other hand, for both signs of γ_{00}/β_0, there may be additional contributions to ρ, like dipole terms $b(g)\kappa^2\delta'(k^2)$, which could be responsible for quark-confinement within the framework of the potential picture.*

4. PROJECTED PROPAGATORS

Let me now turn to projected gauge field propagators.[1,2] We consider, at first, projections into linear subspaces $\mathcal{V}_+ \subset \mathcal{V}$

* See Ref. 4 for a list of references.

with positive metric in a generic form. In order to preserve renormalization group invariance, we assume that the projection does not introduce a new, dimensionful parameter. We also want to retain Poincare covariance. Under these circumstances, the structure function of the projected, transverse gauge field propagator

$$\langle 0 | T A^a_{\mu\nu}(x) P(\vartheta_+) A^b_{\rho\sigma}(x') | 0 \rangle \tag{4.1}$$

is again the boundary value of an analytic function $D_+(k^2, \kappa^2, g)$ which is regular in the cut k^2-plane. We restict our discussion to the Landau gauge, and define the dimensionless function

$$R_+\left(\frac{k^2}{\kappa^2}, g\right) \equiv -k^2 D_+(k^2, \kappa^2, g) . \tag{4.2}$$

Retaining the normalization R(1. g) = 1 for the full propagator, it is important to note that R_+ (1 g) is not determined, a priori. The function R_+ satisfies the same renormalization group equation as R in eq. (2.4), and correspondingly we also get the relation

$$R_+\left(\frac{k^2}{\kappa^2}, g\right) = R\left(\left|\frac{k^2}{\kappa^2}\right|, g\right) R_+\left(e^{i\phi}, \bar{g}\left(\left|\frac{k^2}{\kappa^2}\right|, g\right)\right) \tag{4.3}$$

with $k^2 = -|k^2| e^{i\phi}$, $|\phi| \le \pi$. We see that the asymptotic behavior of R_+ for $k^2 \to \infty$ depends upon the limit of R_+ $(k^2/\kappa^2, g)$ for $g^2 \to +0$ and k^2-fixed, given the known asymptotic expression (3.1) for R. Unless R_+ increases faster than any power of exp $(1/g^2)$ for $g^2 \to +0$, k^2 fixed, we have polynomial boundedness for $k^2 \to \infty$, g^2 fixed and can write an appropriate dispersion representation for D_+.

On the other hand, for $\gamma_{00}/\beta_0 > 0$, unless R_+ diverges at least like

$$R_+\left(\frac{k^2}{\kappa^2}, g\right) \geq O\left((g^2)^{-\gamma_{00}/\beta_0}\right) , \tag{4.4}$$

we find that D_+ is superconvergent, so that we would have

$$\int_{-0}^{\infty} dk^2 \rho_+(k^2, \kappa^2, g) = 0 . \tag{4.5}$$

Since ρ_+ is a positive measure, eq. (4.5) implies $\rho_+ \equiv 0$ and hence $D_+ \equiv 0$. In contrast, for $\gamma_{00}/\beta_0 < 0$, R_+ can be bounded for $g^2 \to +0$ and fixed k^2 without imposing drastic constraints on the structure function D_+.

In the following we consider some special projections.

a) Total Projection.[2)] It is of interest to consider a special subspace $\mathcal{V}_+$ which contains all states $A(k)|0\rangle$, as defined in eq. (3.11), which have positive norm. We call this space $\mathcal{V}_T^+$. It is sufficient for our purpose to define it for the states $A(k)|0\rangle$. Then we have

$$\rho_T^+ = \begin{cases} \rho & \text{if } \rho \; 0 \\ 0 & \text{if } \rho \leq 0 \end{cases} , \tag{4.6}$$

where the discontinuity ρ_T^+ is related to the norm by

$$\begin{aligned} &\langle 0| \bar{A}(k') P(\mathcal{V}_T^+) A(k) |0\rangle \\ &= 2\pi C^2 \delta(k'-k) \Theta(k^0) \rho_T^+(k^2) , \end{aligned} \tag{4.7}$$

with C^2 as defined in eq. (3.12).

In contrast to the situation for $\gamma_{00}/\beta_0 < 0$, where $\rho > 0$ for $k^2 > K^2(\kappa^2, g)$ and hence $\rho_T^+ = \rho$ there, we have $\rho_T^+ = 0$ for $k^2 \geq K^2$ if $\gamma_{00}/\beta_0 > 0$ ($N_F \leq 9$ for QCD). As long as $D_T^+(k^2, \kappa^2, g)$ does not diverge stronger than any power of g^2 for $g^2 \to +0$, we have for $\gamma_{00}/\beta_0 > 0$ the dispersion formula

$$D_T^+(k^2) = \int_{-0}^{K^2} dk'^2 \, \frac{\rho_T^+(k'^2)}{k'^2 - k^2} , \tag{4.8}$$

which can be written in the form

$$D_T^+(k^2) = -\frac{1}{k^2}\int_{-0}^{K^2} dk'^2 \rho_T^+(k'^2) + \frac{1}{k^2}\int_{-0}^{K^2} dk'^2 \frac{k'^2 \rho_T^+(k'^2)}{k'^2 - k^2}, \tag{4.9}$$

where $K^2(\kappa^2, g)$ is as given in eqs. (3.9) and (3.10), vanishing exponentially for $g^2 \to +0$. Using the renormalization group equations, we find that the first term in eq. (4.9) diverges for $g^2 \to +0$ like $(g^2)^{-\gamma_{00}/\beta_0}$, while the second term is proportional to $\exp(1/\beta_0 g^2)$, vanishing exponentially for $g^2 \to +0$. None of the terms in D_T^+ contributes to the perturbative power series expansion of the full propagator function D. This situation for $\gamma_{00}/\beta_0 > 0$ is in sharp contrast to the non-vanishing of ρ_T^+ for large k^2 if $\gamma_{00}/\beta_0 < 0$, where D_T^+ contains the full perturbative expansion of D. In the latter case, the situation is quite consistent with the existence of physical transverse gluons, whereas in the case $\gamma_{00}/\beta_0 > 0$ ($N_F \leq 9$) the properties of D_T^+ are indicative of confinement.

In the case $\gamma_{00}/\beta_0 > 0$, instead of allowing the divergence of $D_T^+(k^2, \kappa^2, g)$ for $g^2 \to +0$ in accordance with eq. (4.4), we might consider the trivial solution $D_T^+ \equiv 0$ of the renormalization group equations (4.3). However, since $\mathcal{V}_T^+$ contains <u>all</u> of the state $A(k)|0\rangle$ with positive norm, the relation $\rho_T^+ \equiv 0$, together with the superconvergence relation (3.4) for the full discontinuity ρ, would imply that $\rho \equiv 0$ except for possible multipole terms of the form

$$\sum_j \sum_{n=1}^{n_j} c_{nj}\, \delta^{(n)}(k^2 - K_j^2), \tag{4.10}$$

which do not contribute to the superconvergence relation. These requirements for ρ are not consistent with the asymptotic form (3.5) and with perturbation theory in the weak-coupling limit $g^2 \to +0$.

b) <u>Physical Subspace</u>

With the total projection we have considered the complete set of all positive norm states of the form $A(k)|0\rangle$, and we found that this set cannot be empty. Furthermore, for $\gamma_{00}/\beta_0 > 0$, the positive norm of these states was seen to diverge for $g^2 \to +0$ at least like $(g^2)^{-\gamma_{00}/\beta_0}$.

An important subspace of $\mathcal{V}$ with positive norm is the physical subspace $\mathcal{V}_P$. Only states $|\Psi\rangle \in \mathcal{V}_P$ contribute as intermediate states in the unitarity relation for the S-matrix. The subspace $\mathcal{V}_P$ can be defined in terms of the Becchi-Rouet-Stora algebra[9] and its representations,[10,11] but here we consider only the possible set of states $A(k)|0\rangle$, as defined in eq. (3.11), which may belong to $\mathcal{V}_P$.* Since the projection $P(\mathcal{V}_P)A(k)|0\rangle$ does not introduce a new dimensionful parameter, the corresponding structure function D_P and its discontinuity ρ_P satisfy the renormalization group relations corresponding to eq. (4.3).

Following the discussion below eq. (4.3) for the generic projection, we see that there are two possibilities for the structure function D_P in the case $\gamma_{00}/\beta_0 > 0$ ($N_F \leq 9$ for QCD):

1. We can allow $R_P \equiv 0$ and hence

$$\rho_P(k^2, \varkappa^2, g) \equiv 0 , \tag{4.11}$$

so that none of the states $A(k)|0\rangle$ belongs to the physical subspace. Then transverse gluons, and related states generated by the Heisenberg operator $A^{\mu\nu}$, are unphysical, they are not in the physical subspace and hence they do not contribute to S-matrix unitarity. Transverse gluons are <u>confined</u> in this sense. Since all positive norm states $A(k)|0\rangle$ are unphysical, the divergence of their norm ρ_T^+ and of D_T^+ is not problematic.

2. We may try to avoid the solution (4.11) by allowing D_P and ρ_P to diverge at least like $(g^2)^{-\gamma_{00}/\beta_0}$ for $g^2 \to +0$. However, since

---

* Within the framework it is used here, the question of relativistic invariance of $\mathcal{V}_P$ requires a detailed analysis and will be dicussed elsewhere. In Ref. 3, Nishijima has give a derivation using the BRS-singlet space in a special Lorentz frame.

we are now dealing with states which are supposed to be physical, we can exclude this possibility for the following reasons:

(a) For physical gluons, one expects that the structure functions $R_P = -k^{-2}D_P$ is bounded for $g^2 \to +0$ and fixed k^2.[5,1)]

(b) We have shown, in collaboration with W. Zimmermann, that the divergence of ρ_P for $g^2 \to +0$ leads to contradictions in unitarity relation for Wightman functions involving BRS- and gauge-invariant operators.[12)]

We conclude that the structure analysis of the transverse gauge field propagator allows us to argue that gluons are confined provided $\gamma_{00}/\beta_0 > 0$ ($N_F \leq 9$ for QCD). If the number of matter fields is sufficiently large so that $\gamma_{00}/\beta_0 < 0$, but $\beta_0 < 0$ ($10 \leq N_F \leq 16$ for QCD), we obtain no corresponding restrictions, and confined as well as unconfined phases appear to be possible, a priori. A theory with unconfined gluons looks quite natural in this case.

Acknowledgement

It is a pleasure to thank Professor Keiji Igi, and the members of the Particle Theory Groups of the University of Tokyo, for their very kind hospitality. I am also grateful for a Fellowship of the Japan Society for the Promotion of Science.

For discussions and helpful conversations, I am indebted to Peter Freund, Hironari Miyazawa, Yoichiro Nambu, Kazuhiko Nishijima, Klaus Sibold, Wolfhart Zimmermann and Myung-Hoon Chung.

This work has been supported in part by the National Science Foundation, PHY-85-21588.

References

1. Oehme, R. and Zimmermann, W., Phys. Rev. D21, 471 (1980); Phys. Lett. 79B, 93 (1978).
2. Oehme, R. and Zimmermann, W., Phys. Rev. D21, 1661 (1980).
3. Nishijima, K., Prog. Theor. Phys. 77, 1035 (1987).
4. Oehme, R., "On the Confinement of Gluons and Quarks", in Festschrift for Val Telagdi, edited by K. Winter (North Holland, Amsterdam, 1988).
5. Nishijima, K., Prog. Theor. Phys. 75, 1221 (1986); Prog. Theor. Phys. Suppl. 86, 253 (1986).
6. Oehme, R., Phys. Lett. 195B, 60 (1987); Fermi Institute Report EFI 86-24 (1986).
7. Weinberg, S., Phys. Rev. D8, 3497 (1973); 't Hooft, G., Nucl. Phys. B61, 455 (1983); Collins, C. and Mac-Farlane, A. J., Phys. Rev. D10, 1201, (1974),
8. Oehme, R. and Zimmermann, W. (to be published).
9. Becchi, C., Rouet, A. and Stora, R., Ann. Phys. (N. Y.), 98, 287 (1976); Tyutin, I. V., Labedev Report FIAN No. 39 (1975).
10. Kugo, T. and Ojima, I., Prog. Theor. Phys. Suppl. No. 66 (1979): Curci, G. and Ferrari, R., Nuovo Cimento 35A, 276, 464 (1976).
11. Nishijima, K., Nucl. Phys. B238, 601 (1984); Prog. Theor. Phys. 72, 1214 (1984).
12. Oehme, R. and Zimmermann, W., Phys. Lett. 108B, 416 (1982); and (to be published).

GAUGE FREEDOM IN THE REPRESENTATIONS OF BRS ALGEBRA

Kazuhiko Nishijima
Research Institute for Fundamental Physics
Kyoto University, Sakyo-ku, Kyoto 606
JAPAN

ABSTRACT

The study of the representations of BRS algebra plays an important rôle in the physical interpretation of non-abelian gauge theories. The entire state vector space is decomposed into a direct sum of the singlet, daughter and parent subspaces by exploiting BRS algebra. This decomposition is not unique, however, and an inquiry is made into the nature of this arbitrariness.

1. REPRESENTATIONS OF BRS ALGEBRA

The BRS transformation[1] plays an essential rôle in the canonical formulation of non-abelian gauge theories.[2] Its generator called the BRS charge Q_B is hermitian and nilpotent.

$$Q_B^\dagger = Q_B, \quad Q_B^2 = 0. \tag{1}$$

These properties of Q_B force us to introduce indefinite metric, and the study of the representations of the graded Lie algebra (1) leads us to two irreducible representations, namely, the BRS singlet and doublet representations.[3] The entire state vector space V is decomposed accordingly as

$$V = V_S \oplus V_D , \tag{2}$$

where V_S denotes the collection of BRS singlet states and V_D that of BRS doublet states, and these two subspaces are orthogonal to one another. The BRS doublet subspace V_D is further decomposed into a sum of the parent and daughter subspaces denoted as V_p and V_d, respectively.

$$V_D = V_p \oplus V_d\,, \tag{3}$$

and hence

$$V = V_S \oplus V_p \oplus V_d\,. \tag{4}$$

It so happens that the above decomposition (4) is not unique, and a systematic way of relating one decomposition to others is investigated in the present article. This arbitrariness is very akin to gauge freedom and hence we have the present title. In studying this problem it is instructive to recall that the following subspaces are uniquely defined:

$$V_d = \{Q_B|x>|\ |x>\in V\}, \tag{5}$$

$$V_{phys} = \{|x>|\ |x>\in V,\ Q_B|x>=0\}. \tag{6}$$

From the properties of various subspaces we can infer the following structure:

$$V_{phys} = V_S \oplus V_d\,. \tag{7}$$

Now we have a decreasing chain of state vector spaces:

$$V \supset V_{phys} \supset V_d\,. \tag{8}$$

These three state vector spaces are uniquely defined, but the differences between two succeeding spaces, namely, V_p and V_S are not unique as we shall see in what follows.

In order to study this problem we shall introduce a complete set of basis in each subspace and shall specify their characteristic features:

$$\{|s_i>\} \in V_S \,, \quad \{|d_i>\} \in V_d \,, \quad \{|p^i>\} \in V_p \,. \tag{9}$$

These states are orthonormalized by

$$\begin{aligned} &< s_i|s_j> = \delta_{ij} \,, < s_i|d_j> = < s_i|p^j> = 0, \\ &< d_i|p^j> = \delta_{ij} \,, < p^i|p^j> = < d_i|d_j> = 0. \end{aligned} \tag{10}$$

It is worth remarking that all the states in V_d or V_p are of zero norm. Furthermore, we have

$$Q_B|p^i> \in V_d \,, \quad Q_B|s_i> = Q_B|d_i> = 0. \tag{11}$$

Eq.(7) is a consequence of Eq.(11) and the definition (6).

Now we shall try to find the most general form of the complete sets of basis preserving Eqs.(10) and (11). Such a change of basis is realized by a unitary transformation U of which both V_{phys} and V_d are invariant subspaces of V.

$$U|d_i> = |d_i'>, \quad U|s_i> = |s_i'> \,, U|p^i> = |p^{i'}> \,. \tag{12}$$

First, since V_d is unique, we may write

$$|d_i'> = \sum_j |d_j> v_{ji} \,. \tag{13}$$

Next, since V_{phys} is unique, we may write

$$|s_i'> = \sum_j \Big(|s_j> + \sum_k |d_k> c_{kj}\Big) u_{ji} \,. \tag{14}$$

The normalization condition (10) requires u to be unitary.

$$u^\dagger u = u u^\dagger = 1. \tag{15}$$

Finally, V_p can be mixed with both V_S and V_d without violating Eq.(11).

$$|p^{k'}> = \sum_k \Big(|p^j> + \sum_i |s_i> a_{ij} + \sum |d_i> b_{ij}\Big) w_{jk} \,. \tag{16}$$

The coefficients, however, have to satisfy three conditions:

(1) The inner product between $|p'>$ and $|d'>$ should satisfy Eq.(10), and this amounts to

$$wv^\dagger = v^\dagger w = 1 \tag{17}$$

(2) $|p'>$ should be orthogonal to $|s'>$, and this requires

$$a + c^\dagger = 0. \tag{18}$$

(3) $|p'>$ must be of zero norm, and we have

$$a^\dagger a + b + b^\dagger = 0 \tag{19}$$

The transformations represented by u, v and w do not mix subspaces, but those represented by a, b and c do mix them indicating the arbitrariness in the decomposition (4), and the latter transformations are very similar to gauge transformations in the sense that physically relevant quantities are invariant against them. Physically important unitary transformations such as the S matrix and Poincaré transformations of state vectors are BRS invariant, of which both V_{phys} and V_d are invariant subspaces. Hence, both V_{phys} and V_d are Poincaré-invariant, but V_S is not. This is a serious problem since V_S is supposed to represent unconfined observable states. The Poincaré invariance of the theory is recovered, however, when we confine ourselves to BRS-invariant quantities. Assume that T is a BRS-invariant operator, then we have

$$< s' + d'|T|s + d > = < s'|T|s > , \tag{20}$$

where $|s>$ and $|s'>$ belongs to V_S and $|d>$ and $|d'>$ and to V_d. Thus, state vectors of the form $|s> + |d>$, where $|s>$ is fixed and $|d>$ is an arbitrary element of V_d, form an equivalent class denoted by $|s\gg$, and we have

$$< s' + d'|T|s + d > = \ll s'|T|s \gg . \tag{21}$$

Thus the collection of such equivalent classes forms a space H_{phys}:

$$H_{phys} = V_{phys}/V_d , \tag{22}$$

which is Poincaré-invariant, although V_S is not.

2. PROJECTION OPERATORS

In the preceding section the representations of unitary transformations have been studied, but its BRS invariance has not been taken into account, and its consequences will be studied in this section. For this purpose, we shall introduce a pair of BRS doublets by

$$Q_B|p^i>=|d^i>, \quad Q_B|p_i>=|d_i> \ . \tag{23}$$

Then the hermiticity of Q_B allows us to normalize them by

$$<p^i|d_j>=<d^i|p_j>=\delta_{ij} \ . \tag{24}$$

Such a pair of doublets can be identified with the Kugo-Ojima quartet.[2)]

Now assume that U is unitary and BRS-invariant, and

$$U|p>=|p'>, \quad U|d>=|d'> \ , \quad Q_B|p>=|d> \ , \tag{25}$$

then

$$Q_B|p'>=Q_BU|p>=UQ_B|p>=U|d>=|d'> \ . \tag{26}$$

In what follows we identify $|p^k>$ with $|p>$ and $|d^k>$ with $|d>$, then

$$Q_B|p^{k'}>=\sum_j|d^j>w_{jk}=\sum_{i,j}|d_i><p^i|d^j>w_{jk} \tag{27}$$

must be equal to

$$|d^{k'}>=\sum_j|d'_j><p^j|d^k>=\sum_{i,j}|d_i>v_{ij}<p^j|d^k> \ . \tag{28}$$

Now let us put

$$<p^j|d^k>=<d^j|p^k>=g^{jk} \ , \tag{29}$$

then the equality reduces to

$$gw = vg \; , \tag{30}$$

or combining this equation with Eq.(17), we find

$$vgv^\dagger = g \tag{31}$$

This constraint on v is a consequence of BRS-invariance of U.

Now the BRS charge Q_B satisfying Eq.(23) may be expressed as

$$Q_B = \sum_j |d^j >< d_j| = \sum_j |d_j >< d^j| \; . \tag{32}$$

In ref.3 the representation of Q_B was denoted by q, and we may write*

$$q = \sum_j |d^j >< d_j| = \sum_j |d_j >< d^j| \; , \tag{33}$$

$$q^\dagger = \sum_j |p_j >< p^j| = \sum_j |p^j >< p_j| \; . \tag{34}$$

We can show that Q_B or q is invariant under a BRS-invariant unitary transformation, for instance, under Poincaré transformations,

$$\begin{aligned} q \to \sum_j |d^{j'} >< d'_j| &= \sum_{i,j,k} |d^i > (g^{-1}vg)_{ij} v^\dagger_{jk} < d_k| \\ &= \sum_i |d^i >< d_i| = q \; , \end{aligned} \tag{35}$$

where use has been made of Eq.(31), but $q^\dagger$ is not invariant.

$$qq^\dagger = \sum_j |d_j >< p^j| = \sum_j |d^j >< p_j| = P(V_d) \; , \tag{36}$$

$\star$ To be more precise the representation matrix q_{ij}, is given for a given complete set of basis $\{|e_i>\}$ by $q|e_j> = \sum_i |e_i> q_{ij}$.

$$q^\dagger q = \sum_j |p^j >< d_j| = \sum_j |p_j >< d^j| = P(V_p) \; . \tag{37}$$

These two operators are the projection operators to the subspaces V_d and V_p, respectively. It is worth emphasizing that the projection operator $P(V_d)$ is not Poincaré-invariant although V_d itself is. Also,

$$qq^\dagger + q^\dagger q = P(V_D) \tag{38}$$

is the projection operator to the BRS doublet subspace as has already been pointed out.[4)]

The projection operator

$$\sum_j |s_j >< s_j| + \sum_j |d_j >< p^j| = P(V_{phys}) \tag{39}$$

certainly satisfies the condition

$$P^2(V_{phys}) = P(V_{phys}) \; , \tag{40}$$

but it is not hermitian, namely.

$$\sum_j |s_j >< s_j| + \sum_j |p^j >< d_j| = P^\dagger(V_{phys}) \; . \tag{41}$$

Furthermore, we have

$$P(V_{phys})P^\dagger(V_{phys}) = P^\dagger(V_{phys})P(V_{phys}) = P(V_S) \; , \tag{42}$$

where

$$P(V_S) = \sum_j |s_j >< s_j| \; . \tag{43}$$

All the projection operators are not Poincaré-invariant, although some of the subspaces are invariant.

Suppose that both $|f>$ and $|g>$ belong to V_{phys}, then as a consequence of Eq.(42) we have

$$<f|g>=<f|P(V_S)|g> \ . \tag{44}$$

The definition of V_{phys} is unique, but it is not the case with V_S. This formula is valid, however, for any choice of V_S, so that the left-hand-side is independent of the choice of the Lorentz frame. Indeed, this expression is a typical example of Eq.(20) and depends only on two classes of states.

The contents of this article were borne out of discussions with Reinhard Oehme on one of his papers.[5] The author is indebted to him for stimulating discussions.

This article is dedicated to Professor Hironari Miyazawa on the occasion of his sixtieth birthday. I cherish our enduring friendship and wish him many happy returns for his direct and indirect contributions to the community of physicists.

REFERENCES

1. Becchi, C., Rouet, A., and Stora, R., Ann. of Phys. **98**, 287 (1976).
2. Kugo, T. and Ojima, I., Phys. Lett. **73B**, 459 (1978); Suppl. Prog. Theor. Phys. No.**66**, 1 (1979). Curci, G. and Ferrari, R., Nuovo Cim. **35A**, 278, 474 (1976).
3. Nishijima, K., Nucl. Phys. **B238**, 601 (1984).
4. Nishijima, K., Prog. Theor. Phys. **72**, 1214 (1984).
5. Oehme, R., Phys. Lett. **195B**, 60 (1987).

Global Gauge Anomaly of Classical Groups in Even Dimension

Susumu Okubo and Huazhong Zhang
Department of Physics and Astronomy
University of Rochester
Rochester, NY 14627 U.S.A.

ABSTRACT

Explicit expression of global gauge anomaly coefficients $A(\omega)$ of locally anomaly-free representation ω of classical groups SU(N), Sp(2N) and SO(N) have been calculated in even dimensional space-time by uses of group theory and homotopy theory. As a by-product, we will prove some modular relations involving the n-th Dynkin indices $Q_n(\omega)$ of these groups.

1. INTRODUCTION AND SUMMARY OF MAIN RESULTS

One complication facing any quantum gauge field theory in spaces of even dimension D=2n is the consistency requirement that it must have no local (or perturbative) as well as global (or non-preturbative) anomalies. Let H be the gauge group under consideration, and let ω denote the representation of H for all fundamental fermion fields in the theory. The absence of the local gauge anomaly in D=2n is now well-known[1] to be satisfied if the representation ω obeys the condition

$$\mathrm{Tr}^{(\omega)}X^{n+1} = 0 \tag{1.1}$$

where X stands for the generic matrix representation of a generic element of the Lie algebra L of H in the representation ω. It may be readily noted that this condition is automatically valid for any self-contragradient representation ω when n is even, i.e. D=0(mod 4). Especially, if H is any one of the following groups:

$$SU(2)\ ,\ Sp(2N)\ ,\ SO(2N+1)\ ,\ SO(4N)\ ,\ G_2\ ,\ F_4\ ,\ E_7\ ,\ E_8\ , \tag{1.2}$$

then any representation ω of these groups possesses[2] no local anomaly at all in dimension D=0(mod 4).

However, the case of the global anomaly is more complicated. Bismut and Freed[3], following Witten[4], have given a general formula for the global anomaly, including the gravitation, as a holonomy of a determinant bundle. If we are only interested in the pure gauge anomaly in flat space or in sphere, then we need not consider the gravitational and mixed anomalies. In that case, we have shown elsewhere[5],[6] the following proposition.

Proposition (1)

(i) The global gauge anomaly coefficient $A(\omega)$ is given by

$$A(\omega) = (-1)^{\text{ind}\ D_{2n}} \tag{1.3}$$

for any locally anomaly-free representation ω satisfying Eq. (1.1). Here, ind D_{2n} stands for the Attiya-Singer index of Dirac operator in 2n-dimensional space. Especially, the global anomaly is at most of Z_2-type for any locally anomaly-free representation ω of any group H.

(ii) Any self-contragradient representation ω has no global anomaly in dimension D=2(mod 4), provided that ω is free of local anomaly. Especially, any locally anomaly-free representation ω of any gauge group listed in Eq. (1.2) has no global anomaly in D=2(mod 4).

In spite of these general statements, it is in general very difficult to compute explicit values of the anomaly coefficient $A(\omega)$ from the formula given in Eq. (1.3). However, there exists a complementary method for computing the purely gauge global anomaly coefficient $A(\omega)$. It has been initiated by Witten[7], as well as by Elitzur and Nair[8] whose formula has been subsequently studied for some special cases of SU(N) groups by several authors[9]. The method has been generalized in the references 5 and 6 which will be hereafter referred to as (I) and (II), respectively. Moreover, the mathematical aspect of the problem has been further elaborated by Lundell and Tosa[10]. Our method displays a rather interesting interplay between group-theoretical consideration and algebraic topology in contrast to that of the ref. 3 which is more topological in nature.

The purpose of this note is to investigate the problem in more details with some new results. Especially, we will show the following results for classical groups.

1) <u>Sp(2N) (N=rank of the group) (N≥1)</u>

i) Any representation ω of Sp(2N) has no global as well as local anomalies in dimension D=0(mod 8).

ii) Any locally anomaly-free representation ω of Sp(2N) has no global anomaly in D=2 or 6(mod 8).

iii) The global anomaly coefficient $A(\omega)$ of Sp(2N) in dimension D=4(mod 8) is given by

$$A(\omega) = \exp[i\pi Q_2(\omega)] \quad , \tag{1.4}$$

where $Q_2(\omega)$ is the 2nd Dynkin index[2],11],12], normalized to $Q_2(\square)=1$ for the 2N-dimensional basic representation $\square$ of Sp(2N). The explicit formula of $Q_2(\omega)$ will be given in section 3.

2) <u>SO(N) (N≥7)</u>

i) Any tensor representation ω, excluding cases of self-dual tensors for N=even, has no global as well as local anomaly in D=4(mod 8). The absence of the global anomaly also holds valid in dimension D=2 or 6(mod 8), assuming in addition that ω is free of local anomaly.

ii) The global anomaly coefficient $A(\omega)$ of non-self-dual tensor representation ω in D=0(mod 8) is given for N≥9 by the formula Eq. (1.4) in terms of the 2nd Dynkin index $Q_2(\omega)$ of the SO(N) group with normalization $Q_2(\square)=1$ where $\square$ is the N-dimensional basic representation of SO(N).

iii) The cases of spinor and self-dual tensor representations are difficult to be dealt with by our method, except for the case of H=SO(2n+1) with n=even and hence D=2n=0(mod 4). However, we will not go into this case since it is discussed in (II).

3) <u>SU(N) (N≥2)</u>

i) SU(N) has no global anomaly for N≥n+1 in D=2n.

ii) Any locally anomaly-free representation ω of SU(n-1) in any D=2n as well as of SU(n) in D=2n (n=odd) is free of global anomaly.

iii) Any locally anomaly-free representation of SU(3) has no global anomaly in D=2(mod 4).

iv) SU(2) has no global anomaly in D=0(mod 8) as well as in D=2 or 6(mod 8), provided that ω is free of local anomaly for the latter cases. For D=4(mod 8), any irreducible representation ω of SU(2) with dimension 2J+1 (J=0, 1/2, 1, 3/2, ...) has no global anomaly for J=integer and for $J = 1/2(4\ell+3) = 3/2, 7/2, \ldots$, while the anomaly coefficient $A(\omega)$ assumes the value -1 for $J = 1/2(4\ell+1) = 1/2, 5/2, \ldots$

For proofs of some results stated above, the following property of the n-th order Dynkin index[2],[12] $Q_n(\omega)$ is crucial. Let den B_n ($n\geq 1$) be the denominator of the n-th Bernoulli number B_n. More accurately, it is defined as product of all prime numbers p which divides n-1. Then for any representation ω of SU(N), and Sp(2N) as well as for non-self-dual tensor representation of SO(N) ($N\geq 7$), we have

$$Q_n(\omega) - Q_m(\omega) = k \text{ den } B_{n-m} \tag{1.5}$$

for integer k, where n and m are any two integers satisfying $n>m\geq 2$ such that the group under the consideration possess fundamental n-th and m-th Casimir invariants[2],[12]. Moreover, if $m\geq 3$ and n-m=even, then k in Eq. (1.5) is an even integer. When we note that den B_n is always even, then Eq. (1.5) especially gives

$$Q_n(\omega) = Q_2(\omega) (\text{mod } 2) \quad , \tag{1.6}$$

for values of n obeying $2\leq n\leq N$ for SU(N), $2\leq n\leq 2N$ (n=even) for Sp(2N) and SO(2N+1), and $2\leq n\leq 2(N-1)$ (n=even) for SO(2N). Especially, this reproduces the even-odd rule of ref. 13,

$$Q_3(\omega) = Q_2(\omega) (\text{mod } 2) \tag{1.7}$$

for SU(N) ($N\geq 3$), which is important for determination of global anomaly of SU(N) group in D=4 dimension. It may also be noted that Eq. (1.7) is the relation needed to link the result of refs. 3 and 4 to that of refs. 7 and 8 in D=4.

We organize our paper as follows. We will present the basic mathematical formulas in the next section, while explicit computations of global anomaly coefficients will be given in section 3. In the Appendix, we will give a proof of Eq. (1.5) as well as related trace identities involing TrX^{n+1}.

2. MATHEMATICAL FORMULATIONS

Let H be the gauge group under consideration. If the homotopy group $\pi_{2n}(H)$ is identically zero, then H has, of course, no global anomaly in D=2n. Also, since any theory with local anomaly is not self-consistent, we implicitly assume that the representation ω of H is free of local anomaly, i.e. it satisfies Eq. (1.1) unless it is stated otherwise or it is unnecessary. Now, the method of refs. 7 and 8 as well as of (I) and (II) is to reduce the problem of computing the global anomaly coefficient $\Lambda(\omega)$ of H to that of computing local anomaly of a larger group G in a representation $\tilde{\omega}$ as follows. For a given ω and H, suppose that we can find a group G and its representation $\tilde{\omega}$ satisfying conditions;

1) $G \supset H$, and $\pi_{2n}(G) = 0$,
2) $\tilde{\omega}$ of G reduces to a direct sum of ω and singlets of H when we restrict G to H.

We call these two conditions the representation conditions. Note that $\pi_{2n}(G) = 0$ implies that G has no global anomaly in D=2n. The second condition (2) cannot be in general satisfied, unless we allow $\tilde{\omega}$ with negative multiplicity coefficients, corresponding to the negative chirality of fermion fields. The same remark also applies to ω satisfying Eq. (1.1) in general, unless it is a self-contragradient representation of H in D=0(mod 4). As we demonstrated in (II), the representation conditions are always satisfied by suitable choices of G and $\tilde{\omega}$ for H=SU(N) or Sp(2N) or G_2, or F_4 as well as non-self-dual tensor representations of SO(N).

Hereafter, we assume that both G and H are compact, connected, simple Lie groups, and consider the exact homotopic sequence[14],

$$\pi_{2n+1}(G) \xrightarrow{p^*} \pi_{2n+1}(G/H) \xrightarrow{\Delta^*} \pi_{2n}(H) \xrightarrow{i^*} \pi_{2n}(G) = 0 \quad . \quad (2.1)$$

We have proved the following Propositions 2 and 3 in (II).

Proposition 2

Suppose that the representation conditions as well as Eq. (1.1) are satisfied. If $\pi_{2n+1}(G)$ or $\pi_{2n+1}(G/H)$ is zero or finite group, then the representation ω of H is free of global anomaly in D=2n.

Proposition 3

We assume again that the representation conditions and Eq. (1.1) are satisfied. Suppose that we have $\pi_{2n+1}(G) = Z$ and $\pi_{2n+1}(G/H) = Z \oplus T$, where T is a finite torsion group. Let $x \in \pi_{2n+1}(G)$ and $y \in \pi_{2n+1}(G/H)$ be generating elements of infinite cyclic groups Z contained in these groups. They must satisfy

$$p^*(x) = dy + t \tag{2.2}$$

where $t \in T$ and d is a integer. Then, the basic global anomaly coefficient $A(\omega)$ of H is given by

$$A(\omega) = \exp\{2\pi i Q_{n+1}(\tilde{\omega})/d\} \tag{2.3}$$

where the integral number $Q_{n+1}(\tilde{\omega})$ is the n+1-th order Dynkin index[2],12] of the representation $\tilde{\omega}$ of G.

Remark (1)

If G possesses no n+1-th fundamental Casimir invariant, then we set $Q_{n+1}(\tilde{\omega}) = 0$ in Eq. (2.3) so that we have $A(\omega) = 1$, i.e. no global anomaly. Otherwise, we normalize $Q_{n+1}(\tilde{\omega})$ by $Q_{n+1}(\square) = 1$ for the basic (i.e. the lowest dimensional) representation $\square$ of G.

Remark (2)

Strictly speaking, the formula Eq. (2.3) must be slightly modified for the case of G = SO(2n+2) with n=odd by reasons explained in (II). However, for this case with H = SO(2n+1) , the Proposition 1 implies no global anomaly at all.

Proposition 4

Under the same assumptions $\pi_{2n+1}(G) = Z$ and $\pi_{2n+1}(G/H) = Z \oplus T$, the value of d defined by Eq. (2.2) is given by

$$d = |\pi_{2n}(H)|/|T| \tag{2.4}$$

where $|\pi_{2n}(H)|$ and $|T|$ are orders (i.e. numbers of distinct elements) of groups $\pi_{2n}(H)$ and T, respectively, with understanding that $|T|=1$ when T is empty. Also, $\pi_{2n}(H) \simeq Z_{\ell d} \oplus T/Z_{\ell}$ for a positive integer ℓ.

Remark (3)

The proof of the Proposition 4 will be presented elsewhere[15]. Note that $|\pi_{2n}(H)|$ is always finite by a theorem proved by Serre[16]. If $\pi_{2n+1}(G/H) = Z$ and $\pi_{2n}(H) = Z_m$ $(2\leq m<\infty)$, then the Proposition 4 gives d=m in accordance with the result of (II).

The following proposition has been also noted in (II) and ref. 10.

Proposition 5

For H = SU(n-k) $(0\leq k\leq n-2)$, we can choose G = SU(n+1). Then, the value of d is given by

$$d = n!/W(n+1,k+1) \tag{2.5}$$

where W(n+1,k+1) is the James number[17] of the complex Stiefel manifold SU(n+1)/SU(n-k). Therefore, we calculate

$$A(\omega) = \exp\{2\pi i \frac{W(n+1,k+1)}{n!} Q_{n+1}(\tilde{\omega})\} \tag{2.6}$$

for any locally anomaly-free representation ω of SU(n-k) group, where $Q_{n+1}(\tilde{\omega})$ is the n+1-th Dynkin index of G = SU(n+1).

Remark (4)

For k=0, we have W(n+1,1) = 1 so that the formula Eq. (2.6) reproduces the one given by Elitzur and Nair[8] for H = SU(n). The cases of k=0 and k=1 have been discussed in (I), while cases k=n-2 and n-3 will be analyzed in section 3.

3. COMPUTATIONS OF $A(\omega)$

3.1 Sp(2N) (N=Rank of the Group)

The Bott's periodicity theorem[18],[19] implies

$$\pi_j[Sp(2N)] = \begin{cases} Z , & j=3,7(\text{mod } 8) \\ Z_2 , & j=4,5(\text{mod } 8) \\ 0 , & j=0,1,2,6(\text{mod } 8) \end{cases} \tag{3.1}$$

for the stable region $4N\geq j-1$. Especially, we see $\pi_{2n}(H) = 0$ for H = Sp(2N) with $4N\geq D = 2n$ for D=0, 2, 6(mod 8). Therefore, any Sp(2N) $(4N\geq D)$ has no global anomaly in D=0, 2, 6(mod 8). Note that this conclusion is independent of whether ω satisfies the locally anomaly-free condition Eq. (1.1) or not. Actually, we can remove the restriction D≤4N as follows. For the case of D=0(mod 8), we can choose G = Sp($2\tilde{N}$) with any $\tilde{N}$ satisfying $\tilde{N}$≥Max(N,D/4) for any H = Sp(2N). As we noted in (II), we can always find $\tilde{\omega}$ which satisfies the representation conditions. But in D=0(mod 8), we see $\pi_{2n+1}(G) = 0$ from Eq. (3.1) so that the Proposition 2 assures the absence of the global anomaly for any value of N as we already noted in (II). On the other hand, suppose D=2, 6(mod 8). Then, we appeal to the Proposition 1 to conclude the absence of the global anomaly again, provided now that ω is required to satisfy the local anomaly-free condition.

For the remaining case of D=4(mod 8), we cannot choose G = Sp($2\tilde{N}$) since $\pi_{2n}(G) = Z_2$ in general. However, we note the Bott's periodicity theorem for SU($\tilde{N}$),

$$\pi_j[SU(\tilde{N})] = \begin{cases} 0 \;, & j=\text{even} \\ Z \;, & j=\text{odd} \end{cases} \tag{3.2}$$

for the stable region $\tilde{N}$≥1/2(j+1). Therefore, for a given H = Sp(2N), we can choose G = SU($\tilde{N}$) for any $\tilde{N}$ satisfying $\tilde{N}$≥Max(2N,n+1) in order to have $\pi_{2n}(G) = 0$. As in (II), we can find then $\tilde{\omega}$ which obeys the representation conditions. Unfortunately, the homotopy group $\pi_{2n}(G/H)$ is not known for the general case. We consider first the case of 2N≥n+1 in D=2n with a choice of G = SU(2N), where we note[18],[20]

$$\pi_j[SU(2N)/Sp(2N)] = \pi_{j+2}[SO] = \begin{cases} Z \;, & j=1,5(\text{mod } 8) \\ Z_2 \;, & j=6,7(\text{mod } 8) \\ 0 \;, & j=0,2,3,4(\text{mod } 8) \end{cases} \tag{3.3}$$

for j<4N. Especially, for D=2(mod 4), this gives

$$\pi_{2n+1}(G/H) = \begin{cases} 0 \;, & D=2(\text{mod } 8) \\ Z_2 \;, & D=6(\text{mod } 8) \end{cases}$$

so that we have no global anomaly again by the Proposition 2 for 4N≥D+2 in accordance with the Proposition 1 as well as $\pi_{2n}(H) = 0$. However for D=4(mod 8), i.e. n=2(mod 4), we find

$$\pi_{2n+1}(G/H) = Z \quad , \quad \pi_{2n}(H) = Z_2 \quad .$$

Therefore, Propositions 3 and 4 lead to d=2 with

$$A(\omega) = \exp\{\pi i Q_{n+1}(\tilde{\omega})\} \qquad (3.4)$$

for 4N≥D+2, where $Q_{n+1}(\tilde{\omega})$ is the n+1-th order Dynkin index of G = SU(2N). For the remaining case of 4N≤D, the same formula Eq. (3.4) is still valid with the choice of G = SU(n+1) by the following reason. For H = Sp(2N) (4N≤D=2n), Lundell and Tosa[10] choose G = SU(n+1) and studied a exact double homotopic sequence. Their result can be rewritten as Eq. (3.4) again.

Now, we can rewrite Eq. (3.4) further as follows. Consider first the case of 4N≥D+1 with G = SU(2N). Then, by Eq. (1.6), we see

$$Q_{n+1}(\tilde{\omega}) = Q_2(\tilde{\omega}) \pmod 2 \quad .$$

We now restrict G = SU(2N) to the sub-group H = Sp(2N), where the branching index sum rule[12] can be shown to give[21]

$$Q_2(\tilde{\omega}) = Q_2(\omega) \quad .$$

Here, $Q_2(\omega)$ is the 2nd order Dynkin index of H = Sp(2N), and we utilized the fact that $\tilde{\omega}$ reduces to a direct sum of ω and singlets of H. Therefore, we find

$$A(\omega) = \exp[\pi i Q_2(\omega)] \qquad (3.5)$$

for H = Sp(2N) in D=4(mod 8) when we have 4N>D. Note that any reference to the auxiliary quantities G and $\tilde{\omega}$ has disappeared in this form. For the remaining case of 4N≤D, we consider the branching of G = SU(n+1) to H = Sp(2N) via a chain

$$SU(n+1) \supset Sp(n) \supset Sp(n-2) \supset \ldots \supset Sp(2N) \quad .$$

Then, we similarly recover the validity of Eq. (3.5), and we conclude that it holds for all cases of Sp(2N) in D=4(mod 8). Since any ω is free of local anomaly in this case, we can restrict ourselves to the case of ω being irreducible. Then, the calculation of $Q_2(\omega)$ is easy from the formula given in ref. 12 or from the definition

$$Q_2(\omega) = d(\omega)I_2(\omega)/d(\square)I_2(\square) \tag{3.6}$$

where $d(\omega)$ and $I_2(\omega)$ are dimensions and 2nd-order Casimir invariant of the irreducible representation ω and where $\square$ is the 2N-dimensional basic rep. of Sp(2N). Explicit expressions for $d(\omega)$ and $I_2(\omega)$ are found in ref. 22. Let $\Gamma = \langle f_1, f_2, \ldots, f_N\rangle$ satisfying $f_1 \geq f_2 \geq \ldots \geq f_N \geq 0$ be the Young tableau corresponding to ω, and set

$$\ell_j = f_j + N - j + 1 \quad , \quad \ell_j^{(o)} = N - j + 1 \quad , \quad (1 \leq j \leq N) \quad . \tag{3.7}$$

Then, we calculate

$$Q_2(\omega) = \frac{d(\omega)}{2N(2N+1)} \sum_{j=1}^{N} \{(\ell_j)^2 - (\ell_j^{(o)})^2\} \quad . \tag{3.8}$$

Especially for SU(2) = Sp(2) with N=1, we set $f_1 = 2J$ (J=0, 1/2, 1, 3/2,...) to find

$$Q_2(\omega) = \frac{2}{3} J(J+1)(2J+1) \tag{3.9}$$

as in (II), which is odd only for J = 1/2 $(4\ell+1)$ = 1/2, 5/2, ...

3.2 SO(N)

The Bott's periodicity theorem is now

$$\pi_j[SO(N)] = \begin{cases} Z & , \quad j=3,7 \pmod 8 \\ Z_2 & , \quad j=0,1 \pmod 8 \\ 0 & , \quad j=2,4,5,6 \pmod 8 \end{cases} \tag{3.10}$$

for the stable region $N \geq j+2$. Especially, we see $\pi_{2n}[SO(N)] = 0$ for $N \geq 2n+2$ in D=2n=2, 4, 6(mod 8). Hence, we conclude that any representation ω of H=SO(N) in $N \geq D+2$ has no global anomaly in D=2, 4, 6 (mod 8). For D=2, 6(mod 8), the Proposition 1 assures that tensor representations of SO(N) have no global anomaly still in $N \leq D+1$, provided that ω is assumed now to be free of local anomaly. For the case of D=4(mod 8), we can also remove the restriction $N \geq D+2$, if we assume ω to be non-self-dual tensor representations of SO(N). To prove it, let $\tilde{N}$ be any integer satisfying $\tilde{N} \geq \mathrm{Max}(N, D+2)$ and choose $G=SO(\tilde{N})$. Then, Eq. (3.10) shows $\pi_{2n+1}(G) = 0$ in D=4(mod 8) so that the Proposition 2 assures the absence of global anomaly in this case also.

For the remaining instance of D=0(mod 8), we cannot choose $G=SO(\tilde{N})$ in general since $\pi_{2n}(G) = Z_2 \neq 0$ at least in the stable region. However, we can choose $G=SU(\tilde{N})$ for any $\tilde{N}$ satisfying $\tilde{N} \geq \mathrm{Max}(N,(D+1)/2)$. In order to satisfy the representation condition, we have to restrict ourselves to non-self-dual tensor representations ω of SO(N) by the same reason given in (II). We first consider the case of $N \geq D+2$ with the choice of G=SU(N) for H=SO(N), where we have

$$\pi_{2n+1}(G) = \pi_{2n}(H) = Z \quad , \quad \text{and} \quad \pi_{2n+1}(G/H) = Z_2 \quad .$$

Therefore, from Propositions 3 and 4, we find

$$A(\omega) = \exp[i\pi Q_{n+1}(\tilde{\omega})] \tag{3.11}$$

where $Q_{n+1}(\tilde{\omega})$ is the n+1-th order Dynkin index of $G \equiv SU(N)$. However just as in the case of H=Sp(2N), we can rewrite this also as

$$A(\omega) = \exp[i\pi Q_2(\omega)] \quad . \tag{3.12}$$

Next, consider the case of $9 \leq N \leq D+2$, where we can choose G=SU(D+3) for H=SO(N). Then, Lundell and Tosa[10] were able to essentially derive the same formula Eq. (3.11) again by considering a double exact homotopic sequence. Then, Eq. (3.12) is also valid in this case, and we conclude that the global anomaly coefficient $A(\omega)$ for H=SO(N) ($N \geq 9$) in D=0(mod 8) is given by Eq. (3.12). The numerical value of $Q_2(\omega)$ can be computed as follows. Let $\Gamma = \langle f_1, f_2, .., f_m \rangle$ satisfying $f_1 \geq f_2 \geq .. \geq f_m \geq 0$ with m=[N/2] be the Young tableau corresponding to the tensor representation ω of SO(N) (N=2m or 2m+1). Since we are excluding the case of self-dual tensors for N=2m=even, we have to set $f_m=0$ for N=2m (but not necessarily for N=2m+1). Setting now

$$\ell_j = f_j + N/2 - j \quad , \quad \ell_j^{(o)} = N/2 - j \quad (1 \leq j \leq m) \quad , \tag{3.13}$$

we calculate[22]

$$Q_2(\omega) = \frac{d(\omega)}{N(N-1)} \sum_{j=1}^{m} \{(\ell_j)^2 - (\ell_j^{(o)})^2\} \quad , \tag{3.14}$$

where $d(\omega)$ is the dimension of ω for the SO(N) group.

3.3 SU(N)

For this case, we base our calculation upon the formula Eq. (2.6). Unfortunately, numerical values of the James number are known only for few cases. For N=n or n-1, the calculation has been done in (I) and we will not repeat the argument. For H=SU(2), the James number has been calculated by Walker[23] and we find[10]

$$A(\omega) = \begin{cases} \exp\left[\frac{2\pi i}{\text{den } B_{n-1}} Q_{n+1}(\tilde{\omega})\right] , & n=3(\text{mod } 4) \\ \exp\left[\frac{4\pi i}{\text{den } B_{n-1}} Q_{n+1}(\tilde{\omega})\right] , & n=1(\text{mod } 4) \end{cases} \tag{3.15}$$

and

$$A(\omega) = \begin{cases} \exp[2\pi i Q_{n+1}(\tilde{\omega})] = 1 , & n=0(\text{mod } 4) \\ \exp[\pi i Q_{n+1}(\tilde{\omega})] , & n=2(\text{mod } 4) \end{cases} \quad . \tag{3.16}$$

Therefore, we conclude that SU(2) in D=0(mod 8) has no global anomaly while the anomaly coefficient for D=4(mod 8) is rewritten as

$$A(\omega) = \exp[i\pi Q_2(\omega)]$$

again by the same reasoning used as before on the basis of Eq. (1.6). These results reproduce, of course, the known cases of Sp(2) ≃ SU(2). We shall postpone for a while discussions of Eq. (3.15).

Finally, consider the case of H=SU(3) in D=2(mod 4) where the James number is known[23] to give

$$A(\omega) = \exp\left[\frac{2\pi i}{\text{den } B_{n-1}} Q_{n+1}(\tilde{\omega})\right] . \tag{3.17}$$

In the Appendix, we will prove that $A(\omega)$ given by Eqs. (3.15) and (3.17) are always one, provided that the representation ω is free of local anomaly, i.e. it satisfies

$$\text{Tr}^{(\omega)} X^{n+1} = 0 \quad , \quad (n=\text{odd}) \quad . \tag{3.18}$$

Therefore, locally anomaly-free representation ω of both SU(2) and SU(3) has no global anomaly in D=2(mod 4). Although this fact for SU(2) is a special case of the Proposition 1, the case of SU(3) is not covered by it. It is curious to note that we have found no example of a case having global anomaly in D=2(mod 4). Hence, we conjecture that any locally anomaly-free representation ω of any simple group has no global anomaly in D=2(mod 4), irrespective of whether ω is self-contragradient or not.

Acknowledgement

This work is supported in part by the U.S. Department of Energy Contract No. DE-AC02-76ER13065. Also, we would like to thank Prof. A. Lundell and Dr. Y. Tosa for having kindly sent their preprint to us.

Appendix

Here, we will prove Eq. (1.5) as well as a related trace identity. Consider first the case of SU(N) group where N-1 fundamental weight system will be denoted as $\Lambda_1, \Lambda_2, \ldots, \Lambda_{N-1}$. From the formula given in the second paper of ref. 12, we know then

$$Q_n(\Lambda_f)-Q_m(\Lambda_f) = -\sum_{\ell=0}^{f-1}(-1)^{f-\ell}\{(f-\ell)^{n-1}-(f-\ell)^{m-1}\}\frac{N!}{\ell!(N-\ell)!}\ , \quad (1\leq f\leq N-1) \tag{A.1}$$

for any two integers n and m satisfying $2\leq m\leq n\leq N$. Let p be a prime number such that p-1 divides n-m. Now it is easy to show

$$x^{n-1} - x^{m-1} = 0(\mathrm{mod}\ p) \tag{A.2}$$

for any integer x. Therefore, we find

$$x^{n-1} - x^{m-1} = k\ \mathrm{den}\ B_{n-m}\ , \quad (n>m\geq 2) \tag{A.3}$$

for some integer k. Also, if n-m=even>0 and $m\geq 3$, we will show soon that k is an even integer. From Eqs. (A.1) and (A.3) with $x=f-\ell$, we find Eq. (1.5) for the special case $\omega=\Lambda_f$ $(1\leq f\leq N-1)$ of SU(N), i.e.

$$Q_n(\Lambda_f) - Q_m(\Lambda_f) = k\ \mathrm{den}\ B_{n-m}\ . \tag{A.4}$$

Before going further, we will now prove Eq. (A.2). If x is an integer multiple of p, then it is trivial. If (x,p)=1, and n>m, then we know $x^{n-m}=1(\mathrm{mod}\ p)$ by the Fermat's theorem so that Eq. (A.2) is still valid. Also, if $n>m\geq 3$ and n-m=even we can easily verify $x^{n-1}-x^{m-1} = 0$ (mod 4) and hence k in Eqs. (A.3) and (A.4) must be even.

Now, let ω be a general representation of SU(N). Then, Eq. (1.5) can be readily derived from special cases of Eq. (A.4) if we use either the general formula for $Q_n(\omega)$ of ref. 12 or the induction method of ref. 13. The cases for Sp(2N) and for non-self-dual tensor representations of SO(N) can be similarly proved. For some special cases, we can say more. Actually, we can show for instance

$$Q_6(\omega) = Q_2(\omega) \pmod{30} \tag{A.5}$$

not only for SU(N)(N≥6) and Sp(2N)(N≥3) with den B_4=30 but also for G_2 as well as for any representation ω of SO(N)(N≥7), including spinor and self-dual tensor representations for the latter. Also, we find

$$Q_6(\omega) = Q_2(\omega) \begin{cases} \pmod{10} & \text{for } F_4 \text{ and } E_6 \\ \pmod{5} & \text{for } E_7 \end{cases} . \tag{A.6}$$

Next, let L be a simple Lie algebra and let I_3, I_+ and I_- satisfying $[I_3, I_\pm] = \pm I_\pm$, and $[I_+, I_-] = 2I_3$ be its su(2) sub-Lie algebra. Let n and m be arbitrary positive even integers. Then, for any representation ω of L, we will prove

$$\mathrm{Tr}^{(\omega)}(2I_3)^n - \mathrm{Tr}^{(\omega)}(2I_3)^m = 4k' \text{ den } B_{n-m} \tag{A.7}$$

for a integer k′. Moreover, k′ is even, if ω is a real representation of L. To prove these, decompose the representation ω of L into a direct sum of irreducible representations Λ_{2J} with dimension 2J+1 (J=0, 1/2, 1, 3/2, ...) of su(2) by restricting L to su(2) as in

$$\omega \downarrow \sum_J \oplus\, g_J\{\Lambda_{2J}\} \tag{A.8}$$

where g_J is the multiplicity coefficient. Therefore, we calculate

$$\mathrm{Tr}^{(\omega)}(2I_3)^n - \mathrm{Tr}^{(\omega)}(2I_3)^m = \sum_J g_J \sum_{M=-J}^{J} \{(2M)^n - (2M)^m\}$$

$$= 2\sum_J g_J \sum_{M>0}^{J} \{(2M)^n - (2M)^m\}$$

for positive even integers n amd m, where M assumes 2J+1 values J,

J-1, ..., -J. Since 2M are integers and n-m is even with $m \geq 2$, k in Eq. (A.3) is even and hence Eq. (A.3) gives Eq. (A.7). Moreover, if ω is a real representation of L, then we have proved in ref. 13, that the multiplicity coefficient g_J for half-integral value of J is always even. This implies that k′ in Eq. (A.7) is a even number in that case and hence especially for G_2 and F_4.

Now, we can prove the assertion made at the end of the section 3. Let ω be a locally anomaly-free representation of L, so that it satisfies Eq. (1.1) for arbitrary X. We choose $X=2I_3$. Then, Eq. (A.7) gives

$$2\mathrm{Tr}^{(\omega)}(I_3)^2 = -2k' \text{ den } B_{n-1} \quad (n=\text{odd}) \quad , \tag{A.9}$$

where we have set m=2 and replaced n by n+1. On the other hand, we have shown elsewhere[13] the validity of

$$2\mathrm{Tr}^{(\omega)}(I_3)^2 = \theta\, Q_2(\omega)\, \tilde{N}_2(\square) \quad , \tag{A.10}$$

where

$$\tilde{N}_2(\square) = \begin{cases} 1 & \text{for} \quad SU(N) \text{ and } Sp(2N) \\ 2 & \text{for} \quad SO(N)(N\geq 7) \text{ and } G_2 \\ 6 & \text{for} \quad F_4 \text{ and } E_6 \\ 12 & \text{for} \quad E_7 \\ 60 & \text{for} \quad E_8 \end{cases} \tag{A.11}$$

for a integer value of θ which is independent of ω. However, we can always choose the su(2) sub-algebra to be the minimum regular embedding[13] of L with $\theta=1$. Hence from Eqs. (A.9) and (A.10), we find

$$Q_2(\omega) = -2k' \text{ den } B_{n-1} \tag{A.12}$$

for locally anomaly-free representation ω of H=SU(N) and Sp(2N) in D=2n(n=odd). But we also have

$$Q_{n+1}(\tilde{\omega}) - Q_2(\tilde{\omega}) = k \text{ den } B_{n-1} \tag{A.13}$$

from Eq. (A.4). Moreover, if we reduce G=SU(n+1) to SU(3) or SU(2), we also have $Q_2(\tilde{\omega}) = Q_2(\omega)$ by the branching index sum rule[12],[21] when we note that $\tilde{\omega}$ reduces to a direct sum of ω and singlets of SU(3) or SU(2). Combining these facts, we see

$$Q_{n+1}(\tilde{\omega}) = \tilde{k} \text{ den } B_{n-1} \quad (n=\text{odd})$$

for a integer $\tilde{k}$. This proves $A(\omega)=1$ for Eqs. (3.15) and (3.17) in D=2(mod 4) as has been stated in section 3.

References

1. Bardeen, A. and White, A. R.; (editors) Symposium on anomalies, geometry, and topology, (World Scientific, Singapore, 1986).
2. e.g. see Okubo, S.; "General Dynkin indices and their applications", in Symmetries in Science II, edited by B. Gruber and R. Lenczewski (Plenum Pub., NY 1986), p. 419.
3. Bismut, J. M. and Freed, D. S.; Commun. Math. Phys. 106, 159 (1986), ibid, 107, 103 (1986), Freed, D. S.; ibid 107, 483 (1986), Freed, D. S. and Vafa, C.; ibid 110, 349 (1987), Della Pietra, S., Della Pietra, V. and Alvarez-Gaume, L.; ibid 109, 691 (1987), 110, 573 (1987).
4. Witten, E.; Phys. Lett. B117 324 (1982), Comm. Math. Phys. 100 197 (1985).
5. Okubo, S., Zhang, H., Tosa, Y., and Marshak, R. E.; Phys. Rev. D in press. This paper will be referred to as (I).
6. Zhang, H., Okubo, S., and Tosa, Y.; to appear in Phys. Rev. D. This paper will be referred to as (II) hereafter.
7. Witten, E.; Nucl. Phys. B223, 422 and 433 (1983).
8. Elitzur, S. and Nair, V. P.; Nucl. Phys. B243, 205 (1984).
9. Holman, R. and Kephart, T. W.; Phys. Lett. 167B, 417 (1986), Kiritsis, E.; Phys. Lett. 178B, 53 (1986). Erratum, ibid, 181B, 416 (1986), Braden, H. W.; Univeristy of North Carolina preprint IFP-296-UNC (1987).
10. Lundell, A. and Tosa, Y. "Global Anomalies and Algebraic Topology" University of Colorado Report, COLO-HEP-167, Dec. 1987.
11. Dynkin, E. B.; Mat. USSR Sb. 30 349 (1952) [Am. Math. Soc. Trans. Ser. 2, 6, 111 (1957)].

12. Okubo, S.; Jour. Math. Phys. 23, 8 (1982), Okubo, S. and Patera, J.; Jour. Math. Phys. 24, 2772 (1983), and 25; 219 (1984).

13. Geng, C., Marshak, R. E., Zhao, Z., and Okubo, S.; Phys. Rev. D36, 1953 (1987), Okubo, S., Geng, C., Marhsak, R. E., and Zhao, Z.; Phys. Rev. D36 3268 (1987).

14. Hu, S. T.; "Homotopy Theory" (Academic Press, NY 1959), Steenrod, N. "The topology of fiber bundle" (Princeton University Press, Princeton, NJ 1951); Whitehead, G. W. "Elements of Homotopy theory" (Spring-Verlag, New York, 1978).

15. Zhang, H. and Okubo, S.; University of Rochester Report in preparation.

16. Serre, J. P.; Ann. of Math. 58, 258 (1953), Borel, A.; Bull. Amer. Math. Soc. 61, 397 (1955).

17. James, I. M.; Proc. London Math. Soc. 8 536 (1958), and "The topology of Stiefel Manifolds" (London Mathematical Society Lecture Note Series Vol. 24, Cambridge University Press, Cambridge 1976).

18. Bott, R.; Ann. Math. 70, 313 (1959). See also, Milnor, J.; "Morse theory" (Princeton Univ. Press, Princeton, 1973).

19. Iyanaga, S., and Kawada, Y.; editors, "Encyclopedic dictionary of mathematics, vol. II" (M.I.T. Press, Cambridge, 1977). In Table 6 VI for real Stiefel manifold in Appendix A, n(m) should be, however, the column (row) variable.

20. Harris, B.; Trans. Amer. Math. Soc. 106, 174 (1963), Mimura, M.; C. R. Acad. Sci. Paris Ser A-B 262, A20 (1966).

21. Okubo, S.; Jour. Math. Phys. 26, 2127 (1985).

22. Okubo, S.; Jour. Math. Phys. 18, 2382 (1977), ibid 23, 8, (1982).

23. Walker, G.; Quart. J. Math. Oxford, 32, 467 (1981).

BRIEF REVIEW OF THE NEW LOCAL SUPERSYMMETRY IN THE VIERBEIN FORMALISM OF EINSTEIN GRAVITY*

Noboru Nakanishi
Research Institute for Mathematical Sciences
Kyoto University, Kyoto 606
JAPAN

ABSTRACT

A new supersymmetric extension of the local Lorentz symmetry of the vierbein formalism is briefly reviewed. The proposed symmetry is local $OSp(N,2;\mathbb{C})$, and is totally independent of spacetime symmetry at the Lagrangian level. The theory is quite a natural extension of the manifestly covariant canonical formalism of quantum gravity. There is no difficulty concerning probabilistic interpretability, though the bosonic part of $OSp(N,2;\mathbb{C})$ is non-compact. The theory requires chiral symmetry to be $SO(N)$, N being unrestricted.

1. INTRODUCTION

In spite of the great successes of the standard theory of particle physics and the Einstein theory of gravity, physicists are still unsatisfied with them because the fundamental theory of physics is expected to provide us a more systematic and unified picture. There have been proposed a number of interesting attempts such as grand unification, supersymmetry, supergravity, Kaluza-Klein theory, superstring theory, etc., but none of them has even the slightest experimental support. It seems more desirable, therefore, to seek for new possibilities of unification rather than concentrating one's attention to one of those attempts.

When no helps are available from experiments, the best thing which theorists can do may be to extend the theories which have been already successful in such a way that the extended theory is both physically and mathematically natural. The superstring theory may, in some sense, have achieved unification of the standard theory and the Einstein gravity perturbatively, but, unfortunately, it can become physically sensible only if a very special assumption is introduced by hand in order to reduce dimensionality. I believe that the fundamental theory should not be crucially based on such artificiality. I therefore attach more importance to the naturalness of the extension than immediate achievement of unification. It is too hasty to wish to achieve the ultimate theory at the present stage.

It is usually believed that supergravity is the local version of supersymmetry, but it should be noted that the local Poincaré symmetry involved in supergravity is <u>not</u> spacetime symmetry but internal one. Indeed, it is <u>impossible</u> to realize the global translation generators in the superalgebra of supergravity. Although supergravity is formulated in such a way that it becomes a generalization of the Einstein gravity, this fact is not the consequence of symmetry principle. Indeed, in the vierbein formalism of the Einstein gravity, the general-coordinate (or Einstein) symmetry is <u>totally independent</u> of the local Lorentz symmetry, that is, there is no Poincaré symmetry structure. Therefore, I belive that supergravity is not a natural extension of the Einstein gravity.

The Poincaré symmetry of particle physics and its absence in the Einstein gravity are reconciled in the manifestly covariant canonical formalism of quantum gravity[1-3] in the following way. By gauge fixing, the Einstein symmetry and the local Lorentz symmetry become spacetime affine symmetry (i.e., translation plus general linear symmetry) and global internal Lorentz symmetry, respectively, and both of them are spontaneously broken. But a particular combination (depending explicitly on the way of spontaneous breakdown) of the Lorentz part of the general linear symmetry and the global internal Lorentz symmetry remains unbroken. This unbroken symmetry turns out to be the spacetime Lorentz symmetry, which, together with translation, gives us the Poincaré symmetry of particle physics. This story suggests that the Poincaré symmetry should not be regarded as a first principle when gravity is taken into account.

Since the most beautiful quantum version of the Einstein gravity is the manifestly covariant canonical formalism of quantum gravity,[2,3] I believe that attempts at unifying gravity and particle physics should be made so as to be consistent with it. This theory has the sixteen-dimensional Poincaré-like superalgebra (inhomogeneous OSp(8,8)),[4,5] which contains both spacetime and internal symmetries in complete harmony without requiring any Kaluza-Klein-type _ad hoc_ procedure. Since this supersymmetry is so remarkable that, I believe, any extension of spacetime symmetry should not destroy it. In particular, I conclude that extensions of spacetime symmetry should be made at the level of _quantum_ Lagrangian density.[6] But attempts in this direction is not yet quite satisfactory, though not impossible.[7]

Since the vierbein formalism of the Einstein gravity has the local Lorentz symmetry, I now turn my attention to its possible extension, that is, I seek for its supersymmetric version. Because this symmetry is _purely Lorentz but not Poincaré_ as emphasized above, the no-go theorem[8] is evaded. Since the Lorentz algebra is equivalent to $SL(2,\mathbb{C})$, one has only to seek for supersymmetric extensions of $SL(2)$. Possible extensions are thus $OSp(N,2;\mathbb{C})$, $SL(N,2;\mathbb{C})$, and three exceptional Lie superalgebras.[9] I take the first possibility;[10] others seem not to be appropriate for the present purpose. The bosonic part of $OSp(N,2;\mathbb{C})$ is $SL(2,\mathbb{C})\oplus SO(N,\mathbb{C})$. It should be emphasized that this type of extension becomes impossible if the dimensionality of spacetime is higher than six[11] because then the Lorentz algebra does not have spinor form. Thus to consider $OSp(N,2;\mathbb{C})$ is in conformity with my belief that higher-dimensional spacetimes should be excluded by some intrinsic reasons.

2. OSp(N,2;ℂ) AND ITS REPRESENTATIONS

It is convenient to employ spinor notation for describing the internal Lorentz part. Let $M_{AB}(=M_{BA})$ and $\bar{M}_{\dot{A}\dot{B}}[=(M_{AB})^{\dagger}]$ be Lorentz generators in spinor notation. Spinor indices are denoted by capital Roman letters which take values 1 and 2. The spinor-space metric is $\epsilon_{AB}(=-\epsilon_{BA})$; $\epsilon_{12}=\epsilon^{12}=1$. The SO(N,ℂ) generators are denoted by $X^{\ell m}(=-X^{m\ell})$ and $\bar{X}^{\dot{\ell}\dot{m}}[=(X^{\ell m})^{\dagger}]$, where lower-case Roman letter indices take values 1,2,...,N. The fermionic generators are denoted by $Q_A{}^{\ell}$ and $\bar{Q}_{\dot{A}}{}^{\dot{\ell}}[=(Q_A{}^{\ell})^{\dagger}]$. Then the Lie superalgebra OSp(N,2;ℂ) is given by

$$i[M_{AB},M_{CD}] = \epsilon_{AC}M_{BD}+\epsilon_{BC}M_{AD}+(C\leftrightarrow D), \tag{1}$$

$$i[M_{AB},Q_C{}^{\ell}] = \epsilon_{AC}Q_B{}^{\ell}+\epsilon_{BC}Q_A{}^{\ell}, \tag{2}$$

$$[M_{AB},X^{\ell m}] = 0, \tag{3}$$

$$\{Q_A{}^{\ell},Q_B{}^{m}\} = \delta^{\ell m}M_{AB}+\epsilon_{AB}X^{\ell m}, \tag{4}$$

$$i[Q_A{}^{\ell},X^{mn}] = \delta^{\ell m}Q_A{}^{n}-\delta^{\ell n}Q_A{}^{m}, \tag{5}$$

$$i[X^{\ell m},X^{pq}] = \delta^{mp}X^{\ell q}-\delta^{\ell p}X^{mq}-(p\leftrightarrow q), \tag{6}$$

and likewise for dotted generators; any undotted generator (anti-) commutes with any dotted one.

Now, I consider the representations of the above superalgebra. Because of complete parallelism, I usually omit writing dotted quantities unless necessary. Since all quantities considered below transform manifestly covariantly under the bosonic generators, it suffices to consider the transformation properties under $Q_A{}^{\ell}$. A supermultiplet $\{F_{BC}(=F_{CB}),\ \Psi_B{}^{m},\ K^{mn}(=-K^{nm})\}$ constitutes an adjoint representation if they transform as the generators do, that is, if

$$i[Q_A{}^{\ell},\ F_{BC}] = \epsilon_{AB}\Psi_C{}^{\ell}+\epsilon_{AC}\Psi_B{}^{\ell}, \tag{7}$$

$$\{Q_A{}^{\ell},\ \Psi_B{}^{m}\} = \delta^{\ell m}F_{AB}+\epsilon_{AB}K^{\ell m}, \tag{8}$$

$$i[Q_A{}^{\ell},\ K^{mn}] = \delta^{\ell m}\Psi_A{}^{n}-\delta^{\ell n}\Psi_A{}^{m}. \tag{9}$$

The super-invariant inner product of two supermultiplets of adjoint representation is given by

$$F'^{BC}F_{BC}+2i\Psi'^{Bm}\Psi_B{}^{m}-K'^{mn}K^{mn}. \tag{10}$$

Likewise, a supermultiplet $\{\Phi_B, R^m\}$ constitutes a fundamental representation if

$$\{Q_A{}^{\ell},\ \Phi_B\} = \epsilon_{AB}R^{\ell}, \tag{11}$$

$$i[Q_A{}^{\ell},\ R^{m}] = \delta^{\ell m}\Phi_A. \tag{12}$$

The super-invariant inner product of two supermultiplets of fundamental representation is given by

$$i\Phi'^{B}\Phi_B-R'^{m}R^{m}. \tag{13}$$

The adjoint representation can be formed as a product representation of two fundamental representations.

3. ξ-FIELD REALIZATION

In the spinor notation, the vierbein field is written as $\hat{h}_{\mu\dot{B}C}$, where the Greek script is, of course, a spacetime index (μ=0,1,2,3). It should be the first member of a supermultiplet which behaves like $\{\bar{\Phi}_{\dot{B}}\Phi_C,\ \bar{\Phi}_{\dot{B}}R^n,\ \Phi_C\bar{R}^{\dot{m}},\ \bar{R}^{\dot{m}}R^n\}$. Here, however, in order to avoid complexity of introducing independent superpartners,* I make a very crucial

---

* Note that $Q_A{}^{\ell}$ is dimensionless in contrast to the ordinary supersymmetry whence superpartners necessarily have a common dimension in contradiction to the usual difference of dimensions between boson and fermion fields.

assumption that all superpartners can be formed by using a particular fermion field $\xi_B{}^m$ and its dotted counterpart.[10,12] With

$$\{Q_A{}^\ell,\ \xi_B{}^m\} = -\delta^{\ell m}\epsilon_{AB} - i\xi_A{}^m\xi_B{}^\ell, \tag{14}$$

it is easy to see that both $\{\phi_B,\ -i\xi^{Cm}\phi_C\}$ and $\{-\xi_B{}^n r^n, r^m\}$ constitute a fundamental representation and that they are mutually orthogonal with respect to the inner product (13). They are nonlinear realizations different from the usual one except for N=1. Any generic fundamental representation can be expressed as a sum of the above two types of nonlinear realizations. Furthermore, as product representations, I can construct three types of nonlinear realizations of adjoint representation, which are mutually orthogonal with respect to the inner product (10) and complete in the sense that any generic adjoint representation can be decomposed into a sum of those three types.[13)]

Now, I return to the vierbein. From the above consideration, I should set

$$i[Q_A{}^\ell,\ \hat{h}_{\mu\dot{B}C}] = \epsilon_{AC}\xi^{D\ell}\hat{h}_{\mu\dot{B}D}. \tag{15}$$

After constructing all superpartners in terms of $\hat{h}_{\mu\dot{B}C}$, $\xi_B{}^m$ and $\bar{\xi}_{\dot{B}}{}^{\dot{m}}$, it is easy to see from (13) that $\Xi\bar{\Xi}\hat{h}_\mu{}^{\dot{B}C}\hat{h}_{\mu\dot{B}C}$ is super-invariant, where

$$\Xi \equiv 1+\frac{1}{2}i\xi^{Bm}\xi_B{}^m, \tag{16}$$

$\bar{\Xi}$ being its dotted counterpart. Hence, if one redefines the vierbein field $h_{\mu BC}$ by[12)]

$$h_{\mu\dot{B}C} \equiv \Xi^{1/2}\bar{\Xi}^{1/2}\hat{h}_{\mu\dot{B}C}, \tag{17}$$

then the metric tensor, defined by

$$g_{\mu\nu} \equiv \frac{1}{2}h_\mu{}^{\dot{B}C}h_{\nu\dot{B}C}, \tag{18}$$

is super-invariant. This $OSp(N,2;\mathbb{C})$ invariance of $g_{\mu\nu}$ is the fundamental requirement that spacetime symmetry remains untouched.[10)] Since (14) yields

$$i[Q_A{}^{\ell}, \Xi] = -\Xi\xi_A{}^{\ell}, \tag{19}$$

(15) implies that*

$$\begin{aligned} i[Q_A{}^{\ell}, h_{\mu\dot{B}C}] &= \epsilon_{AC}\xi^{D\ell}h_{\mu\dot{B}D} - \tfrac{1}{2}\epsilon_{AD}\xi^{D\ell}h_{\mu\dot{B}C} \\ &= \tfrac{1}{2}\xi^{D\ell}(\epsilon_{AC}h_{\mu\dot{B}D}+\epsilon_{DC}h_{\mu\dot{B}A}), \end{aligned} \tag{20}$$

that is,

$$[Q_A{}^{\ell}, h_{\mu\dot{B}C}] = \tfrac{1}{2}\xi^{D\ell}[M_{AD}, h_{\mu\dot{B}C}]. \tag{21}$$

4. HIDDEN $GL(N,\mathbb{C})$ SYMMETRY

The relation (21) is not accidental. To see this, it is necessary to introduce the generators $Y^{\ell}{}_m$ of a $GL(N,\mathbb{C})$ symmetry,[14)] though it is not a genuine symmetry. They satisfy

$$i[Y^{\ell}{}_m, Y^p{}_q] = \delta^p{}_m Y^{\ell}{}_q - \delta^{\ell}{}_q Y^p{}_m. \tag{22}$$

Any quantity $\mathcal{O}$ which does not explicitly depend on $\xi_B{}^m$ transforms under $GL(N,\mathbb{C})$ like

$$\begin{aligned} i[Y^{\ell}{}_m, \mathcal{O}^{p_1\ldots p_r}{}_{q_1\ldots q_s}] &= (s-r)\delta^{\ell}{}_m \mathcal{O}^{p_1\ldots p_r}{}_{q_1\ldots q_s} \\ &+ \sum_{k=1}^{r}\delta^{p_k}{}_m \mathcal{O}^{p_1\ldots\ell\ldots p_r}{}_{q_1\ldots q_s} - \sum_{k=1}^{s}\delta^{\ell}{}_{q_k}\mathcal{O}^{p_1\ldots p_r}{}_{q_1\ldots m\ldots q_s}. \end{aligned} \tag{23}$$

---

* Note that any quantity antisymmetric in A and B is proportional to ϵ_{AB}.

Then the following general formula holds:

$$[Q_A{}^{\ell}, \mathcal{O}]_{\mp} = \frac{1}{2}\xi^{B\ell}[M_{AB}, \mathcal{O}] + \xi_A{}^m[Y^{\ell}{}_m, \mathcal{O}]. \tag{24}$$

Of course, $Y^{\ell}{}_m$ is related to $X^{\ell m}$ through

$$X^{\ell m} = Y^{\ell}{}_m - Y^m{}_{\ell}\ . \tag{25}$$

Although $\xi_B{}^m$ itself does not have an appropriate transformation property uner $GL(N,\mathbb{C})$, the contravariant and covariant components of the $GL(N,\mathbb{C})$ metric tensor, γ^{mn} and γ_{mn}, inverse to each other, can be constructed in the following way:*

$$\gamma^{mn} \equiv \Xi^2(\delta^{mn} - i\xi^{Bm}\xi_B{}^n) = \gamma^{nm}, \tag{26}$$

$$\gamma_{mn} \equiv \Xi^{-2}(\delta^{mn} + i\Xi^{-1}\xi^{Bm}\xi_B{}^n) = \gamma_{nm}. \tag{27}$$

Indeed, (26) and (27) are consistent with (24) together with (23). Furthermore, (19) and (24) imply that

$$i[Y^{\ell}{}_m, \Xi] = -\delta^{\ell}{}_m\Xi, \tag{28}$$

that is, Ξ is a $GL(N,\mathbb{C})$ scalar density.

As is well known, in order to formulate the Dirac theory in a manifestly generally-covariant way, it is necessary to introduce the vierbein field. Quite analogously, in order to formulate the Dirac theory in a manifestly $SO(N,\mathbb{C})$-covaraint way, it is necessary to introduce the $GL(N,\mathbb{C})$ "vielbein" field.[14] The contravariant and covariant components of the $GL(N,\mathbb{C})$ vielbein field, $e^{\alpha m}$ and $e^{\alpha}{}_m$, inverse to each other, satisfy the relations

---

* The factors $\Xi^{\pm 2}$ in (26) and (27) are due to the convention of introducing the first term in the rhs of (23) but have no intrinsic meaning.

$$\gamma^{mn} = e^{\alpha m} e^{\alpha n}, \tag{29}$$

$$\gamma_{mn} = e^{\alpha}{}_{m} e^{\alpha}{}_{n}. \tag{30}$$

Here the new index α runs over $1,2,\ldots,N$ but it is independent of the $SO(N,\mathbb{C})$ symmetry. Furthermore, it is "real" in the sense that $(e^{\alpha m})^{\dagger} = e^{-\alpha\dot{m}}$, i.e., $\dot{\alpha} \equiv \alpha$. Just as in the case of the ordinary vierbein, the relations (29) and (30) are invariant under a new local $SO(N)$ symmetry, which turns out to be nothing but chiral symmetry, that is, <u>chiral symmetry should be SO(N).</u>

5. CONNECTION SUPERMULTIPLET

As is well known, the spin connection $\omega_{\mu BC}(=\omega_{\mu CB})$ can be defined by the requirement that the covariant derivative of the vierbein field vanishes:

$$\partial_{\mu} h_{\nu\dot{B}C} - \Gamma_{\mu\nu}{}^{\lambda} h_{\lambda\dot{B}C} - \frac{1}{2}\omega_{\mu C}{}^{D} h_{\nu\dot{B}D} - \frac{1}{2}\bar{\omega}_{\mu\dot{B}}{}^{\dot{E}} h_{\nu\dot{E}C} = 0, \tag{31}$$

where $\Gamma_{\mu\nu}{}^{\lambda}$ stands for the affine connection. Since the internal Lorentz symmetry has now been extended to the $OSp(N,2;\mathbb{C})$ symmetry, it is natural to define the connection supermultiplet $\{\mathbb{F}_{\mu BC}(=\mathbb{F}_{\mu CB}), \Psi_{\mu B}{}^{m}, \mathbb{K}_{\mu}{}^{mn}(=-\mathbb{K}_{\mu}{}^{nm})\}$ by requiring that both vierbein supermultiplet and vielbein one have vanishing covariant derivatives with respect to spacetime and $OSp(N,2;\mathbb{C})$ symmetries.[12] The expressions for $\mathbb{F}_{\mu BC}$, $\Psi_{\mu B}{}^{m}$ and $\mathbb{K}_{\mu}{}^{mn}$ can be found explicitly in terms of $\omega_{\mu BC}$, $\xi_{B}{}^{m}$ and $e^{\alpha}{}_{m}$, but I refrain from reproducing them here. Under <u>global</u> transformations, $\{\mathbb{F}_{\mu BC}, \Psi_{\mu B}, \mathbb{K}_{\mu}{}^{mn}\}$ constitutes an adjoint representation given by (7)-(9). Under <u>local</u> transformations, they have the following correct transformation property as the connection:

$$\text{L.P.}[M_{AB}, \mathbb{F}_{\mu CD}] = 2i(\epsilon_{AC}\epsilon_{BD} + \epsilon_{AD}\epsilon_{BC})\partial_{\mu}(\cdot), \tag{32}$$

$$\text{L.P.}\{Q_{A}{}^{\ell}, \Psi_{\mu B}{}^{m}\} = -2\epsilon_{AB}\delta^{\ell m}\partial_{\mu}(\cdot), \tag{33}$$

$$L.P.[X^{\ell m}, \mathbb{K}_{\mu}{}^{pq}] = -2i(\delta^{\ell p}\delta^{mq} - \delta^{mp}\delta^{\ell q})\partial_{\mu}(\cdot), \tag{34}$$

and other commutators have no $\partial_{\mu}(\cdot)$ part. Here L.P. means that the terms corresponding to global transformations are omitted, and $\partial_{\mu}(\cdot)$ denotes the simple derivative of an infinitesimal transformation function.

It is now easy matter to construct locally $OSp(N,2;\mathbb{C})$-invariant Lagrangian densities, e.g., the locally super-invariant Dirac Lagrangian density.[14] On the other hand, the Einstein Lagrangian density is evidently trivially intact; in particular, $\Gamma_{\mu\nu}{}^{\lambda}$ does not acquire torsion in contrast to supergravity.

6. BRS INVARIANCE

In order to quantize the theory, it is necessary to introduce gauge fixing and FP-ghost Lagrangian densities so as to be BRS invariant. In the manifestly covariant canonical formalism of quantum gravity, the local-Lorentz gauge fixing is given by a scalar density formed by a product of the spin connection and the simple derivative of the local-Lorentz B-field.[15] Its straightforward extension is achieved by considering a scalar density formed by a super-invariant inner product (10) of the connection supermultiplet $\{\mathbb{F}_{\mu BC}, \Psi_{\mu B}{}^{m}, \mathbb{K}_{\mu}{}^{mn}\}$ and the simple derivative of the B-field supermultiplet, $\{S_{BC}, \Sigma_{B}{}^{m}, B^{mn}\}$, constituting a generic adjoint representation.[12] Then the gauge-fixing action is general-coordinate invariant and globally super-invariant. It should be noted that since $\mathbb{F}_{\mu BC}$, $\Psi_{\mu B}{}^{m}$ and $\mathbb{K}_{\mu}{}^{mn}$ already contain time derivatives, the B-fields must be regarded as canonical variables, that is, they are <u>not</u> multiplier fields.

In order to construct the FP-ghost Lagrangian density, the FP ghost supermultiplet $\{T_{BC}, \mathcal{T}_{B}{}^{m}, C^{mn}\}$ and the FP anti-ghost one $\{U_{BC}, \Upsilon_{B}{}^{m}, D^{mn}\}$ are introduced. They both should constitute a generic adjoint representation, but since they must obey "wrong" statistics, (7)-(9) would have to be modified. In order to avoid this trouble, the FP (anti-)ghosts are assumed to obey the following unconventional statistics: The usual spin-statistics rule is reversed only when both members are FP (anit-)ghosts.[16] Such statistics is admissible because of the superselection rules of the FP-ghost number and spinor even-oddness, though, of course, then FP (anti-)ghosts are neither purely bosonic nor purely fermionic. Thanks to this unconventional statistics, the FP (anti-)ghost supermultiplets can obey (7)-(9) and the BRS transformation is definable so as to commute, rather than anticommute, with the super-transformation induced by $Q_A{}^{\ell}$.

The BRS transformation $\underline{\delta}$ for any $\mathcal{O}$ involving no FP (anti-) ghosts is defined by[16]

$$\underline{\delta}\mathcal{O} \equiv -\frac{i}{2}T^{AB}[M_{AB}, \mathcal{O}] + \mathcal{J}^{A\ell}[Q_A{}^{\ell}, \mathcal{O}]_{\mp} + \frac{i}{2}C^{\ell m}[X^{\ell m}, \mathcal{O}]$$

$$+ \text{dotted counterpart,} \tag{35}$$

where if the (anti-)commutators involve $\partial_\mu(\cdot)$, it should be understood to act on the corresponding FP ghost. The BRS transforms of the FP ghosts are just one half of what are obtained from the right-hand side of (35) by substituting the FP ghosts into $\mathcal{O}$. The BRS transforms of the FP anti-ghosts are, of course, i times the B-fields. Then the BRS transformation commutes with M_{AB}, $Q_A{}^{\ell}$ and $X^{\ell m}$ and is nilpotent. The FP-ghost Lagrangian density is given by a scalar density formed by a super-invariant inner product of the super-covariant derivative of the FP ghost supermultiplet and the simple derivative of the FP anti-ghost supermultiplet.[12] In this way, one obtains the BRS-invariant Lagrangian density, which is globally super-invariant; this result should be compared with the fact that globally super-invariant gauge fixing seems to be impossible in supergravity.

It is now straightforward, at least in principle, to carry out canonical quantization. Abe has explicitly worked it out in the N=1 case.[17] The thory is quite a natural extension of the manifestly covariant canonical formalism of quantum gravity in the sense that the former reduces to the latter (vierbein formalism) if the ξ field is absent and that all symmetry properties (such as the extended BRS) of the latter survive in the former.[12)]

7. REMARKS

I make a comment on the ghost counting.[16)] As usual, the number of the B-fields, that of the FP ghosts and that of the FP anti-ghosts are all equal to the number of the generators,

$$F \equiv 2[3+2N+\frac{1}{2}N(N-1)]. \quad (36)$$

As for the numbers of the (field) degrees of freedom related to $OSp(N,2;\mathbb{C})$ other than the above are as follows: 6 for the local Lorentz part of $h_{\mu\dot{B}C}$, $2\cdot 2N$ for $\xi_B{}^m$ and $\bar{\xi}_{\dot{B}}{}^{\dot{m}}$, and $2\cdot N(N-1)/2$ for $e^{\alpha m}$ and $\bar{e}^{\alpha\dot{m}}$ because of (29). Thus their total sum is just equal to F. Since the B-fields describe independent canonical degrees of freedom as noted already, the number of relevant bosonic degrees of freedom is 2F, which is precisely equal to that of the FP ghosts and anti-ghosts. Thus all those degrees of freedom are unobservable owing to the Kugo-Ojima quartet mechanism.[18)] Therefore the existence of the non-compact symmetry $OSp(N,2;\mathbb{C})$ does not destroy probabilistic interpretability.

The fact that the $OSp(N,2;\mathbb{C})$ symmetry is unobservable is not disappointing. Indeed, what is important in constructing a unified theory is the symmetry principle at the Lagrangian level but not at the physical S-matrix level.

As stated in the Introduction, the internal Lorentz symmetry is spontaneously broken, but a particular combination of it with space-time symmetry is unbroken and becomes the spacetime Lorentz symmetry. Quite analogously, the $SO(N,\mathbb{C})$ symmetry is spontaneously broken because of $\langle 0|e^{\alpha m}|0\rangle \simeq \delta^{\alpha m}$, but a particular combination of its real part with chiral symmetry is unbroken and becomes the physical chiral symmetry. Furthermore, I note that (14) implies the spontaneous breakdown of $Q_A{}^{\ell}$. Thus all of $OSp(N,2,\mathbb{C})$ are spontaneously broken.

Finally, I note that the ξ-field non-linear realization has quite general applicability to the problems of the spontaneous breakdown.

REFERENCES

1. Nakanishi, N., Prog. Theor. Phys. 66, 1843 (1981).
2. Nakanishi, N., Publications Res. Inst. Math. Sci. (Kyoto) 19, 1095 (1983).
3. Nakanishi N. and Ojima, I., "Covariant Operator Formalism of Gauge Theories and Quantum Gravity", World Scientific (Singapore), to be published.
4. Nakanishi, N., Prog. Theor. Phys. 63, 2078 (1980).
5. Nakanishi, N., Prog. Theor. Phys. 64, 639 (1980).
6. Nakanishi, N., Prog. Theor. Phys. Suppl. No.86, 203 (1986).
7. Abe, M. and Nakanishi, N., Prog. Theor. Phys. 74, 881 (1985).
8. Haag, R., Łopuszanski, J.T. and Sohnius, M., Nucl. Phys. B88, 257 (1975).
9. Kac V.G., Adv. Math. 26, 8 (1977).
10. Nakanishi, N., Prog. Theor. Phys. 77, 1533 (1987).
11. Lukierski, J. and Nowicki, A., Phys. Lett. 151B, 382 (1985).
12. Abe, M. and Nakanishi, N., preprint RIMS-609.
13. Abe, M. and Nakanishi, N., Prog. Theor. Phys. 79, 227 (1988).
14. Abe, M. and Nakanishi, N., Prog. Theor. Phys. 78, 704 (1987).
15. Nakanishi, N., Prog. Theor. Phys. 62, 779 (1979).
16. Abe, M. and Nakanishi, N., Prog. Theor. Phys. 79, 240 (1988).
17. Abe, M., preprint RIMS-606.
18. Kugo, T. and Ojima, I., Prog. Theor. Phys. Suppl. No.66, 1 (1979).

Dynamical Symmetry Breaking and the Effective Potential of Local Composite Operators

M. Inoue*), H. Katata**), T. Muta and K. Shimizu

Department of Physics
Hiroshima University
Hiroshima 730

Abstract

The effective potential of local composite operators is employed in order to study the general feature of dynamical symmetry breaking. The ambiguity unique to the effective potential of local composite operators originating from the arbitrariness of the renormalization scheme is pointed out and its consequences are discussed. The relation of the multivaluedness of the Schwinger functional to the shape of the effective potential is clarified.

----

*) Present address: Nautical Science Department, Hiroshima National College of Maritime Technology, Higashino-cho, Toyota-gun, Hiroshima 725-02.

**) Present address: Research Department 4, Central Research Laboratories, Engineering Center, SHARP Corporation, 2613-1 Ichinomoto-cho, Tenri-shi, Nara 632.

§ 1. Introduction

The notion of the spontaneous symmetry breaking (SSB) is considered to be one of the most important ideas in modern quantum field theory. It has been indispensable in constructing a variety of field-theory models such as Glashow-Salam-Weinberg model, grand unification theories and technicolor models.

One of the simplest ways of giving rise to SSB is to introduce Higgs fields, i.e. elementary scalar fields which have nonvanishing vacuum expectation values. This procedure has in fact been employed in Glashow-Salam-Weinberg model and grand unification theories. In this method, however, we meet with a number of uncontrollable free parameters associated with Higgs fields, i.e. their masses and self-coupling constants. We feel this feature unsatisfactory and prefer to investigate the possibility of SSB without any recourse to Higgs fields, i.e. the dynamical symmetry breaking.

In the scheme of the dynamical symmetry breaking, SSB takes place as a result of nonperturbative effects such as the formation of bound states through interactions among fundamental fields. This mechanism was in fact primarily employed in the original proposal of SSB in quantum field theory by Nambu and Jona-Lasinio.[1] One of the most important applications of this mechanism is found in the technicolor model[2] where the Higgs fields in Glashow-Salam-Weinberg model are replaced by dynamical bound states of subquarks.

A useful tool to investigate SSB is the effective potential[3] which is interpreted as a ground-state energy in the presence of an external source. We regard the effective potential as a function of the vacuum expectation value of the operator relevant to SSB. In the case of the dynamical symmetry breaking the relevant operator is a composite operator made of fundamental fields.

There are two kinds of definition of the composite operator: the local and nonlocal (bilocal) composite operator. It is nowadays common to use the bilocal composite operator as a dynamical variable following Cornwall, Jackiw and Tomboulis.[4] The effective potential for the bilocal composite operator turns out to be a functional of the vacuum expectation value of this operator; the stationary condition for the effective potential is known to be the Dyson-Schwinger equation for Green functions.[4] We have to solve the Dyson-Schwinger equation to derive an information on the ground state. The equation, however, is usually intractable. On the other hand the effective potential for the local composite operator is an ordinary function of the vacuum expectation value of this operator and hence the stationary condition reduces to an algebraic equation which may be solved without any difficulty. Since the local composite operator consists of the product of operators at the same space-time point, it develops an infinity due to the short-distance singularity. This infinity has to be removed by the renormalization procedure for the composite operators so that the corresponding counter term is necessarily included in the Lagrangian. It has been pointed out that the introduction of the above counter term creates an ambiguity which may spoil the stability criterion for the ground state.[5,6]

In spite of the above bothering situation we reconsider, in the present article, an advantage of using the effective potential for local composite operators. This is partially motivated by the fact that the ambiguity of the similar nature has been found recently in the effective potential for bilocal composite operators[7,8] and the stationary condition is conspicuously simple in the case of local composite operators rather than in the case of bilocal ones. We discuss the renormalization of the effective potential for local composite operators and find that we need the counter term which is of the form of polynomial of the source function. The arbitrariness of the renormalization condition leads us to the ambiguity of the effective potential. We argue how to see the stability of the stationary point of such effective potentials with ambiguity.

Yet another important issue related to the effective potential for local composite operators is the multivaluedness of the Schwinger functional when SSB is present. Here the effective action/potential is obtained from the Schwinger functional through the Legendre transformation. We use the Gross-Neveu model[9] in 3 space-time dimensions to explain the multivalued property of the Schwinger functional.

§ 2. Renormalization of the effective potential of local composite operators

We shall discuss an ambiguity of the effective potential of local composite fields which originates from extra counter terms needed to cancel divergences in vacuum diagrams. In order to explain our idea we use the solvable model, the Gross-Neveu model,[9] which is characterized by the following Lagrangian in two space-time dimensions,

$$\mathcal{L} = \bar{\psi}_a (i\gamma^\mu \partial_\mu) \psi_a + (g^2/2)(\bar{\psi}_a \psi_a)^2, \tag{2.1}$$

where ψ_a is an N-component massless fermion field with $a = 1, 2, \ldots, N$ (Repeated indices in Eq.(2.1) are summed) and g which is chosen to be positive is a bare coupling constant.

The Lagrangian (2.1) exhibits the chiral symmetry which may be broken spontaneously. This symmetry breaking is dynamical in the sense that it is signaled by the composite operator $g\bar{\psi}_a\psi_a$ instead of an elementary field. We shall investigate the dynamical chiral-symmetry breaking by using the effective potential as a function of the vacuum expectation value of $g\bar{\psi}_a\psi_a$ in the presence of the external source.

The effective action $\Gamma[\phi_c]$ is given as a Legendre transform of the Schwinger functional W[J],

$$\Gamma[\phi_c] = W[J] - \int d^Dx \, \frac{\delta W}{\delta J} J, \tag{2.2}$$

where we remain in D-dimensional space-time for later convenience, the Schwinger functional is defined by

$$\exp(iW[J]) = \int [d\psi d\bar{\psi}]\exp\{i\int d^Dx(\mathcal{L} + g\bar{\psi}_a\psi_a J)\}, \tag{2.3}$$

and the new variable ϕ_c is given by

$$\phi_c = \frac{\delta W}{\delta J} = \langle g\bar{\psi}_a\psi_a\rangle_J, \tag{2.4}$$

which is regarded as a vacuum expectation value of $g\bar{\psi}_a\psi_a$ in the presence of the external source J. From the above definition it follows that

$$\frac{\delta \Gamma}{\delta \phi_c} = -J. \tag{2.5}$$

By letting the source function J(x) to be independent of x we define the Schwinger function G(J) and effective potential $V(\phi_c)$,

$$G(J) = W[J]/\Omega_D \, . \tag{2.6}$$

$$V(\phi_c) = -\Gamma[\phi_c]/\Omega_D, \tag{2.7}$$

where Ω_D is the D-dimensional space-time volume.

Before going into the discussion of renormalization we introduce a different definition of the Schwinger functional and derive the relation between two different Schwinger functionals. The new definition of the Schwinger functional consists in the use of an auxiliary field σ,[9]

$$\exp(i\hat{W}[J]) = \int [d\psi d\bar{\psi} d\sigma] \exp\{i\int d^Dx(\hat{\mathcal{L}} + \sigma J)\}, \tag{2.8}$$

where

$$\hat{\mathcal{L}} = \bar{\psi}_a(i\gamma^\mu\partial_\mu)\psi_a - \sigma^2/2 + g\bar{\psi}_a\psi_a\sigma. \tag{2.9}$$

Without the source term, i.e. for J = 0, the new Schwinger functional $\hat{W}[J]$ coincides with the old one W[J]. In the presence of the source term these two Schwinger functionals are different. In fact the relation

$$\hat{W}[J] = W[J] + (1/2)\int d^Dx J(x)^2, \tag{2.10}$$

can be derived by noting that Eq.(2.3) may be multiplied by the constant factor,

$$\int [d\sigma] \exp\{-(i/2)\int d^Dx(\sigma - J - g\bar{\psi}_a\psi_a)^2\}, \tag{2.11}$$

and the integration variable σ can be shifted.[7] By setting J(x) to be constant we obtain

$$\hat{G}(J) = G(J) + J^2/2, \tag{2.12}$$

where $\hat{G}(J)$ is defined from $\hat{W}[J]$ in a similar manner as in Eq.(2.6). Hence we have

$$\sigma_c = \phi_c + J, \tag{2.13}$$

$$\hat{V}(\sigma_c) = V(\phi_c) + J^2/2, \tag{2.14}$$

where σ_c is given by

$$\sigma_c = \frac{\delta \hat{W}[J]}{\delta J} = \langle \sigma \rangle_J, \tag{2.15}$$

and $\hat{V}(\sigma_c)$ is defined through an equation similar to Eq.(2.7). It is obviously seen in Eqs.(2.13) and (2.14) that $\sigma_c = \phi_c$ and $\hat{V}(\sigma_c) = V(\phi_c)$ for $J = 0$. Thus the above two effective potentials are stationary at the same point and acquire the same value at the stationary point.

We then turn our attention to the renomalization of the effective potential of local composite operators. We expand $W[J]$ and $\hat{W}[J]$ in powers of J,

$$W[J] = \langle 0 \mid 0 \rangle_c + \int \langle 0 \mid g\bar{\psi}\psi \mid 0 \rangle_c J$$

$$+ (1/2) \int \int \langle 0 \mid T(g\bar{\psi}\psi)(g\bar{\psi}\psi) \mid 0 \rangle_c J^2 + \ldots , \tag{2.16}$$

$$\hat{W}[J] = \langle 0 \mid 0 \rangle_c + \int \langle 0 \mid \sigma \mid 0 \rangle_c J$$

$$+ (1/2) \int \int \langle 0 \mid T\sigma\sigma \mid 0 \rangle_c J^2 + \ldots , \tag{2.17}$$

where the space-time variable x is suppressed wherever it appears, the subscript c indicates connected Green functions and T represents the time ordering. The Green functions appearing in each coefficient of the expansion (2.17) is readily known to be renormalizable. This is because the Schwinger functional $\hat{W}[J]$ generates ordinary Green functions for elementary scalar field σ interacting with fermions ψ through Yukawa coupling. For the Schwinger functional W[J], however, we need an extra counter term which is foreign to the Lagrangian (2.9). In fact we realize in perturbative expansion in g that all diagrams corresponding to the Schwinger functionals W[J] and $\hat{W}$[J] are the same except for the one corresponding to the coefficient of J^2. Expressing these coefficients in terms of diagramatic expansion, we observe that for $\langle 0 | T(g\bar{\psi}\psi)(g\bar{\psi}\psi) | 0\rangle_c$ we have no graph that corresponds to the tree diagram for $\langle 0 | T(\sigma\sigma) | 0\rangle_c$ (see Fig.1). The origin of this disparty is nothing but the difference term $(1/2)\int J^2$ in Eq.(2.10). Usually the counter term contained in this tree diagram cancels the divergence of the 1-loop graphs of the order g^2. On the other hand, since W[J] does not have such term as $(1/2)\int J^2$, it is not renormalizable without an extra counter term $BJ^2/2$ where B is the divergent constant to be determined by a suitable renormalization condition. Adding this term to $\mathcal{L}$ we obtain the renormalizable Lagrangian density $\mathcal{L}_R$,

$$\mathcal{L}_R = \mathcal{L} + \frac{1}{2} BJ^2$$
$$= \bar{\psi}(i\gamma^\mu \partial_\mu)\psi + \frac{1}{2} g^2(\bar{\psi}\psi)^2 + \frac{1}{2} BJ^2. \qquad (2.18)$$

Obviously B is renormalization-scheme dependent and hence is arbitrary. We shall call this arbitrariness the J^2 ambiguity.

According to the renormalizable Lagrangian $\mathcal{L}_R$ the Schwinger functional $W_R[J]$ takes the form,

$$W_R[J] = W[J] + (1/2)\int BJ^2. \qquad (2.19)$$

The corresponding effective potential $V_R(\phi_c)$ may be calculated in a standard manner and reads

$$V_R(\phi_c) = V(\phi_c) + BJ^2/2, \qquad (2.20)$$

where J should be expressed as a function of ϕ_c by solving the equation like Eq.(2.4) for $W_R[J]$. A comparison of Eqs.(2.19) and (2.20) with Eqs.(2.10) and (2.14) suggests that $\widehat{W}[J]$ and $\widehat{V}(\sigma_c)$ which are defined through the auxiliary field method can be viewed as renormalized $W[J]$ and $V(\phi_c)$ in a specific renormalization scheme.

It is easy to see that the above phenomenon is not specific to the Gross-Neveu model but is more general. In fact one needs extra counter terms as introduced above in any calculation of the Schwinger functional for local composite operators. Since the choice of renormalization schemes is arbitrary, an ambiguity emerges in the expression of the Schwinger functional and effective potential.

We shall see in the following that this ambiguity will give rise to the crucial difficulty in arguing the stability of the stationary point for the effective potential. In order to discuss the stability one usually observe the sign of the curvature of the effective potential at the stationary point. On the other hand we see that

$$\frac{d^2V}{d\phi_c^2} = \frac{dJ}{d\phi_c} = \left(\frac{d^2G(J)}{dJ^2}\right)^{-1} . \tag{2.21}$$

Hence the value of the curvature of $V[\phi_c]$ at the stationary point where $J = 0$ depends on the finite part of B which is arbitrary. Thus the sign of the curvature may be flipped by suitably changing the renormalization scheme and the stability argument based on the curvature is spoiled.

Although the curvature of the effective potential at the stationary point loses its physical meaning, the stationary point of the effective potential and its value at the stationary point are unique and physically significant. Hence we can at least compare values of the effective potential at several stationary points and find out the lowest one which is a candidate for the stable vacuum.

§ 3. Multivaluedness of the Schwinger functional W[J]

In this section we discuss the general property of the Schwinger functional W[J] whose source function J(x) couples to a field ϕ which may be a local elementary field or a local composite field. We calculate the effective action by following the Legendre transformation,

$$\Gamma [\phi_c] = W[J] - \int d^D x \frac{\delta W}{\delta J} J, \tag{3.1}$$

with ϕ_c defined by

$$\phi_c = \frac{\delta W}{\delta J} . \tag{3.2}$$

It is straightforward to show that

$$\frac{\delta \Gamma}{\delta \phi_c} = -J . \tag{3.3}$$

These relations clearly shows that, if SSB occurs, i.e. if

$$\frac{\delta \Gamma [\phi_c]}{\delta \phi_c} = 0 \tag{3.4}$$

has several solutions, the Schwinger functional W[J] becomes multivalued for J = 0. We discuss the interrelation between the multivaluedness of the Schwinger functional at J = 0 and the structure of the effective action when SSB takes place in the Gross-Neveu model in 3 space-time dimensions.

We calculate the effective potential $\hat{V}[\sigma_c]$ corresponding to the Schwinger functional $\hat{W}[J]$ in Eq.(2.8) in the leading order of the 1/N expansion. The effective potential in D space-time dimension is give by

$$\frac{dV(\sigma)}{d\sigma} = \sigma - \frac{2^{[d/2]}}{(4\pi)^{D/2}} Ng^D \Gamma(1-D/2)\sigma^{D-1} , \tag{3.5}$$

where [ ] is the Gauss' notation, Γ is the gamma function and the subscript c is simply neglected.

By setting D = 2 and subtracting the divergence we recover the result of Gross and Neveu. It is seen that there are two vacua ($\sigma = 0$ and $\sigma \neq 0$) and the broken vacuum has lower energy than that of the symmetric one. The following relations hold for the Schwinger function $\hat{G}(J)$ and the effective potential $\hat{V}(\sigma)$

$$\sigma = \frac{d\hat{G}(J)}{dJ} , \tag{3.6}$$

$$\hat{V}(\sigma) = -\hat{G}(J) + \frac{d\hat{G}(J)}{dJ} J , \tag{3.7}$$

$$\frac{d\hat{V}(\sigma)}{d\sigma} = J . \tag{3.8}$$

Using Eqs.(3.6), (3.7) and (3.8) we find

$$\hat{G}(J) = -\hat{V}(\sigma) + \sigma \frac{d\hat{V}(\sigma)}{d\sigma} . \tag{3.9}$$

In principle the relation between σ and J can be derived from Eq.(3.5) with Eq.(3.8). For even-dimensional space-time, however, we cannot express σ explicitly as a function of J because of the logarithmic function contained in the effective potential and hence an explicit expression for $\hat{G}(J)$ can not be derived.

When the space-time dimension is set equal to three, the simplest relation between σ and J results. Equation (3.5) with D equal to 3 reads

$$\frac{d\hat{V}(\sigma)}{d\sigma} = \sigma + \frac{\sigma^2}{4a} , \tag{3.10}$$

where a is the constant defined by

$$a = \frac{\pi}{2Ng^3} . \tag{3.11}$$

With the normalization condition

$$\hat{V}(0) = 0, \tag{3.12}$$

Eq.(3.10) is solved to give

$$\hat{V}(\sigma) = \frac{1}{2}\sigma^2 + \frac{\sigma^3}{12a} . \tag{3.13}$$

We realize that Eq.(3.13) is nothing but a concrete example of the case considered by Woodhouse.[10]

This potential tells us that there are two vacua, i.e., the symmetric vacuum ($\sigma = 0$) and the broken one ($\sigma = -4a$), and the symmetric vacuum is lower than the broken one (see Fig. 2). We are interested in the Schwinger function corresponding to this effective potential.

Using Eq.(3.10) in Eq.(3.8) we obtain σ as a function of J,

$$\sigma(J) = 2a[-1 \pm \sqrt{1 + J/a}]. \tag{3.14}$$

For the symmetric vacuum we have

$$\sigma(0) = 0, \tag{3.15}$$

and hence we have to choose the upper sign in Eq.(3.14). The lower sign corresponds to the broken vacuum. Taking into account Eq.(3.6) and denoting "symmetric" and "broken" by S and B respectively, we obtain

$$\frac{d\hat{G}^S(J)}{dJ} = \sigma^S(J) = 2a[-1 + \sqrt{1 + J/a}], \qquad (3.16a)$$

$$\frac{d\hat{G}^B(J)}{dJ} = \sigma^B(J) = 2a[-1 - \sqrt{1 + J/a}]. \qquad (3.16b)$$

We fix constants of integration by using Eq.(3.7) with J set equal to zero, i.e.,

$$\hat{G}^S(0) = -\hat{V}(\sigma)|_{\sigma=0} = 0, \qquad (3.17a)$$

$$\hat{G}^B(0) = -\hat{V}(\sigma)|_{\sigma=-4a} = -\frac{8a^2}{3} \quad . \qquad (3.17b)$$

It then follows that

$$\hat{G}^S(J) = 2a[-J + \frac{2a}{3}(1 + \frac{J}{a})^{\frac{3}{2}} - \frac{2a}{3}] , \qquad (3.18a)$$

$$\hat{G}^B(J) = 2a[-J - \frac{2a}{3}(1 + \frac{J}{a})^{\frac{3}{2}} - \frac{2a}{3}] . \qquad (3.18b)$$

Let us now invert our argument and trace back to the origin of the multivaluedness of $\hat{G}(J)$. Suppose that we compute $\hat{G}(J)$ directly through its definition. In this calculation we have to choose the symmetric vacuum: the generating function which can be obtained by this procedure is $\hat{G}^s(J)$. When there are many vacua, this function may have several branch cuts. Eq.(3.18a) is a typical example. It develops a branch cut through which we may continue $\hat{G}^s(J)$ analytically in J to the next Riemann sheet. After this continuation the sign of the second term in the square brackets is changed. Thus $\hat{G}^s(J)$ reduces to $\hat{G}^B(J)$ by analytic continuation (see Eq.(3.18b) and Fig. 3). From these generating functions the expectation value of σ is obtained as a double-valued function of J(see Eq.(3.14)). For J = 0 we have two solutions $\sigma^S(0)$ and $\sigma^B(0)$ one of which corresponds to the lower energy. By Eq.(3.17) it is apparent that the symmetric vacuum gives the ground-state energy so that $\sigma^S(0)$ is the right one. The effective potential is given by Eq.(3.7). Here it should be noted that the Legendre transform of $\hat{G}^S$ and $\hat{G}^B$ agree with each other but the region of σ is separated by the condition,

$$\sigma^S > -2a , \quad \sigma^B < -2a .$$

(see Fig.4)

So far we have been considering the auxiliary field Lagrangian density $\hat{\mathcal{L}}$. We now come back to $\mathcal{L}$ defined in Eq.(2.1) and compute the Schwinger function G and the effective potential V. Using Eq.(2.12) we have

$$G^S(J) = 2a[-J + \frac{2a}{3}(1 + \frac{J}{a})^{\frac{3}{2}} - \frac{2a}{3}] - \frac{1}{2}J^2, \tag{3.19a}$$

$$G^B(J) = 2a[-J - \frac{2a}{3}(1 + \frac{J}{a})^{\frac{3}{2}} - \frac{2a}{3}] - \frac{1}{2}J^2. \tag{3.19b}$$

Differentiation of Eq.(3.19) gives $\Sigma \equiv \langle g\bar{\psi}_a \psi_a \rangle_J$ just as σ was given in Eq.(3.16):

$$\Sigma^S(J) = 2a[-1 + \sqrt{1 + J/a}] - J, \tag{3.20a}$$

$$\Sigma^B(J) = 2a[-1 - \sqrt{1 + J/a}] - J. \tag{3.20b}$$

Solving the above equations with respect to J we obtain the following relation

$$J_{\pm} = \frac{dV_{\pm}}{d\Sigma} = -\Sigma \pm 2a\sqrt{\frac{-\Sigma}{a}} , \tag{3.21}$$

where the subscript $\pm$ is to remind us of the signs in the right-hand side. Integration of Eq.(3.21) with the condition

$$V_{\pm}(0) = 0, \tag{3.22}$$

leads us to the effective potential for the theory under $\mathcal{L}$ with the source term $J\bar{\psi}\psi$,

$$V_{\pm}(\Sigma) = -\frac{1}{2}\Sigma^2 \mp \frac{4}{3}a^2\left(-\frac{\Sigma}{a}\right)^{\frac{3}{2}}. \tag{3.23}$$

We derived G and V in the 3-dimensional Gross-Neveu model up to the one-loop level by use of the difference between the original theory and the auxiliary field theory. This is an indirect procedure which is valid only for this special example. Though the stationary points of the effective potential are still equal to 0 and -4a, its form is quite different from $\hat{V}(\sigma)$. Not only the Schwinger function G(J) but also the effective potential $V(\Sigma)$ is double-valued though $\hat{V}(\sigma)$ is single-valued.

Acknowledgements

One of the authors (T.M.) expresses his sincere gratitude to Professor Hironari Miyazawa for the excellent guidance while T.M. was a graduate student in University of Tokyo.

References

1) Y. Nambu and G. Jona-Lasinio, Phys. Rev. **122** (1961) 345; **124** (1961) 246.

2) S. Weinberg, Phys. Rev. **D13** (1976) 974; **D19** (1979) 1277. L. Susskind, Phys. Rev. **20** (1979) 2619.

3) G. Jona-Lasinio, Nuovo Cim. **34** (1964) 1790. See also W. Heisenberg and H. Euler, Z. Physik **98** (1936) 714; J. Schwinger, Phys. Rev. **82** (1951) 664.

4) J.M. Cornwall, R. Jackiw and E. Tomboulis, Phys. Rev. **D10** (1974) 2428.

5) S. Coleman, Point-like Structures inside and outside Hadrons (ed. A. Zichichi, Plenum Press. New York, 1981).

6) T. Banks and S. Raby, Phys. Rev. **D14** (1976) 2182.

7) R.W. Haymaker, T. Matsuki and F. Cooper, Phys. Rev. **D35** (1987) 2567.

8) M. Inoue, H. Katata, T. Muta and K. Shimizu, Prog. Theor. Phys. **79** (1988) 519.

9) D.J. Gross and A. Neveu, Phys. Rev. **D10** (1974) 3235.

10) K.A. Woodhouse, Nucl. Phys. **B113** (1976) 303.

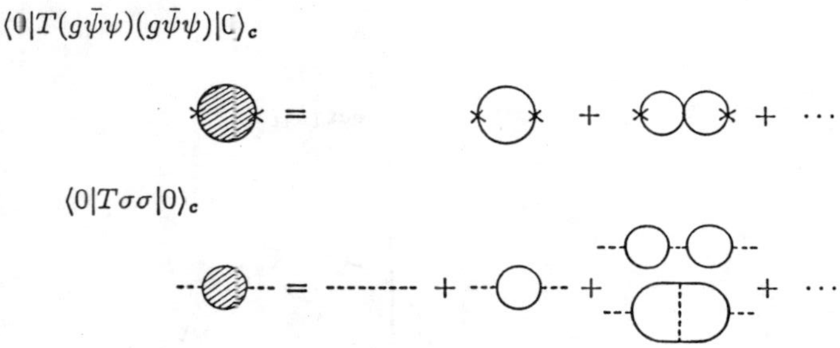

Fig. 1

Diagrammatic expression of the perturbative expansion of connected two-point Green functions corresponding to W[J] and $\hat{W}$[J] respectively

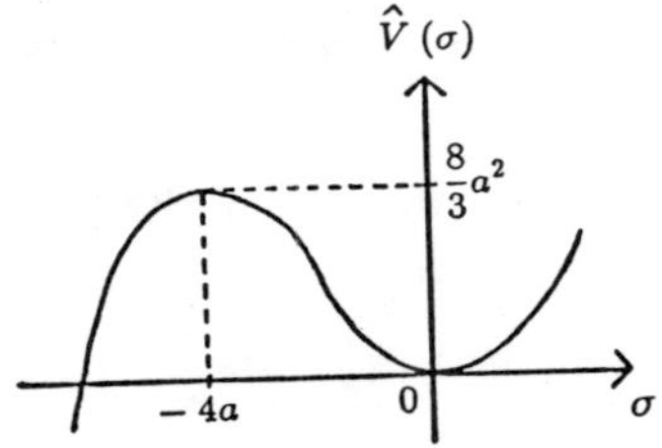

Fig. 2

Effective potential for Gross-Neveu model in three space-time dimensions to one-loop order where σ is defined by Eg.(3.6).

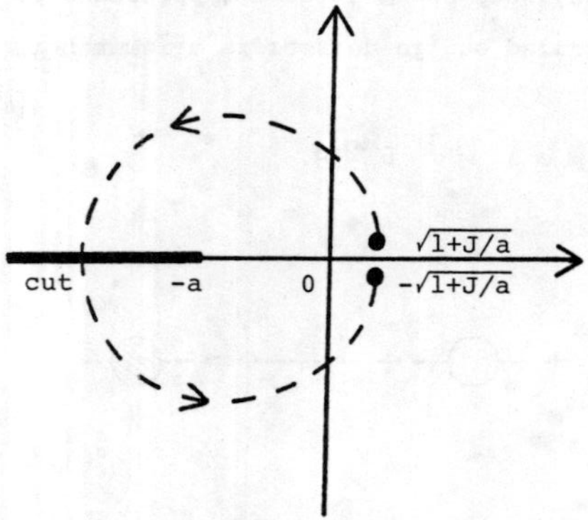

Fig. 3

Path of the analytic continuation on the complex J-plane connecting $\hat{G}^S(J)$ to $\hat{G}^B(J)$.

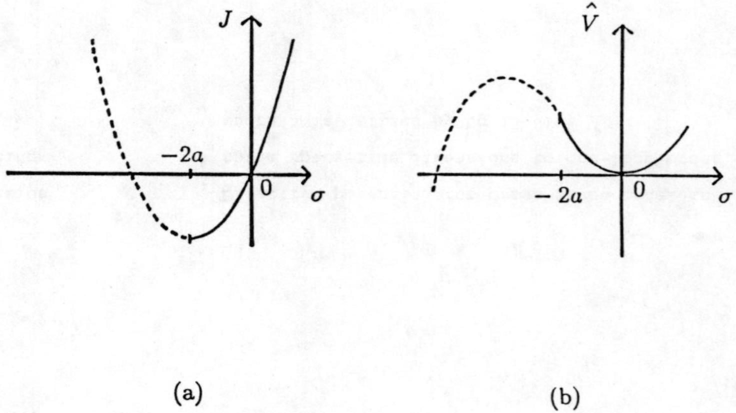

Fig. 4

(a) Diagrammatic expression of Eqs.(3.16a) and (3.16b)

(b) Effective potential $\hat{V}(\sigma)$ when it is separated into two parts $\hat{V}(\sigma^S)$ and $\hat{V}(\sigma^B)$.

How to Calculate One–Loop Diagrams

Y. Oyanagi [§], T. Kaneko [*], T. Sasaki [‡], S. Kawabata and Y. Shimizu

§ Institute of Information Sciences, University of Tsukuba
Tsukuba, Ibaraki, 305 Japan

* Meiji Gakuin University
Totsuka, Yokohama, 244 Japan

‡ The Institute of Physical and Chemical Research
Wako, Saitama, 351 Japan

KEK, National Laboratory for High Energy Physics
Tsukuba, Ibaraki, 305 Japan

Abstract

A numerical method to estimate the Feynman integrals for one–loop diagram with on–shell internal particles is described in terms of an antithetic cancellation technique using an adaptive Monte Carlo integration. This method enables us to estimate necessary integrals of loop diagrams for physical processes without analytic calculation.

1. Introduction

The radiative corrections to e^+e^- annihilation processes have been recognized to play an important role to test the standard electroweak theory quantitatively at TRISTAN, SLC and LEP energy regions [1]. Especially around the Z^o-pole, the knowledge of the radiative corrections is indispensable for precise measurement of the peak. In the $O(\alpha)$ correction we have to calculate a lot of one-loop diagrams as well as real photon emission cross sections with some experimental cuts. The latter can be estimated by integrating the differential cross sections over the phase space by making use of multi-dimensional adaptive Monte Carlo integration routines, such as VEGAS or BASES [2]. The calculation of the former, which amounts to multiple-integration on Feynman parameters, cannot be done by computer in a naive way but only by hand, because in these diagrams the internal particles can be on the mass shell in the physical region, in other words, the integrand of the Feynman integral has poles on the real axis. Although in the literature [3] all the formulas for one-loop 3- and 4-points diagrams, vertex and box, have been given for general mass combinations, they are sometimes not suitable for numerical calculation for special mass combination because we encounter cancellation of huge numbers among many terms in complicated way. Thus it is clever to calculate diagram by diagram by taking advantage of special values of masses from the first (e.g. smallness of the electron mass compared with energy or Z^o mass and so on).

In general, however, the integrands which we have to deal with in realistic problems are very complicated so that the analytic integration becomes extremely difficult and tedious. Further, if we proceed to 5-point one-loop diagrams or multi-loop diagrams appearing in higher order correction, it is almost impossible to perform integration by hand. Therefore we cannot help to invent a completely different method to solve the problem. In this note we would like to propose one possible method which uses numerical Monte Carlo integration together with the antithetic sampling.

One can see a naive stochastic method is not applicable to such a singular integration that needs to evaluate principal values. Even the adaptive Monte Carlo integration program which is powerful to estimate very sharp peak precisely, fails to converge when we apply it directly to such an integral,

$$R = \mathrm{Re} \lim_{\varepsilon \to 0} \int_S \frac{f(x)}{x - i\varepsilon} \mathrm{d}x = \mathrm{P} \int_S \frac{f(x)}{x} \mathrm{d}x, \tag{1}$$

since the integrand is divergent at $x = 0$.

In order to estimate the principal part correctly, we impose a kind of antithetic sampling, that is, we rewrite the integrand as

$$\frac{1}{2}\left\{\frac{f(x)}{x} - \frac{f(-x)}{x}\right\}.$$

Since the function value at x is cancelled with that at its mirror point $-x$, this function is no longer singular at $x=0$ and can easily be integrated numerically. If the range of integration S is not symmetric with respect to $x=0$, we must divide S into two parts, S_1 and S_2, in such a way that any point in S_1 has its mirror point in S_1 and any point S_2 has its mirror point nowhere in S. Then the integation (1) can be reformulated as

$$R = \int_{S_1} \frac{1}{2}\left\{\frac{f(x)}{x} - \frac{f(-x)}{x}\right\} \mathrm{d}x + \int_{S_2} \frac{f(x)}{x} dx. \tag{2}$$

We note that $x=0$ belongs to S_1.

In the next section we will show how the integral for the triangle graph is obtained using this antithetic technique. In section 3, the extention of this method to box or one-loop graphs with more external lines is discussed. The last section is devoted to the conclusion.

2. Triangular Graph

In general the amplitude for a triangular Feynman graph is given in terms of the integral with respect to the so-called Feynman parameters,

$$A = \iint_S \frac{f(x,y)}{\mathfrak{D}(x,y) - i\varepsilon} \mathrm{d}x\mathrm{d}y, \tag{3}$$

where the denominator $\mathfrak{D}(x,y)$ is a quadratic function of x and y, and the range S is the traiangular region $0<x$, $0<y$, $x+y<1$.

Let us consider the simplest diagram in Fig. 1. In this case the denominator is,

$$\mathfrak{D}(x,y) = -xys + (x+y)^2m^2 + (1-x-y)M^2. \tag{4}$$

If s is greater than $4m^2$, $\mathfrak{D}$ has a zero line in S. In order to see the structure of the zero line more clearly, we first transform $\mathfrak{D}$ into a standard form by changing variables,

$$x = x_0+(x'+y')/\sqrt{2}, \quad y = y_0+(x'-y')/\sqrt{2}, \tag{5}$$

with $x_0=y_0=-M^2/(s-4m^2)$. Then we have,

$$\mathfrak{D} = -(\frac{s-4m^2}{2})x'^2 + \frac{s}{2}y'^2 + M^2(1+\frac{M^2}{s-4m^2}). \tag{6}$$

We can now see clearly that the zero line $\mathfrak{D}=0$ is a hyperbola whose foci are on the x'-axis.

The zero on a line $y'=const$ locates at

$$x' = \pm x_1 = \pm\sqrt{s(s-4m^2)y'^2+2M^2(s+M^2-4m^2)}/(s-4m^2). \tag{7}$$

We will define the coordinate x'' of the mirror point of (x',y') in such a way as

$$x'' = \begin{cases} 2x_1 - x' & (x'>0) \\ -2x_1 - x' & (x'<0) \end{cases}$$

The point (x'',y') is accepted as a mirror point only if

a) x'' has the same sign as x',

b) (x'',y') is in the integral region S.

If these two criteria are not satisfied, the point (x',y') has no mirror point, i.e. it belongs to S_2.

The integration for the real part R is performed numerically. Given a point (x,y), we first determine whether it belongs to S_1 or S_2, i.e. its mirror point $(\tilde{x},\tilde{y})$ exist or not. The integrand is

$$\frac{1}{2}\{\frac{f(x,y)}{\mathfrak{D}(x,y)}+\frac{f(\tilde{x},\tilde{y})}{\mathfrak{D}(\tilde{x},\tilde{y})}\} \qquad if\ (x,y)\in S_1,$$

$$\frac{f(x,y)}{\mathfrak{D}(x,y)} \qquad if \ (x,y)\in S_2. \tag{8}$$

Although the singularity at $\mathfrak{D}(x,y)=0$ is cancelled, the integrand is discontinuous at the interface between S_1 and S_2, so that such numerical integration methods as Gauss or double exponential methods are not applicable. We used an adaptive Monte Carlo program BASES[2] developed by one of the present authors.

The imaginary part I is given

$$\begin{aligned} I &= \pi\int\int f(x,y)\delta(\mathfrak{D}(x,y))\mathrm{d}x\mathrm{d}y, \\ &= \pi\int\frac{f(x,y(x))}{\frac{\partial\mathfrak{D}}{\partial y}\Big|_{y=y(x)}}\mathrm{d}x, \end{aligned} \tag{9}$$

where $y(x)$ is the y-coordinate of the zero line at x.

The numerical results are given in Table 1 for $m=40$ (top quark), $M=93$ (Z^0) and s=4000−10000. The numerator $f(x,y)$ is assumed to be unity. The parameters for BASES are:

ITMX1 = 15,

ITMX2 = 20,

NTRIAL = 20.

The results obtained by an analytic integration [4] are also given in this table. The corresponding analytic formula is given by

$$\begin{aligned} I(s) &= \int_0^1 dx\int_0^{1-x}dy\frac{1}{-xys+(x+y)^2m^2+(1-x-y)M^2}, \\ &= (-1)^l\cdot\frac{2}{s}\frac{1}{\sqrt{c}}\frac{1}{\eta_0\eta_+}\{ln(\omega_2)^2-ln(\omega_1)^2-ln(\frac{\omega_1}{\omega_2})\cdot(ln(f)-ln(-f)) \\ &\quad -Sp(\frac{\omega_1}{\alpha_+})+Sp(\frac{\omega_2}{\alpha_+})-Sp(\frac{\omega_1}{\alpha_-})+Sp(\frac{\omega_2}{\alpha_-})-Sp(\frac{\beta_+}{\omega_1})+Sp(\frac{\beta_+}{\omega_2})-Sp(\frac{\beta_-}{\omega_1})+Sp(\frac{\beta_-}{\omega_2})\}, \end{aligned} \tag{10}$$

where

$$a=1+\frac{4m^2}{-s}, \qquad b=-\frac{4M^2}{s}, \qquad c=-\frac{4M^2}{s},$$

$$\eta_+=\frac{-b+\sqrt{b^2-4ac}}{2c},$$

$$\eta_-=\frac{-b-\sqrt{b^2-4ac}}{2c}, \qquad \eta_0=\sqrt{\eta_-/\eta_+},$$

$$\omega_1=\frac{1-\eta_0}{1+\eta_0}\,, \qquad \omega_2=\frac{\sqrt{1-\eta_-}-\eta_0\sqrt{1-\eta_+}}{\sqrt{1-\eta_+}-\eta_0\sqrt{1-\eta_+}}\,,$$

$$d=\eta_0^2-1-\eta_0\sqrt{(b^2-4ac)/c}\,,$$

$$e=2(\eta_0^2+1),$$

$$f=\eta_0^2-1+\eta_0\sqrt{(b^2-4ac)/c}\,,$$

$$\alpha_+=\frac{-e+\sqrt{e^2-4fd}}{2d}\,, \qquad \beta_+=\frac{d}{f}\alpha_+,$$

$$\alpha_-=\frac{-e-\sqrt{e^2-4fd}}{2d}\,, \qquad \beta_-=\frac{d}{f}\alpha_-,$$

and $l=1$ for $s\geq 4m^2$, $l=0$ for $s<4m^2$. The complex Spence function is defined as ,

$$Sp(z)= \quad -\int_0^z \frac{dt}{t}ln(1-t).$$

The agreement is satisfactory except for the imaginary part near the threshold; the reason for this is not yet clear. In Fig.2 we also show how the cancellation of principal integral is achieved by this method.

The extention to general trianglar graphs with arbitrary masses is straightforward.

3. Box Diagram

The amplitude for a box diagram with four external lines is given by

$$A = \lim_{\varepsilon\to 0} \int\int\int_S \frac{f(x,y,z)}{(\mathfrak{D}(x,y,z)-i\varepsilon)^2}dxdy\ dz, \tag{11}$$

where S is now a tetragonal region, $0<x$, $0<y$, $0<z$, $x+y+z<1$. $\mathfrak{D}(x,y,z)$ is again a quadratic function of x,y and z. Here the point is how to estimate the *principal* part of the integral with double zero in the denominator.

Let us consider the single dimensional case,

$$A = \int \frac{f(x)}{(x-i\varepsilon)^2}\mathrm{d}x,$$

$$= \int \frac{(x^2-\varepsilon^2)+2x\varepsilon i}{(x^2+\varepsilon^2)^2} f(x)\mathrm{d}x. \tag{12}$$

From this we can see that the cancellation of the real part around $x=0$ is very delicate and cannot be estimated numerically. We will therfore integrate by part, assuming the differentiability of $f(x)$,

$$A = \lim_{\varepsilon\to 0}\int_a^b \frac{f(x)}{(x-i\varepsilon)^2}\mathrm{d}x,$$

$$= -\left[\frac{f(x)}{x}\right]_a^b + \quad \mathrm{P}\int_a^b \frac{f'(x)}{x}\mathrm{d}x + \pi i f'(0). \tag{13}$$

The second term on the righthand side can be estimated numerically as in the previous section if the derivative $f'(x)$ is given.

Thus the same method is also applicable to the box amplitude (11). The result will be expressed in three terms: surface integral, principal part and the integral on the zero surface (imaginary part). The detail will be published in a forthcoming paper. It can in principle be generalized to any one-loop Feynman amplitude with internal particles on the mass shell.

4. Conclusion

We have shown that the principal part of singular integrals of a Feynman diagram with internal particles on the mass shell can be estimated numerically in terms of an antithetic cancellation technique using an adaptive Monte Carlo integration package BASES. The results were compared with an analytic calculation for the simplest case. The advantage of this method is that it is applicable to Feynman integrals with any numerator functions $f(x,y,...)$. More interested cases would be estimation of two or higher order loop diagram in electroweak theory.

Acknowledgement

We thank members of TRISTAN Theory Group for valuable discussions. Particularly we are indebed to Dr. J. Fujimoto for providing us the results of analytic formula for triangle diagram.

References

[1] For example, see papers published by TRISTAN Radiative Correction Group: Nucl. Phys. B263, 347 (1986), *ibid* B288, 301 (1987), Prog. Theor. Phys. 75, 905 (1986), *ibid* 77, 118 (1987).

[2] G. P. Lepage, J. Comp. Phys. 27, 192 (1978) and "VEGAS Adaptive Multi-dimensional Integration Program", CLNS-80/447, March 1980.
S. Kawabata, Comp. Phys. Commun. 41, 127 (1986).

[3] G. 'tHooft and M. Veltman, Nucl. Phys. B153, 365 (1979).

[4] J. Fujimoto, Thesis for doctorate, Nagoya University (1988).
J. Fujimoto and Y. Shimizu, Mod. Phys. Lett., in press.

s	real part		imag. part	
	numerical	analytic	numerical	analytic
5000	2.3036±0.0004	2.3033	–	0.0
6000	2.7457±0.0004	2.7456	–	0.0
7000	3.1835±0.0021	3.1832	0.9932±0.0017	1.0282
8000	2.7701±0.0022	2.7753	1.4689±0.0034	1.4905
9000	2.4465±0.0019	2.4480	1.7018±0.0039	1.7070
10000	2.1808±0.0018	2.1796	1.8218±0.0031	1.8221
11000	1.9542±0.0016	1.9557	1.8778±0.0058	1.8835
12000	1.7678±0.0015	1.7662	1.9072±0.0045	1.9133
13000	1.6042±0.0015	1.6037	1.9199±0.0045	1.9233
14000	1.4606±0.0014	1.4631	1.9230±0.0044	1.9205

Table 1. The numerical results for the graph in Fig. 1.
The analytic results are also given.

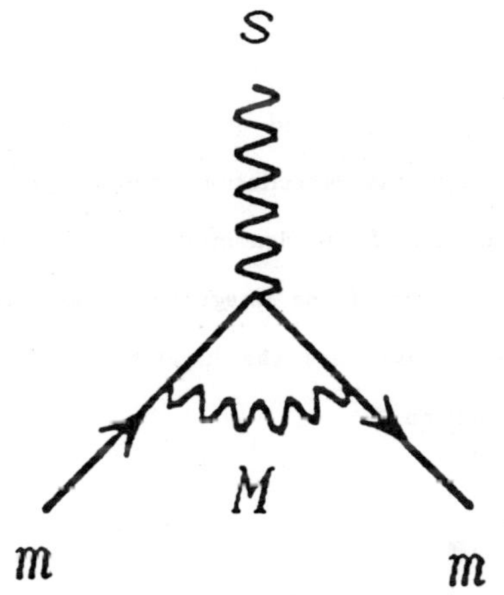

Fig. 1 The triangular Feynman diagram.

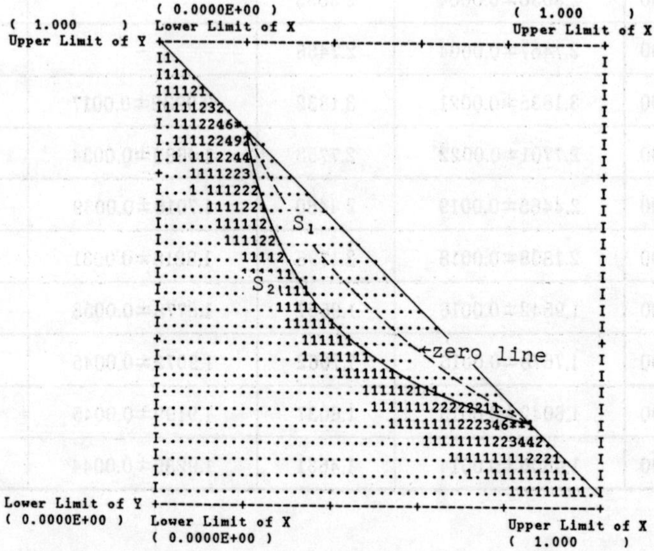

Fig. 2 The map of the integrand onto the integral region, $0<x,y<1$ and $x+y<1$. This shows how the cancellation between pair of mirror points is realized around zero line of the denominator. The numbers in this map indicate the relative height of the integrand. The dots are those points less than 1 and the asterisks are the highest ones. Two regions S_1 and S_2 are defined in section 2.

MANY HILBERT SPACES IN MACROSCOPIC SYSTEM

R. Fukuda

Department of Physics, Faculty of Science and Technology, Keio University, Yokohama 223, Japan

ABSTRACT

The macroscopic system has necessarily the structure of many Hilbert spaces. This explains various quantum mechanical puzzling problems inherent in the macroscopic system. Discussions are given about the measurement problems, the increase of the entropy, the violation of the time reversal invariance, the connection between the spin and statistics, long time average and ensemble average and the time dependent phase transition.

Quantum theory is nowadays thought to be applicable to the macroscopic system just as we have applied it to a microscopic system. There are many macroscopic quantum effects in various field of physics. They are usually related to the phenomena where the wave function describing the macroscopic system has a well defined phase. This is called the macroscopic coherence and the phenomenology of this case is very far from a classical view point and leads to amazing effects.

There is however another limit of the macroscopic quantum system. Here the phase of the wave function is not well defined and only its magnitude is the object of our study. The physical picture, therefore, resembles the classical (statistical) one and we can indeed think of a lot of phenomena which are essentially related to the dephasing of the wave function of the macroscopic body. Our observation in this paper is that the complete loss of the phase coherence of the macroscopic system is a consequence of the many Hilbert space structure. We list below several important issues which are believed to be related to the above stated loss of the information about the phase of the wave function:

1. Measurement theory in quantum mechanics
2. Increase of the entropy of the isolated system
3. The time reversal non-invariance for the macroscopic system
4. Spin and statistics of the macroscopic body
5. Long time average and ensemble average
6. The deterministic equation of motion of the macroscopic variables

These are puzzling problems if we admit that the time evolution of the system is a unitary transformation and therefore a pure state cannot evolve into a mixed state. However once we decided to apply quantum mechanics to a macroscopic system, we immediately encounter and have to solve the above difficult problems.

The purpose of this paper is to apply general observations presented in refs. 1) and 2). There we have discussed the measurement problem in quantum theory by applying the quantum mechanics to the detector -- a macroscopic system. The essence is the following:

For any macroscopic system with the individual degree of freedom q_i ($i=1\sim N$), we divide the variables into the extensive variables (the total momentum, for instance) and the intensive variables. The latter have two classes. The class I contains variables averaged over a macroscopic extent (the center of mass co-ordinate) and the class II consists of other microscopic variables (relative co-ordinates). Let the class I intensive variables be ϕ_k ($k=1\sim n$). In the macroscopic limit $N\to\infty$, these ϕ_k's lose fluctuation and become c-numbers and each continuous c-number value now specifies separate Hilbert space. The class II intensive operators are the only operator variables and they define (or operate on) this Hilbert space. Extensive operators including the Hamiltonian or the Lagrangian do not exist as operators in the macroscopic limit. Since the above c-number values change with the time, we have a picture of the Hilbert space moving continuously in time. This is the same phenomenon as the time dependent phase transition in various systems and actually our arguments are a generalization of the theory of the static phase transition to the time dependent case. The Hilbert spaces of the different phases are known to be disconnected and there does not exist any operator which shifts the value of the order parameter. It is easy to see that the order parameter belongs to class I and the fluctuation vanishes in the macroscopic limit.

The equation of motion governing the time evolution of the c-number is to be supplied since the Hamiltonian of the whole system does not exist. It is most conveniently given by the effective action density $\Gamma[\phi]$ which is the generalization of the classical action and includes the quantum effects. The equation we are looking for is

$$\delta\Gamma[\phi]/\delta\phi(t) = 0. \tag{1}$$

Here ϕ is the c-number for $N=\infty$ and is not the expectation value of the operator.

Now we come to the discussion of the problems listed above.

1. Measurement theory in quantum mechanics is discussed first. The detector is a macroscopic system with the Hilbert space structure specified above. The variable used for measurement (pointer position, for instance) belongs to class I since the variable to be used to measure a microscopic object should not have fluctuations (in an ideal macroscopic limit of course) otherwise we cannot say what is measured. Let the detector variable be ϕ and let its c-number value before the interaction with the object be c.

The wave function of the detector is denoted by ϕ_c where c specifies the Hilbert space. We are going to measure the operator O of the object. Now the object wave function is written in the form $\Sigma a_n \phi_n$ where ϕ_n is the eigen-function of the operator O with the eigen-value λ_n. The total wave function is

$$(\sum_n a_n \phi_n)\phi_c = \sum_n (a_n \phi_n \phi_c). \qquad (2)$$

The detector is a device reacting differently for different n in such a way that by the interaction with the object $\phi_n \phi_c \to \phi_n \phi_{c_n}$ where $c_n \neq c_m$ if $n \neq m$. The pointer position is different for different states n, for instance. Thus after measurement we have

$$\sum_n a_n \phi_n \phi_{c_n} . \qquad (3)$$

Since ϕ_{c_n} belongs to different Hilbert space for different c_n, eq. (3) represents the mixed state whereas eq. (2) is the wave function for the pure state. The interference term between different c_n vanishes as $1/\sqrt{N}$ for large N. Both the magnitude of N and the difference $c_n - c_m$ for $n \neq m$ determine the efficiency of the detector. The change from (2) to (3) by measuring process has been calculated by using several examples.[2] Since the informations contained in the wave function have been lost, the entropy production has taken place between (2) and (3), namely from zero to $-\sum_n |c_n|^2 \ln |c_n|^2$.

2. The increase of the entropy for an isolated system is a puzzling problem since the isolated system obeys the law of the quantum mechanics and its time development is governed by the total Hamiltonian and is expected to be unitary. However once the macroscopic nature of the system is seriously taken into account, we have to realize that the class I intensive variables of the system show the deterministic time evolution and the isolated system no longer realizes the unitary time development. We have seen the

entropy increase in the measuring process. Now we can think of a macroscopic body as an assembly of the subsystems which is macroscopic by itself. The structure of the Hilbert space and the time evolution of the whole system become necessarily complicated. There are many c-number variables. The approximate wave function ϕ_{c_n} of the subsystem is specified by the c-number value c_n of the class I intensive variables of the subsystem we are considering. It undergoes the collision with a microscopic degrees of freedom of other subsystems. Before and after the collision the wave function looks like (2) and (3) respectively and the branching of the Hilbert spaces has occurred. The entropy of the isolated system increases with time continuously in this way just like the entropy increase in the measuring process: each subsystem behaves as a detector.

3. The time reversal invariance seems to be violated for any macroscopic system. As is well known, although the law of dynamics is invariant under the time inversion the macroscopic phenomenon is not. This is explained quite naturally in our theory. Since some variables show the c-number time development while others develop as operators, we cannot find any operator transformation for the time inversion. The consequence of this fact is that any processes in which the valves of c-number variables change violate the time reversal invariance. The increase of the entropy of the isolated system and that in the measuring process are examples.

4. The connection between spin and statistics for the macroscopic body is believed to be irrelevant. The explanation of this belief is that the phase of the wave function of the macroscopic system is not well defined and only the magnitude has a meaning. We can make this statement precise by our theory. Consider the exchange of two macroscopic bodies. One says that the two systems are exchanged if the macroscopic variables of each system are ex-

changed. The center of mass co-ordinates are good examples. But they are necessarily class I intensive variables so that in the macroscopic limit they are c-numbers and they specify different Hilbert spaces. There is not any operator that can shift the c-number value so that the interference between two macroscopic system vanishes.

5. The arguments can be extended to the microscopic systems and the microscopic variables. Let us take the long time average q of the microscopic variable q(t); $q=\frac{1}{2T}\int_{-T}^{T} q(t)dt$. It can be shown that in the limit $T\to\infty$, q loses fluctuation and becomes a c-number. The time average is an extension of the average of the individual variable to define the class I intensive variables. Now the above c-number q defines different Hilbert spaces for different values of q. This leads to the conclusion that any pure state evolves after a long time into the mixed state. This may give a hint to the connection between the long time average and the ensemble average.

6. The equation of motion for the class I intensive variables is a deterministic one in the macroscopic limit. It is different from the classical equation of motion in general. The equation of motion (1) is a deterministic one but it contains $\hbar$. In the limit $\hbar\to 0$ the classical deterministic equation is recovered. The center of mass co-ordinate $\mathbb{X}$ of an isolated macroscopic body however obeys the c-number equation of motion $\ddot{\mathbb{X}}=0$ (the quantum diffision is absent). If there are two macroscopic systems 1 and 2, the relative co-ordinate $\mathbb{X}_1-\mathbb{X}_2$, where $\mathbb{X}_1(\mathbb{X}_2)$ is center of mass co-ordinate of the first (second) system, satisfies the deterministic equation which is different from the classical one. Since $\hbar$ is small, it will not be easy to detect the deviation from the classical motion.

Finally we compare our theory of measurement with other approaches. Among others the many world theory of Everett III[3] is an interesting proposal. He pointed out that the probabilistic interpretation of quantum mechanics can be made consistent with quantum mechanical law if the world branches into several worlds in the process of measurement. Our theory realizes just this story when the system develops from (2) to (3). The important difference between our theory and many world theory is that all the "worlds" are our worlds in our case. Each world constitutes a Hilbert space labeled by the c-number value of the class I operators and each world is connected with each other if the parameter N is not strictly infinite. The continuous c-number parameter appearing above is also the realization of the observation by Araki.[4] He showed that the Hilbert space of the macroscopic system should be labeled by a continuous parameters. We have shown that these parametors are the c-number values of the class I variables.

REFERENCES

1. Fukuda, R., Phys. Rev. A35, 8 (1987).
2. Fukuda, R., Phys. Rev. A36, 3023 (1987).
3. Everett III, H., Rev. Mod. Phys. 29, 454 (1957).
4. Araki, H., Prog. Theor. Phys. 64, 719 (1980).

CONSISTENT CHIRAL ANOMALIES FROM COMMUTATOR ANOMALIES

Koichi Seo
Gifu Women's Junior College
2693 Nagara Fukumitsu, Gifu 502
JAPAN

ABSTRACT

Faddeev's commutator anomaly is perturbatively obtained in a fixed-time regularization method in the two-dimensional space-time. The consistent chiral anomaly is derived from commutator anomalies.

§1. INTRODUCTION

It is usually believed that the chiral gauge theory cannot be consistently quantized, since the gauge anomaly breaks unitarity and renormalizability. However, remarkable properties of the chiral anomaly have been revealed since Adler first discovered it[1]. Adler and Bardeen proved that the chiral anomaly is not affected by higher order corrections[2]. Though there is no proof that Adler-Bardeen's theorem holds for the non-Abelian gauge anomaly also, the absence of higher order corrections is strongly suggested by the recently unveiled topological structure of the gauge anomaly[3]. Then the possibility arises that the chiral gauge theory might be consistently quantized in spite of the obstruction by the non-Abelian gauge anomaly.

Among others Faddeev[4] conjectured that the gauge symmetry is realized as a projective representation in chiral gauge theories, and that a non-trivial two-cocycle appears as a commutator anomaly of the Gauss law operator $G^a(x) = \frac{1}{g}(D_i E^i)^a(x) + i\, j_0^a(x)$ as follows:

$$[\, G^a(x),\ G^b(y)\,] = if_{abc} G^c(x)\delta(x-y) + \mathcal{A}^{ab}(x,y) \tag{1}$$

with

$$\mathcal{A}^{ab}(x,y) = \frac{1}{24\pi^2}\,\varepsilon^{0ijk}\mathrm{tr}(\{t^a,t^b\}\partial_i A_j(x))\partial_k\delta(x-y). \tag{2}$$

In order to confirm Faddeev's conjecture, Kobayashi and Sugamoto[5] tried to evaluate the commutator anomaly through the Bjorken-Johnson-Low (BJL) method[6]. They evaluated only the linear term with respect to the gauge field and found that the commutator anomaly of the gauge part of G has the same expression as (2), while the commutator anomaly of G vanishes as a whole. Meanwhile, Jo[7] completed the computation of the commutator anomaly at the one loop level. There appears a non-trivial commutator anomaly consisting of quadratic and cubic terms with respect to the gauge field:

$$\mathcal{A}^{ab}(x,y) = \frac{1}{48\pi^2}\,\varepsilon^{0ijk}\mathrm{tr}\{[t^a,t^b](\partial_i A_j A_k + A_i\partial_j A_k + A_i A_j A_k)$$
$$+t^a\partial_i A_j t^b A_k - t^b\partial_i A_j t^a A_k\}. \tag{3}$$

The apparent difference between expressions (2) and (3) can be absorbed by redefinition of the Gauss law operator or the charge density. This can be achieved by addition of the following counter terms to the Lagrangian,

$$\Delta\mathcal{L} = \frac{i}{48\pi^2}\,\varepsilon^{0ijk}\mathrm{tr}\{A_0(\partial_i A_j A_k + A_i\partial_j A_k + A_i A_j A_k)\}. \tag{4}$$

Thus Faddeev's conjecture was confirmed by the BJL method. Kobayashi, Sugamoto and the present author[8] also arrived at the same expression as Jo's by making use of the Ward identity, and clarified the connection of the commutator anomaly to the current divergence anomaly.

On the other hand, others have tried to evaluate the commutator anomaly directly in the Hamiltonian formalism. Faddeev and others[9] carried out the computation by a generalized point-splitting method, and could not obtain either Faddeev's form (2) or Jo's form (3). Niemi and Semenoff[10] made use of the ζ function regularization method, and claimed that they obtained Faddeev's form (2). Hosono and the present author[11] adopted a gauge covariant regularization method, and gave a gauge covariant commutator anomaly for the Gauss law operator. The same expression was also derived by Ghosh and Banerjee[12] through a conventional point-splitting method. Recently, Semenoff[13] published a controversial paper claiming that the commutator anomaly eventually vanishes at least in (1+1) dimensions. Thus the problem has not been settled to confirm Faddeev's conjecture in the Hamiltonian formalism.

In the present paper, another regularization is proposed, which partly respects the Bose symmetry at the expense of the gauge symmetry. Utilizing this regularization method, this author evaluated the commutator anomaly of the charge density (j-j anomaly) and the commutator anomaly of the electric field and the charge density (E-j anomaly) in (1+1) dimensions. The results are consistent with those obtained by the BJL analysis[7]. The current divergence anomaly is derived from these commutator anomalies, resulting in the so-called consistent anomaly†. If we assume that the electric fields commute with each other, we obtain Maxwell's equation as the Heisenberg equation. Under the same assumption, Faddeev's commutator anomaly is reproduced.

In Sect. 2, a regularization method which partly respects the Bose symmetry is presented, and formal expressions of commutator anomalies are given. In Sect. 3, they are perturbatively evaluated. In Sect. 4, the current divergence anomaly is derived from the commutator anomalies, and the commutator anomaly of the Gauss law operator is given provided that the electric fields commute with each other.

§2. A REGULARIZATION METHOD RESPECTING THE BOSE SYMMETRY

In the Weyl gauge ($A_0^a = 0$), the Hamiltonian for the chiral gauge theory consists of two parts:

$$\mathcal{H} = \mathcal{H}_{\text{gauge}} + \mathcal{H}_{\text{fermion}}. \tag{5}$$

---

†In fact the consistent anomaly and the covariant anomaly have the same tensor structure in the Weyl gauge and the former is one half of the latter.

The gauge part $\mathcal{H}_{\text{gauge}}$ is the usual one and is given by

$$\mathcal{H}_{\text{gauge}} = \int dx(-\frac{1}{2} E_i^a E^{ia} + \frac{1}{4} F_{ij}^a F^{ij,a}). \tag{6}$$

As for the fermion part $\mathcal{H}_{\text{fermion}}$, we regularize only the interaction term[†] as follows:

$$\mathcal{H}_{\text{fermion}} = \int dx\ \psi^\dagger H\psi, \tag{7}$$

where the Hamilton operator H is iteratively defined by

$$H = H_0 + F(\varepsilon H^2)\ A\ F(\varepsilon H^2), \tag{8}$$

with

$$H_0 = -i\ \gamma^0\gamma^i\ \frac{1+\gamma_5}{2}\partial_i, \tag{9}$$

$$A = -i\ \gamma^0\gamma^i\ \frac{1+\gamma_5}{2}A_i = -ig\ \gamma^0\gamma^i\ \frac{1+\gamma_5}{2}\ A_i^a t^a. \tag{10}$$

Here F(x) is a damping factor with the properties, F(0)=1 and F(∞)=0. For an appropriate differential operator $\mathcal{O}$ and a perturbating operator $\delta\mathcal{O}$, $F(\mathcal{O} + \delta\mathcal{O})$ is assumed to be expanded as

$$F(\mathcal{O} + \delta\mathcal{O}) = F(\mathcal{O}) + F^{(1)}(\mathcal{O}, \delta\mathcal{O}) + F^{(2)}(\mathcal{O}, \delta\mathcal{O}) + \cdots, \tag{11}$$

where $F^{(k)}$ represents the k-th order term in $\delta\mathcal{O}$. The Hamilton operator is expanded with respect to the gauge field as

$$H = H_0 + V_1 + V_2 + \cdots, \tag{12}$$

where V_k stands for the k-th order term in A. Substituting the expansions (11) and (12) into Eq.(8), we can determine V_k iteratively as follows:

$$V_1 = F(\varepsilon H_0^2)\ A\ F(\varepsilon H_0^2), \tag{13}$$

[†]In Ref. 14, Kolokolov and Yelkhovsky applied a point-splitting prescription only to the interaction term and derived a consistent anomaly, but could not reproduce Faddeev's anomaly.

$$V_2 = F^{(1)}(\varepsilon H_0^2,\ \varepsilon(H_0V_1+V_1H_0))\ A\ F(\varepsilon H_0^2) +$$

$$+\ F(\varepsilon H_0^2)\ A\ F^{(1)}(\varepsilon H_0^2,\ \varepsilon(H_0V_1+V_1H_0)),\quad \cdots \tag{14}$$

We define the regularized current by

$$j_\mu^a(x) = \psi^\dagger(x)\ \overleftarrow{F}(\varepsilon H^2)\ \gamma_0\gamma_\mu \frac{1+\gamma_5}{2}\ t^a\ F(\varepsilon H^2)\ \psi(x), \tag{15}$$

where $\psi^\dagger(x)\ \overleftarrow{F}(\varepsilon H^2)$ is the hermitian conjugate of $F(\varepsilon H^2)\ \psi(x)$. Then $\mathcal{H}_{\text{fermion}}$ can be written as

$$\mathcal{H}_{\text{fermion}} = \int dx(-i\psi^\dagger\gamma^0\gamma^k \frac{1+\gamma_5}{2}\partial_k\psi\ -ig\ A_k^a\ j^{ka}). \tag{16}$$

In the Weyl gauge the electric field is regarded as the canonical momentum conjugate to the gauge field, and the following canonical commutation relations are imposed,

$$[\ E_i^a(x),\ A_j^b(y)\ ] = -i\ g_{ij}\ \delta_{ab}\ \delta(x-y). \tag{17}$$

Meanwhile the fermion fields satisfy the anticommutation relations,

$$\{\ \psi(x),\ \psi^\dagger(y)\ \} = \delta(x-y). \tag{18}$$

Niemi and Semenoff have taken the view that the canonical momentum conjugate to the gauge field does not commute with the fermion field and that the electric field commutes with the fermion field[10,13]. Therefore, the electric field differs from the canonical momentum by some operator involving the fermion fields, and the electric fields no longer commute with each other. We take the usual view that the electric fields are independent of the fermion fields and can be regarded as the canonical momentum satisfying the following commutation relations,

$$[\ E_i^a(x),\ \psi(y)\ ] = 0, \tag{19}$$

$$[\, E_i^a(x),\ \psi^\dagger(y)\,] = 0, \tag{20}$$

$$[\, E_i^a(x),\ E_j^b(y)\,] = 0. \tag{21}$$

Now we are able to identify the commutator anomaly of the current as follows:

$$[\, j_0^a(x),\ j_\mu^b(y)\,] = f_{abc}\, j_\mu^c(x)\ \delta(x-y) + \hat{\mathcal{S}}_\mu^{ab}(x,Y), \tag{22}$$

with

$$\hat{\mathcal{S}}_\mu^{ab}(x,y) = \psi^\dagger(x)\ \overleftarrow{F}(\varepsilon H^2)\ t^a\ [\{F(\varepsilon H^2)\}^2 -1]\ \delta(x-y)\gamma_0\gamma_\mu\ t^b\ F(\varepsilon H^2)\ \psi(y)$$

$$-\ (\ x\leftrightarrow y,\ t^a\leftrightarrow \gamma_0\gamma_\mu\ t^b\ ). \tag{23}$$

At least in (1+1) dimensions, it is proved that the operator $\hat{\mathcal{S}}_\mu^{ab}$ can be replaced by its vacuum expectation value in the vanishing limit of ε, and therefore we evaluate the expectation value of $\hat{\mathcal{S}}$ on the fermionic vacuum in the presence of the gauge field. It is formally written as

$$\mathcal{S}_\mu^{ab}(x,y) = \lim_{\varepsilon\to 0}\ {}_A\langle 0|\ \hat{\mathcal{S}}_\mu^{ab}(x,y)\ |0\rangle_A$$

$$= \lim_{\varepsilon\to 0} \mathrm{Tr}[\{F(\varepsilon H^2)\}^2 P_-(y,x)t^a[\{F(\varepsilon H^2)\}^2-1]\delta(x-y)\gamma_0\gamma_\mu t^b$$

$$-\ (\ x\leftrightarrow y,\ t^a\leftrightarrow\gamma_0\gamma_\mu t^b\ )\ ], \tag{24}$$

where $P_-(x,y)$ is the projection operator to the negative energy states.

Next we consider the commutator of the current with the electric field:

$$[\, E_i^a(x),\ j_\mu^b(y)\,] = g\ \hat{\mathcal{T}}_{i\mu}^{ab}(x,y). \tag{25}$$

Again $\hat{\mathcal{T}}_{i\mu}^{ab}$ can be replaced by the vacuum expectation value at least in (1+1) dimensions, which is given by

$$\mathcal{J}^{ab}_{i\mu}(x,y) = \lim_{\varepsilon\to 0} {}_A\langle 0|\hat{\mathcal{J}}^{ab}_{i\mu}(x,y)|0\rangle_A$$

$$= \lim_{\varepsilon\to 0}\lim_{z\to y} \mathrm{Tr}[-\frac{i}{g} t^b\gamma_0\gamma_\mu[\{\frac{\delta}{\delta A^{ia}(x)} F(\varepsilon H^2)\} P_-(y,z) \overleftarrow{F}(\varepsilon H^2)$$

$$+ F(\varepsilon H^2) P_-(y,z)\{\frac{\delta}{\delta A^{ia}(x)} \overleftarrow{F}(\varepsilon H^2)\}]]. \qquad (26)$$

§3. PERTURBATIVE CALCULATION IN (1+1) DIMENSIONS

Now we carry out the perturbative expansion of the commutator anomalies (24) and (26). The Hamilton operator taken in the present paper differs from the one in Ref. 11, but most formulas therein are applicable to the present case with minor modifications.

Since $\mathcal{S}$ and $\mathcal{J}$ have mass dimensions 2 and 1, respectively, and involve a delta function, $\mathcal{S}$ is linear with respect to derivative or the gauge field and $\mathcal{J}$ is essentially constant†. Therefore it is enough to expand the damping factor up to first order in A, and we are able to ignore the terms involving derivatives of the gauge field. Noting that H_0^2 is a unit matrix commuting with the γ matrices and the generator matrices t^a's, we obtain the relevant terms in the expansion as

$$F(\varepsilon H^2) = F(\varepsilon H_0^2) + 2\varepsilon\, A^1 F'(\varepsilon H_0^2)\{F(\varepsilon H_0^2)\}^2\partial_1 + \cdots , \qquad (27)$$

$$\overleftarrow{F}(\varepsilon H^2) = \overleftarrow{F}(\varepsilon H_0^2) - 2\varepsilon\, \overleftarrow{\partial}_1 \overleftarrow{F}'(\varepsilon H_0^2)\{\overleftarrow{F}(\varepsilon H_0^2)\}^2 A^1 + \cdots . \qquad (28)$$

Substituting the expansion (27) into Eq.(24) and retaining the projection operator up to first order in A, we obtain the following result independent of the shape of the damping factor,

$$\mathcal{S}^{00} = \mathcal{S}^{01}$$

$$= \frac{2i}{\pi}\,\mathrm{tr}(t^a t^b)\partial_1\delta(x-y)\int_0^\infty dt\; F'(t)F(t)^3$$

†The gauge field A is always accompanied by the coupling constant g with a mass dimension 1.

$$- \frac{i}{\pi} \mathrm{tr}([t^a,t^b]A^1(x))\delta(x-y)$$

$$\times\int_0^\infty dt[\, F'(t)F(t)^3\{F(t)^2-1\} - F'(t)F(t)^5 \,]$$

$$= - \frac{i}{2\pi} \mathrm{tr}(t^a t^b)\partial_1\delta(x-y) - \frac{i}{4\pi} \mathrm{tr}([t^a,t^b]A^1(x))\delta(x-y). \tag{29}$$

Likewise $\mathcal{J}$ is evaluated as

$$\mathcal{J}^{10} = \mathcal{J}^{11}$$

$$= \frac{1}{\pi} \mathrm{tr}(t^a t^b)\delta(x-y)\int_0^\infty dt\ F'(t)F(t)^3$$

$$= - \frac{1}{4\pi} \mathrm{tr}(t^a t^b)\delta(x-y). \tag{30}$$

These results are in agreement with those obtained by the BJL analysis[7].

§4. CONSISTENT ANOMALY, MAXWELL'S EQUATION AND FADDEEV'S ANOMALY

Now we take the commutator of the charge density with the total Hamiltonian to evaluate the chiral anomaly. The commutator of the charge density with the fermion part of $\mathcal{H}$ is transformed as

$$-i\ [\, j_0^a(x),\ \mathcal{H}_{\text{fermion}} \,] = - (D_1 j^1)^a(x) + \mathcal{O}\!\!\mathcal{L}^a_{\text{spurious}}, \tag{31}$$

with

$$\mathcal{O}\!\!\mathcal{L}^a_{\text{spurious}} = \mathrm{Tr}[t^a\{F(\varepsilon H^2)\ A^1\ F(\varepsilon H^2) - A^1\}P_-(x,y)\{\overleftarrow{F}(\varepsilon H^2)\}^2$$

$$- t^a\{F(\varepsilon H^2)\}^2 P_-(x,y)\{\overleftarrow{F}(\varepsilon H^2)A^1\overleftarrow{F}(\varepsilon H^2) - A^1\}]\big|_{y=x}. \tag{32}$$

Performing the perturbative calculation, we obtain

$$\mathcal{O}\!\!\mathcal{L}^a_{\text{spurious}} = \frac{i}{\pi} \mathrm{tr}[t^a\partial_1 A^1(x)]\int_0^\infty dt\ F'(t)F(t)^3$$

$$= - \frac{i}{4\pi} \mathrm{tr}[t^a\partial_1 A^1(x)]. \tag{33}$$

On the other hand, the commutator of the charge density with the gauge part of $\mathcal{H}$ is known from Eq.(30) as follows:

$$-i\ [\ j_0^a(x),\ \mathcal{H}_{gauge}\ ] = -\frac{i}{4\pi}\,\mathrm{tr}(t^a E^1(x)). \tag{34}$$

Then we have

$$\frac{\partial}{\partial t} j_0^a(x) = -i\ [\ j_0^a(x),\ \mathcal{H}\ ]$$

$$= -\ (D_1 j^1)^a(x) - \frac{i}{4\pi}\,\mathrm{tr}\{t^a(E^1+\partial_1 A^1)(x)\}. \tag{35}$$

As is usually done, $\mathcal{O}\!\mathcal{l}_{spurious}$ can be eliminated by redefinition of the current,

$$\tilde{j}^{1a} = j^{1a} - \frac{i}{8\pi} A^{1a}. \tag{36}$$

It is, however, interesting to note that the spurious anomaly appears with the same coefficient as the Pauli-Villars regularization and the dimensional regularization.

Next we take the commutator of the electric field with the total Hamiltonian and obtain Maxwell's equation as follows:

$$\frac{\partial}{\partial t} E^a = -i\ [\ E^a,\ \mathcal{H}\ ]$$
$$= i\ j^{1a} - \frac{1}{4\pi}\,\mathrm{tr}(t^a A^1) = i\ \tilde{j}^{1a}. \tag{37}$$

Furthermore, we can prove the following relation with the aid of Eqs.(35) and (37),

$$\frac{\partial}{\partial t} G^a = \frac{1}{4\pi}\,\mathrm{tr}(t^a E^1). \tag{38}$$

This relation was first derived by Fujikawa[15] in the path integral method.

Finally the commutator of the Gauss law operator is known from j-j anomaly (29) and E-j anomaly (30) as

$$[\ G^a(x),\ G^b(y)\ ] - i f_{abc} G^c(x)\delta(x-y) = \frac{i}{4\pi}\,\mathrm{tr}([t^a,t^b]A^1)\delta(x-y). \tag{39}$$

Computation in (1+3) dimensions is in progress and will be reported in the future.

ACKNOWLEDGEMENTS

I would like to dedicate this work to Professor Hironari Miyazawa on the occasion of his sixtieth birthday. I was a student unworthy of the gifted professor, but he patiently guided me to theoretical particle physics. Were it not for his generosity, I could not have been able to continue my research life. I express my sincere gratitude to him on this occasion. I wish him many more active and happy years to come.

REFERENCES

1. S.L. Adler, Phys. Rev. 177 (1969) 2426.
2. S.L. Adler and W.A. Bardeen, Phys. Rev. 182 (1969) 1517.
3. B. Zumino, Y.-S. Wu and A. Zee, Nucl. Phys. B239 (1984) 477; L. Alvarez-Gaumé and P. Ginsparg, Nucl. Phys. B243 (1984) 449.
4. L.D. Faddeev, Phys. Lett. 145B (1984) 81; L.D. Faddeev and S.L. Shatashvili, Theor. Math. Phys. 60 (1984) 770.
5. M. Kobayashi and A. Sugamoto, Phys. Lett. 159B (1985) 315.
6. J.D. Bjorken, Phys. Rev. 148 (1966) 1467; K. Johnson and F.E. Low, Prog. Theor. Phys. Suppl. 37-38 (1966) 74.
7. S.-G. Jo, Phys. Lett. 163B (1985) 353.
8. M. Kobayashi, K. Seo and A. Sugamoto, Nucl. Phys. B273 (1986) 607.
9. L.D. Faddeev and S.L. Shatashvili, Phys. Lett. 167B (1986) 225; S.-G. Jo, Phys. Rev. D35 (1987) 3179.
10. A.J. Niemi and G.W. Semenoff, Phys. Rev. Lett. 55 (1985) 925.
11. S. Hosono and K. Seo, to appear in Mod. Phys. Lett. A3 (1988); S. Hosono and K. Seo, Gifu Women's Junior College preprint, GWJC-4 (1987).
12. S. Ghosh and R. Banerjee, Saha Inst. of Nucl. Phys. preprint, SINP-TNP/87-26 (1987).
13. G.W. Semenoff, Phys. Rev. Lett. 60 (1988) 680.
14. I.V. Kolokolov and A.S. Yelkhovsky, Inst. of Nucl. Phys. (USSR) preprint (1987).
15. K. Fujikawa, Phys. Lett. 171B (1986) 424.

LIST OF PUBLICATIONS OF PROFESSOR MIYAZAWA

1. Photo-Meson Production and Nucleon Isobars
   Y. Fujimoto and H. Miyazawa
   Prog. Theor. Phys. **5** (1950) 1052.

2. A Possible Explanation for the Deviations of Nuclear Moments from the Schmidt Lines
   H. Miyazawa
   Prog. Theor. Phys. **6** (1951) 236.

3. A Model for V-Particles
   H. Miyazawa
   Prog. Theor. Phys. **6** (1951) 631.

4. Deviations of Nuclear Magnetic Momemts from the Schmidt Lines
   H. Miyazawa
   Prog. Theor. Phys. **6** (1951) 801.

5. Non-Additivity of Nucleon Magnetic Moment in Deuteron
   H. Miyazawa
   Prog. Theor. Phys. **7** (1952) 207.

6. An Interpretation of the Meso-Disintegration of the Deuteron in Terms of the Strong Coupling Meson Theory
   S. Matsuyama and H. Miyazawa
   Prog. Theor. Phys. **8** (1952) 140.

7. Pi-D Scattering and Strong Coupling Theory
   H. Miyazawa and S. Matsuyama
   Prog. Theor. Phys. **8** (1952) 573.

8. Meson-Deuteron Reaction and Nucleon Isobars
   S. Matsuyama and H. Miyazawa
   Prog. Theor. Phys. **9** (1953) 492.

9. Strong Coupling Theory for the Case of Pseudoscalar Coupling
   H. Miyazawa
   Phys. Rev. **97** (1955) 1413.

10. Sachs' Theorem on the Magnetic Moments of Nucleons
    H. Miyazawa
    Phys. Rev. **97** (1955) 315.

11. Decay of Spin-Zero Mesons into Two Leptons
    H. Miyazawa and R. Oehme
    Phys. Rev. **99** (1955) 315.

12. Feynman Amplitudes and Strong-Coupling Pion Theory
H. Miyazawa
Phys. Rev. **99** (1955) 944.

13. Application of Dispersion Relations to Pion-Nucleon Scattering
M.L. Goldberger, H. Miyazawa and R. Oehme
Phys. Rev. **99** (1955) 986.

14. Anomalous Magnetic Moment of the Nucleon and the Pion-Nucleon Scattering
H. Miyazawa
Phys. Rev. **101** (1956) 1564.

15. Interaction of P- and S-Wave Pions with Fixed Nucleons
H. Miyazawa
Phys. Rev. **104** (1956) 1741.

16. Pion Theory of Three-Body Forces
J. Fujita and H. Miyazawa
Prog. Theor. Phys. **17** (1957) 360.

17. Spin-Orgit Coupling in Heavy Nuclei
J. Fujita and H. Miyazawa
Prog. Theor. Phys. **17** (1957) 366.

18. Electron-Neutron Interaction
F. Ando and H. Miyazawa
Prog. Theor. Phys. **17** (1957) 607.

19. Two-Pion Exchange Potential
M. Konuma, H. Miyazawa and S. Otsuki
Phys. Rev. **107** (1957) 320.

20. Dispersion Relations in Nucleon-Nucleon Scattering
S. Matsuyama and H. Miyazawa
Prog. Theor. Phys. **18** (1957) 328.

21. Two-Pion-Exchange Nuclear Potential
M. Konuma, H. Miyazawa and S. Otsuki
Prog. Theor. Phys. **19** (1958) 127.

22. Dispersion Relations in Nucleon-Nucleon Scattering
S. Matsuyama and H. Miyazawa
Prog. Theor. Phys. **19** (1958) 517.

23. On the Symmetry Law and the Pion Decay
H. Umezawa and H. Miyazawa
Nuovo Cimento **9** (1958) 586.

24. Mesic Correction to the Beta-Decay in a Nucleus
J.I. Fujita, Z. Matsumoto, E. Kuroboshi and H. Miyazawa
Prog. Theor. Phys. **20** (1958) 308.

25. On the Strong Interaction
H. Miyazawa, T. Nakano and H. Umezawa
Prog. Theor. Phys. **22** (1959) 643.

26. S-Wave Pion-Nucleon Scattering
K. Kawarabayashi and h. Miyazawa
Prog. Theor. Phys. **23** (1960) 490.

27. Dispersion Relations in Nucleon-Nucleon Scattering
Y. Hawa and H. Miyazawa
Prog. Theor. Phys. **23** (1960) 490.

28. Application of Nucleon-Nucleon Dispersion Relations to Nuclear Many-Body Problem
I. Hamamoto and H. Miyazawa
Phys. Rev. **123** (1961) 1860.

29. Regge Poles in Nultiple Meson Production
H. Miyazawa and M. Suzuki
Prog. Theor. Phys. **28** (1962) 1102.

30. Bound State Problem and the Determination of Coupling Constants
H. Miyazawa and H. Sugawara
Prog. Theor. Phys. **29** (1963) 735.

31. Pion Dynamics in Nuclear Matter
H. Miyazawa
Jour. Phys. Soc. Japan **19** (1964) 1764.

32. Symmetry SU(9)
H. Miyazawa and H. Sugawara
Prog. Theor. Phys. **33** (1965) 771.

33. Intrinsically Broken U(18) Symmetry
H. Miyazawa and H. Sugawara
Prog. Theor. Phys. **34** (1965) 263.

34. Proton-Neutron Mass Difference
H. Miyazawa, Y. Oi and M. Suzuki
Prog. Theor. Phys. Supplement Extra Number (1965) 436.

35. Static Pion Theory and Sum Rules
H. Miyazawa
Prog. Theor. Phys. Supplement **37** and **38** (1966) 315.

36. Baryon Number Changing Current
H. Miyazawa
Prog. Theor. Phys. **36** (1966) 1266.

37. Spontaneous Symmetry Breaking and Related Problems
H. Miyazawa
Recent Development in Particle Physics, M.J. Moravcsik ed. (1966) 22.

38. Spinor Currents and Symmetries of Baryons and Mesons
H. Miyawaza
Phys. Rev. **170** (1868) 1586.

39. Absorbing Hard Core and Regge Cuts
I. Kimel and H. Miyazawa
Prog. Theor. Phys. **40** (1968) 124.

40. Large Angle Diffraction Scattering
R. Fukuda and H. Miyazawa
Prog. Theor. Phys. **41** (1969) 1247.

41. Weak Interaction as an S-Matrix
A.N. Maheshwari and H. Miyazawa
Prog. Theor. Phys. **44** (1970) 772

42. Finite Energy Sum Rule as a Test of the Pomeranchuk Theorem
K. Igi, M. Kuroda and H. Miyazawa
Phys. Lett. **34b** (1971) 140.

43. Role of Mesons in the Physics of Nuclear Moments
H. Miyazawa
Jour. Phys. Soc. Japan Supplement **34** (1973) 10.

44. High Energy Scattering and Hadron Structure
M. Kuroda and H. Miyazawa
Prog. Theor. Phys. **50** (1973) 569.

45. High Energy Hadron Collisions and Composite Model
M. Kuroda and H. Miyazawa
Prog. Theor. Phys. **51** (1974) 205.

46. Rising Total Cross Sections and Hadron Structure
M. Kuroda, H. Miyazawa and Y. Yamaguchi
Prog. Theor. Phys. **51** (1974) 541.

47. Description of Inclusive Reactions in Quark Model
Z.F. Ezawa, J. Maharana and H. Miyazawa
Prog. Theor. Phys. **51** (1974) 877.

48. Electrostatic Correction to High Energy Hadron Scattering
M. Kuroda and H. Miyazawa
Prog. Theor. Phys. **53** (1974) 1088.

49. Quark Confinement and Forces between Quarks
H. Miyazawa
Proc. Seminaire Franco-Japonais (Inst. for Nuclear Study, Univ. Tokyo, 1977) 265.

50. Energy Dependence of Proton-Proton Cross Section
K. Kato, H. Miyazawa and K. Yoshikawa
Prog. Theor. Phys. **59** (1978) 882.

51. Large/t/ Proton-Proton Scattering and Quark-Quark Scattering
K. Kato and H. Miyazawa
Prog. Theor. Phys. **60** (1978) 1236.

52. Long Range Forces between Hadrons
S. Matsuyama and H. Miyazawa
Prog. Theor. Phys. **61** (1979) 942.

53. Reconnection of Strings and Quark Matter
H. Miyazawa
Phys. Rev. **D20** (1979) 2953.

54. String Flip-Flop, Quark Matter and Multibaryon Resonances
H. Miyazawa
Proc. Hakone Japan-U.S. Seminar (J.S.P.S. 1980) II,224.

55. Quarks and Strings in Dense Matter
H. Miyazawa
Statistical Mechanics of Quarks and Hadrons, H. Satz ed. (1981) 79.

56. Conserved Spinor Quantities
H. Miyazawa
Proc. 4th INS Winter Seminar on Structure of Quarks and Leptons,(Inst. for Nuclear Study, Univ. Tokyo, 1981) 71.

57. Quantum Mechanics in a Multilply Connected Region
H. Miyazawa
Proc. Xth Hawaii Conf. in High Energy Physics (Univ. Hawaii, August, 1985) 441.

58. Bound State Problem for Quarks and Leptons
H. Miyazawa
Supplement, Prog. Theor. Phys. **86** (1986) 65.

59. Pions and Quarks in the Nucleon
H. Miyazawa
Particles and Nuclei, H. Terazawa ed. (1986) 3.

60. Multivalued Wave Functions and the Ehrenberg-Siday-Bohm-Aharonov Effect
H. Miyazawa
To be published.

Symposium of Perspectives on Particle Physics

— In commemoration of the sixtieth birthday of Professor Hironari Miyazawa —

March 29 – 30, 1988
Department of Physics, University of Tokyo

Organizers

J. Arafune

T. Eguchi

Y. Hara

K. Igi

K. Kawarabayashi

S. Matsuda

T. Muta

Y. Nagashima

R. Sasaki

H. Sugawara

S. Yazaki

Program

March 29

Opening talk 13:30 – 13:45 K. Igi

Afternoon session I .... Chairman: K. Igi

Superconformal field theory with extended supersymmetry
13:45 – 14:30 S. Matsuda

Quantum mechanics in macroscopic system and theory of measurement
14:35 – 15:20 R. Fukuda

Afternoon session II .... Chairman: T. Muta

Hadron interactions based on the flip-flop mechanism of strings
15:40 – 16:25 K. Yazaki

Nuclei and pions
16:30 – 17:15 T. Yamazaki

March 30

Morning session I .... Chairman: T. Eguchi

String in background fields
10:00 – 11:00 C.G. Callan

Morning session II .... Chairman: K. Kawarabayashi

Superconvergent propagators
11:15 – 12:15 R. Oehme

Afternoon session I .... Chairman: Y. Nagashima

Compositeness of particles
13:30 – 14:30 M. Suzuki

Cosmic rays and elementary particles
14:35 – 15:20 J. Arafune

Afternoon session II .... Chairman: Y Hara

Field theory of interacting strings
15:40 – 16:25 H. Sugawara

Quantum field theory
16:30 – 17:15 H. Miyazawa

List of Participants

Aihara, H.	Physics Dept., Univ. of Tokyo
Akiba, T.	Physics Dept., Tohoku Univ.
Arafune, J.	Inst. of Cosmic Ray Research, Univ. of Tokyo
Aramaki, S.	Physics Dept., Nagoya Univ.
Eguchi, T.	Physics Dept., Univ. of Tokyo
Ezawa, Z. F.	Physics Dept., Tohoku Univ.
Fujisaki, H.	Physics Dept., Rikkyo Univ.
Fukuda, R.	Physics Dept., Keio Univ.
Fukuma, M.	Physics Dept., Univ. of Tokyo
Furuichi, S.	Physics Dept., Rikkyo Univ.
Hara, Y.	Physics Dept., Tsukuba Univ.
Hatsukade, S.	Faculty of Engineering, Seikei Univ.
Hayano, R.	Physics Dept., Univ. of Tokyo
Hidaka, K.	Physics Dept., Tokyo Gakugei Univ.
Higashijima, K.	KEK
Hosoya, A.	RITP, Hiroshima Univ.
Hotta, Y.	Physics Dept., Univ. of Tokyo
Igi, K.	Physics Dept., Univ. of Tokyo
Ikehata, S.	Physics Dept., Univ. of Tokyo
Imazato, J.	KEK
Inami, T.	RIFP, Kyoto Univ.
Ino, S.	Physics Dept., Univ. of Tokyo
Ishibashi, N.	Physics Dept., Univ. of Tokyo
Iso, A.	Physics Dept., Univ. of Tokyo
Iso, C.	Physics Dept., Tokyo Inst. of Technology
Iwasaki, Y.	Physics Dept., Tsukuba Univ.
Kadono, R.	Meson Sci. Lab., Univ. of Tokyo
Kamae, T.	Physics Dept., Univ. of Tokyo
Kamimura, A.	Physics Dept., Univ. of Tokyo
Kamitsubo, H.	Inst. of Physical and Chemical Research
Kanki, T.	Coll. of Sci. and Technology, Kinki Univ.
Kanbe, T.	Physics Dept., Univ. of Tokyo
Kato, A.	Physics Dept., Univ. of Tokyo
Kato, K.	Physics Dept., Kogakuin Univ.
Kato, M.	Coll. of Arts ans Science, Univ. of Tokyo, Komaba
Kawabe, S.	KEK
Kawarabayashi, K.	Coll. of Arts ans Science, Univ. of Tokyo, Komaba
Kawashima, Y.	Meson Sci. Lab., Univ. of Tokyo

Kikkawa, K.	Physics Dept., Osaka Univ.
Kinoshita, K.	Physics Dept., Kagoshima Univ.
Kitazawa, Y.	Physics Dept., Univ. of Tokyo
Kobayashi, K.	Coll. of Liberal Arts and Science, Kyoto Univ.
Kobayashi, T.	Physics Dept., Univ. of Tokyo
Kobayashi, T.	Physics Dept., Tokyo Metropolitan Univ.
Kohara, Y.	Coll. of Agriculture and Vet. Med., Nihon Univ.
Konuma, M.	Physics Dept.-Hiyoshi, Keio Univ.
Kuroboshi, E.	Physics Dept., Tokyo Women's Christian Univ.
Kuroda, M.	Physics Dept., Meiji Gakuin Univ.
Maki, Z.	RIFP, Kyoto Univ.
Matsuda, S.	Coll. of Liberal Arts and Science, Kyoto Univ.
Matsuo, Y.	Enrico Fermi Inst., Univ. of Chicago, Chicago
Miyake, Y.	Physics Dept., Hokkaido Univ.
Miyazaki, T.	Physics Dept., Science Univ. of Tokyo
Morita, M.	Physics Dept., Osaka Univ.
Muta, T.	Physics Dept., Hiroshima Univ.
Nagamine, K.	Meson Sci. Lab., Univ. of Tokyo
Nagashima, Y.	Physics Dept., Osaka Univ.
Naito, S.	Physics Dept., Osaka City Univ.
Nakai, K.	KEK
Nakamura, K.	KEK
Nakamura, K.	Physics Dept., Meiji Univ., Izumi
Nakamura, S.	Physics Dept., Tokai Univ.
Nakanishi, N.	RIMS, Kyoto Univ.
Nakazawa, N.	Physics Dept., Kogakuin Univ.
Ninomiya, M.	INS, Univ. of Tokyo
Nishijima, K.	RIFP, Kyoto Univ.
Nishioka, A.	Coll. of Arts ans Science, Univ. of Tokyo, Komaba
Nishiyama, K.	Meson Sci. Lab., Univ. of Tokyo
Nomura, T.	INS, Univ. of Tokyo
Ogawa, K.	Physics Dept., Univ. of Tokyo
Ohguri, H.	Physics Dept., Univ. of Tokyo
Okamura, H.	Physics Dept., Kogakuin Univ.
Okawa, M.	KEK
Ono, M.	Physics Dept., Univ. of Tokyo
Ono, S.	Physics Dept., Univ. of Tokyo
Otake, S.	Physics Dept., Univ. of Tokyo
Otokozawa, J.	Physics Dept., Nihon Univ.
Oyanagi, Y.	Inst. of Information Science, Tsukuba Univ.

Sagawa, H.	Physics Dept., Univ. of Tokyo
Sasaki, R.	RITP, Hiroshima Univ.
Sato, H.	Physics Dept., Hyogo Univ. of Teacher Education
Sato, S.	Physics Dept., Rikkyo Univ.
Satô, A.	Physics Dept., Osaka Univ.
Sawada, T.	Physics Dept., Tsukuba Univ.
Seo, K.	Physics Dept., Gifu Women's Junior College
Shima, K.	Mathematics Dept., Science Univ. of Tokyo
Shimizu, K.	Physics Dept., Univ. of Tokyo
Shirai, I.	Coll. of Arts ans Science, Univ. of Tokyo, Komaba
Tonomura, A.	Central Research Lab., Hitachi Ltd.
Suematsu, H.	Inst. of Materials Science, Tsukuba Univ.
Sugamoto, A.	Physics Dept., Ochanomizu Univ.
Sugawara, H.	KEK
Suzuki, M.	Physics Dept., Univ. of California, Berkeley
Suzuki, M.	Physics Dept., Univ. of Tokyo
Tada, T.	Physics Dept., Univ. of Tokyo
Takasugi, E.	College of General Education, Osaka Univ.
Takebe, H.	Physics Dept., Seikei Univ.
Tanabe, K.	Physics Dept., Saitama Univ.
Terazawa, H.	INS, Univ. of Tokyo
Tokuda, N.	Faculty of Engineering, Hokkaido Univ.
Tomiya, M.	INS, Univ. of Tokyo
Uematsu, T.	Coll. of Liberal Arts and Science, Kyoto Univ.
Wada, A.	Physics Dept., Univ. of Tokyo
Wada, M.	Physics Dept., Tokai Univ.
Wada, Y.	Physics Dept., Univ. of Tokyo
Wakabayashi, T.	Physics Dept., Univ. of Tokyo
Watanabe, T.	Physics Dept., Kobe Univ.
Yamada, S.	INS, Univ. of Tokyo
Yamamoto, S.	Physics Dept., Univ. of Tokyo
Yamazaki, T.	INS, Univ. of Tokyo
Yang, S.-K.	RIFP, Kyoto Univ.
Yazaki, K.	Physics Dept., Univ. of Tokyo
Yazaki, S.	Physics Dept., Univ. of Tokyo
Yoneya, T.	Coll. of Arts ans Science, Univ. of Tokyo, Komaba
Yoneyama, K.	Physics Dept., Univ. of Tokyo
Yoshikawa, K.	Syumei High School